Springer Tracts in Advanced Robotics

Volume 144

The Springer Tracts in Advanced Robotics (STAR) publish new developments and advances in the fields of robotics research, rapidly and informally but with a high quality. The intent is to cover all the technical contents, applications, and multidisciplinary aspects of robotics, embedded in the fields of Mechanical Engineering, Computer Science, Electrical Engineering, Mechatronics, Control, and Life Sciences, as well as the methodologies behind them. Within the scope of the series are monographs, lecture notes, selected contributions from specialized conferences and workshops, as well as selected PhD theses.

Special offer: For all clients with a print standing order we offer free access to the electronic volumes of the Series published in the current year.

Indexed by SCOPUS, DBLP, EI Compendex, zbMATH, SCImago.

All books published in the series are submitted for consideration in Web of Science.

More information about this series at https://link.springer.com/bookseries/5208

Bruno Siciliano · Fabio Ruggiero
Editors

Robot Dynamic Manipulation

Perception of Deformable Objects and Nonprehensile Manipulation Control

Springer

Editors
Bruno Siciliano
Department of Electrical Engineering
University of Naples Federico II
Naples, Italy

Fabio Ruggiero
Department of Electrical Engineering
University of Naples Federico II
Naples, Italy

ISSN 1610-7438 ISSN 1610-742X (electronic)
Springer Tracts in Advanced Robotics
ISBN 978-3-030-93289-3 ISBN 978-3-030-93290-9 (eBook)
https://doi.org/10.1007/978-3-030-93290-9

This Springer imprint is published by the registered company Springer Nature Switzerland AG
The registered company address is: Gewerbestrasse 11, 6330 Cham, Switzerland

Foreword

At the dawn of the century's third decade, robotics is reaching an elevated level of maturity and continues to benefit from the advances and innovations in its enabling technologies. These all are contributing to an unprecedented effort to bringing robots to human environment in hospitals and homes, factories, and schools; in the field for robots fighting fires, making goods and products, picking fruits, and watering the farmland, saving time and lives. Robots today hold the promise for making a considerable impact in a wide range of real-world applications from industrial manufacturing to healthcare, transportation, and exploration of the deep space and sea. Tomorrow, robots will become pervasive and touch upon many aspects of modern life.

The Springer Tracts in Advanced Robotics (STAR) is devoted to bringing to the research community the latest advances in the robotics field on the basis of their significance and quality. Through a wide and timely dissemination of critical research developments in robotics, our objective with this series is to promote more exchanges and collaborations among the researchers in the community and contribute to further advancements in this rapidly growing field.

Bruno Siciliano and Fabio Ruggiero present in this book the results and accomplishments in RoDyMan, a European Research Council's project on robotic dynamic manipulation. This project brought a strong team of robotics researchers with a wide range of complementary skills and competencies to work together on the challenges of nonprehensile dynamic manipulation and deformable object manipulation. The Pizza Maker, which was selected as the robot demonstrator, constitutes a compelling robotic benchmark in this project and marks a vivid signature of the project's host city, Naples.

The book is organized in two parts addressing the real-time perception and the planning and control of deformable objects and nonprehensile manipulation challenges. In its nine chapters, this collection offers a comprehensive treatment of dynamic manipulation and establishes significant advances in the theoretical formulation and experimental validation of nonprehensile robotic manipulation systems.

The outcome is a book which is confirmed to be shining in our STAR series!

Stanford, USA
September 2021

Oussama Khatib
STAR Editor

Preface

This book collects the main results of the Advanced Grant project RObotic DYnamic MANipulation (RoDyMan) funded by the European Research Council (ERC) to the first editor from June 2013 to May 2019.

The project started from observing that nonprehensile dynamic manipulation can be reasonably considered the most complex manipulation task. In a nonprehensile manipulation task, the object can still be manipulated by the hand. However, it is possible neither to prevent any infinitesimal motions of the object nor to resist all external wrenches applied to it. Think of an object held by the palm of a human hand: the object is sustained, and it is not dropped; however, it is not possible to resist to a force lifting the object, while it is possible to manipulate it by either moving the hand or breaking the contact by throwing. Both prehensile and nonprehensile manipulation become even more demanding if the handled object is not rigid. Several objects that humans handle every day are subject to both plastic and elastic deformations, such as newspapers, sheets, food, banknotes, tissues, bags, pillows, sponges. Understanding how to manage a deformable object and make it assume the desired shape is something very challenging. It might be argued that all such tasks are still rather far from being fully solved and applied in robotics, reducing the opportunities for broad adoption of robots within human co-habited environments.

As a final demonstrator of the RoDyMan project, we thought to realize a pizza-maker robot. This represents a perfect example to understand the robot challenge, considering the difficulties that every inexperienced person encounters in preparing a pizza having to manipulate a mixture of water and flour, of every varying shape and consistency.

Through RoDyMan, we got the opportunity to merge all the acquired competencies in a blended theoretical and technological challenge, advancing the state of the art in nonprehensile dynamic manipulation, also considering deformable objects. The RoDyMan project contributed to paving the way towards enhancing autonomy and operational capabilities of service robots.

The teamwork produced tangible results during the project. The developed platform often mentioned throughout this book is a humanoid-like robot with a pan-tilt sensorized robotic head, a two-degree-of-freedom (DoF) torso, two seven-DoF arms, and an omnidirectional mobile platform. The pizza-maker demonstrator was realized in all its aspects in simulation, and at 90% in the experimental phase.

From a theoretical point of view, several significant advances in the mathematical formulation of nonprehensile manipulation tasks were produced. Port-Hamiltonian, Lagrangian, and geometric control frameworks were chosen to handle nonprehensile manipulation problems. As concerns with perception, the main modules regarding environmental awareness were successfully developed. It was possible to model and track 3D deformable and texture-less objects subject to both elastic and plastic deformations. As said above, the possibility to take care also of possible fractures of the deformable object was considered. The viscosity of the pizza dough was taken into account by modeling a deformable object through a smoothed-particle hydrodynamic approach.

Unforeseen applications were found in the industrial scenario to glue the sole on the shoe upper. Also, the connection between nonprehensile manipulation and walking gaits of a legged robot is currently carried on in follow-up projects (WELDON, PRINBOT) after the successful experience of the RoDyMan project. Last but not the least, several of the developed control techniques can be applied within the robotic surgery field (MUSHA, PROSCAN, BARTOLO, PACMAN projects), and it is foreseeable to adopt similar methodologies for manipulation of soft tissues, muscles, organs, and the skin.

Regarding the engagement and outreach of the project, the scientific community, the media, and the general public warmly recognized the developed work. Several collaborations with international research groups were established. The RoDyMan project had a wide track record of dissemination to press and general public. The project was covered by media like Scientific American, The Times, The Telegraph, and Mashable. The RoDyMan robot was also exhibited at the Maker Faire European Edition 2017, in Rome, where it was awarded the Maker of Merit recognition.

As stated above, the main results achieved during the RoDyMan project have been collected in this book. Its contents are organized in two parts for a total of nine chapters by nine contributors: Part I deals with perception and includes four chapters on modeling, recognition, motion planning, and tracking of objects undergoing different types of deformation, e.g., elastic, plastic, fractures, collisions. The five chapters in Part II cover several aspects of nonprehensile manipulation planning and control for robotic systems, subject to both holonomic and nonholonomic constraints.

The contents have been duly updated and pedagogically organized in a harmonious way to highlight connections between the already published works. Besides, this book presents newly developed and unpublished results. In particular, Chaps. 3, 4, and 9 contain novelties in the face of the current literature. The notation has been homogenized throughout the book. Each chapter starts with a table containing the local notation employed therein. The symbols in common among the chapters are tabled within the associated list at the beginning of this book. Further, each chapter starts with a brief introduction presenting the addressed problem. A discussion related to achieved results, limitations, open problems, and possible future directions closes the chapter.

The perception and control problems addressed in this book are challenging research topics dealing with nonprehensile dynamic manipulation. Nonetheless, these topics are shared with other research domains such as soft robotics, legged systems, wearable robotics. This is ideally in line with the spirit of the blue-sky research funded by the European Research Council under the Advanced Grant agreement number 320992 which is here gratefully acknowledged. For this reason, this book not only represents the apex of the RoDyMan project, but it can be used as a reference source for undergraduates and Ph.D. students willing to start a research program in the robotic nonprehensile manipulation domain and the related topics, understand the already developed work, and have a view about the associated open challenges.

As a final word, the RoDyMan project culminating with this volume is meant as a tribute to Naples, the hosting city of the project, an avant-garde city in robotics technology, automation, gastronomy, and art culture.

Naples, Italy
August 2021

Bruno Siciliano
Fabio Ruggiero

Contents

Contributors

Alejandro Donaire The University of Newcastle, Callaghan, NSW, Australia

Alejandro Gutierrez-Giles CECAv-UNAM, Ciudad Universitaria, CDMX, Mexico

Jung-Tae Kim MIKNMEK Inc., Gyeongsangbuk-do, Republic of Korea

Vincenzo Lippiello Department of Electrical Engineering and Information Technology, PRISMA Lab, CREATE Consortium & University of Naples Federico II, Naples, Italy

Midhun S. Menon OLA Electric, Bangalore, India

Antoine Petit Therenva, Rennes, France

Fabio Ruggiero Department of Electrical Engineering and Information Technology, PRISMA Lab, CREATE Consortium & University of Naples Federico II, Naples, Italy

Aykut C. Satici Boise State University, Boise, USA

Bruno Siciliano Department of Electrical Engineering and Information Technology, PRISMA Lab, CREATE Consortium & University of Naples Federico II, Naples, Italy

Acronyms

1D	One-dimensional
2D	Two-dimensional
3D	Three-dimensional
AABB	Axis-aligned bounding boxes
AR	Augmented reality
BnB	Ball-and-beam
BRISK	Binary robust invariant scalable keypoints
B-spline	Basis spline
BVP	Boundary value problem
CGAL	Computational geometry algorithms library
CLIK	Closed-loop inverse kinematics
CoM	Center of mass
CPU	Central processing unit
CUDA	Compute unified device architecture
DCC	Differential contact constraint
DC-PBD	Doubly-constrained position-based dynamics
DFSPH	Divergence-free smooth-particle hydrodynamics
DLO	Deformable linear object
DoD	Disk-on-disk
DoF	Degree of freedom
DVH	Diffused vortex hydrodynamics
EFL	Exact feed-forward linearization
FAST	Features from accelerated segment test
FDM	Finite difference method
FEM	Finite element method
FLin	Feedback linearization
FLIP	Fluid implicit particle
FTL	Follow the leader
FVM	Finite volume method
Gb	Gigabyte
GHz	Gigahertz

GMM	Gaussian mixture model
GPU	Graphics processing unit
GSM	Gradient smoothing method
ICP	Iterative closest point
IDA	Interconnection and damping assignment
IISPH	Implicit incompressible smooth-particle hydrodynamics
ISPH	Incompressible smooth-particle hydrodynamics
K-BKZ	Kaye–Bernstein-Kearsley-Zapas
KE-ME	Kinetic energy matching equation
k-NN	k-nearest neighbours
LUT	Lookup table
MAC	Marker-and-cell
MFC	Microsoft foundation class library
MJED	Multiplicative Jacobian energy decomposition
MPM	Material point method
MSER	Maximally stable extremal regions
MuJoCo	Multi-joint dynamics with contact
NURBS	Non-uniform rational B-spline
ODE	Ordinary differential equation
OS	Operating system
PBC	Passivity-based control
PBDM	Position-based dynamics method
PBFM	Position-based fluid method
PCA	Principal component analysis
PCG	Preconditioned conjugate gradient
PCISPH	Predictive-corrective incompressible smooth-particle hydrodynamics
PDE	Partial differential equation
PE-ME	Potential energy matching equation
pH	Port-Hamiltonian
PIC	Particle-in-cell
PID	Proportional–integral–derivative
PSPH	Projection smoothed-particle hydrodynamics
QR	Quick response
RGB	Red Green Blue
RGB-D	Red Green Blue Depth
RMSE	Root-mean-square error
SIFT	Scale-invariant feature transform
SOFA	Simulation Open Framework Architecture
SPH	Smoothed-particle hydrodynamics
S-R-S	Spherical-Revolute-Spherical
SURF	Speeded up robust features
TWMM	Two-wheeled mobile manipulator
WCSPH	Weak compressible smooth-particle hydrodynamics
ZMP	Zero moment point

Common Symbols

$[\mathbf{f}(\mathbf{x}), \mathbf{g}(\mathbf{x})]$	Lie bracket of two vector fields $\mathbf{f}(\mathbf{x})$ and $\mathbf{g}(\mathbf{x})$, with $\mathbf{x} \in \mathbb{R}^n$
$\otimes$	Kronecker product of matrices
∇	Divergence operator
$\|\mathbf{x}\|_{\mathbf{A}} = \mathbf{x}^T\mathbf{A}\mathbf{x}'$	Short way for the quadratic form
$\|\cdot\|$	Cartesian norm
0_n	The zero vector of dimension n
$\mathbf{A} \in \mathbb{R}^{n \times m}$	$\mathbf{A}$ is a $(n \times m)$ matrix of real numbers
$ad_{\mathbf{f}}^k \mathbf{g}(\mathbf{x})$	Short way to indicate the Lie bracket $[\mathbf{f}(\mathbf{x}), ad_{\mathbf{f}}^{k-1}\mathbf{g}(\mathbf{x})]$, with $k > 0$
$diag([a_1 \cdots a_n])$	Definition of a diagonal matrix through a vector specifying its main diagonal
$\mathbf{I}_n \in \mathbb{R}^{n \times n}$	Identity matrix of dimension n
$L_{\mathbf{x}}\lambda(\mathbf{x})$	Lie derivative of a real valued function $\lambda(\mathbf{x})$ and a vector field $\mathbf{f}(\mathbf{x})$, with $\mathbf{x} \in \mathbb{R}^n$
$\mathbf{O}_n$	The square zero matrix of dimension n
$\mathbf{O}_{n \times m}$	The $(n \times m)$ zero matrix
$\mathbb{R}$	Set of real numbers
$SE(3)$	Special Euclidean group of dimension three
$\mathfrak{so}(3)$	Lie algebra of $SO(3)$
$SO(3)$	Special orthogonal group of dimension three
$t \in \mathbb{R}$	Time variable
$\mathcal{W}$	World-fixed inertial frame
$<\mathbf{x}, \mathbf{y}>$	Another way to indicate $\mathbf{x}^T\mathbf{y}$, where $\mathbf{x}, \mathbf{y} \in \mathbb{R}^n$
$\mathbf{x} \times \mathbf{y}$	Cross product between the vectors $\mathbf{x}, \mathbf{y} \in \mathbb{R}^2$ or $\mathbf{x}, \mathbf{y} \in \mathbb{R}^3$
$\mathbf{x} \in \mathbb{R}^n$	$\mathbf{x}$ is a $(n \times 1)$ vector of real numbers

Perception

Deformation Modelling for a Physics-Based Perception System

Antoine Petit, Vincenzo Lippiello, and Bruno Siciliano

Abstract Manipulation tasks require capturing the deformations undergone by the manipulated object. In order to handle various sorts and magnitudes of deformations, physical deformation models have been employed based on FEM. By resorting to linear elasticity with a volumetric formulation, these models simulate large strains while handling volumetric effects. For perception, such modelling provides physically consistent regularization to registered deformations, especially over areas unobserved by the sensing system. Besides elasticity modelling, fracture models are investigated to cope with potential cracks or tears of the manipulated objects as well as collision models to deal with multiple interacting objects or interactions with the environment. Also, these models have proven to be computationally efficient and are thus suitable for dynamic manipulation. This chapter is based on the works presented in [1, 2].

1 Brief Introduction

Unlike vision-based tracking problems with rigid objects, for which a certain maturity has been reached, perception for non-rigid objects is still a challenging problem. It has aroused much interest in recent years in computer vision, computer graphics [3, 4], and robotics community [5]. Many potential applications would indeed be targeted in augmented reality, medical imaging, and robotic manipulation by handling a vast variety of objects: tissues, paper, rubber, viscous fluids, cables, food, organs, and so on. For instance, in the RoDyMan project, a demonstration scenario is the

A. Petit (✉)
Therenva, 74F rue de Paris, 35000 Rennes, France
e-mail: antoine.a.petit@inria.fr

V. Lippiello · B. Siciliano
Department of Electrical Engineering and Information Technology, PRISMA Lab, CREATE Consortium & University of Naples Federico II, Via Claudio 21, 80125 Naples, Italy
e-mail: vincenzo.lippiello@unina.it

B. Siciliano
e-mail: bruno.siciliano@unina.it

B. Siciliano and F. Ruggiero (eds.), *Robot Dynamic Manipulation*,
Springer Tracts in Advanced Robotics 144,
https://doi.org/10.1007/978-3-030-93290-9_1

Table 1 Main symbols used in this chapter

Definition	Symbol
Position of a generic point of the element e in the world frame $\mathcal{W}$	$\mathbf{x} \in \mathbb{R}^3$
Deformation field	$\mathbf{u}_e(\mathbf{x}) \in \mathbb{R}^3$
Initial coordinates of the four vertices of the element e in the world frame $\mathcal{W}$	$\mathbf{x}_{e,0} \in \mathbb{R}^{12}$
Coordinates of the four vertices of the element e in the world frame $\mathcal{W}$	$\mathbf{x}_e \in \mathbb{R}^{12}$
Cauchy's linear symmetric strain tensor using Voigt notation	$\boldsymbol{\epsilon}_e \in \mathbb{R}^6$
Infinitesimal stress tensor using Voigt notation	$\boldsymbol{\sigma}_e \in \mathbb{R}^6$
Internal elastic forces on the four vertices of the element e	$\mathbf{f}_e \in \mathbb{R}^{12}$
Stiffness matrix of the element e	$\mathbf{K}_e \in \mathbb{R}^{12\times12}$
Block-diagonal matrix containing four copies of the rotation matrix in $SO(3)$ corresponding to the rotational component of the deformations of the element e	$\mathbf{R}_e \in \mathbb{R}^{12\times12}$

humanoid dual-arm/hand manipulation of the pizza dough, which requires complex dynamic manipulation tasks of a deformable object. Research has increasingly focused on robots involved in food manipulation tasks in recent years [6–9]. The perception system aims to provide the robot controller accurate, robust, and real-time sensing of the manipulated deformable object. With respect to rigid objects, dealing with deformations poses several additional challenges, such as modelling the considered material properties and fitting this model with the vision and/or range data. This registration problem also involves critical real-time concerns, which are primarily required for robotic dynamic manipulation. Although numerous studies have proposed efficient real-time techniques to handle 3D surfaces (paper, clothes) which undergo isometric or slightly elastic deformations, a large open field remains when considering more significant elastic deformations. Thus, the aim is to propose a real-time tracking system able to handle elastic objects, potentially textureless, by tracking large deformations and fast rigid motions, using visual and range data provided by an RGB-D sensor, as described in the next chapter. Our approach involves physical modelling of the considered object to cope with deformations by relying on a FEM. The considerable progresses recently made within the computer graphics and medical simulation domains have enabled real-time performance for processing such models.

In this chapter, we first introduce in Sect. 2 the various techniques to model deformations, and also topological events such as tearing, fractures, splitting, or merging. These numerical methods have been widely studied in the domains of computer graphics [3, 10] and medical simulation [11].

Based on this background, in Sects. 3, 4, and 5 we describe in more details the elastic, fracture, and collision models we have employed for the RoDyMan project to address, in a physics-based manner, the perception task when these behaviours are encountered.

2 Background on Deformation Models

Modelling deformations can be handled through different approaches: mesh-based heuristic or physically realistic approaches, and mesh-free methods such as particle systems.

2.1 Mesh-Based Approaches

For mesh-based heuristic approaches, potentially relying on simple heuristical physical modeling, can be distinguished free-form geometrical mesh models, such as Bezier curves, deformable splines, active contours [12, 13], or NURBS. These representations implicitly take into account elastic or stretching constraints in an energy function to minimize. Other models are explicitly based on approximate physical properties. In this category, a famous and widespread model is the planar or volumetric mass-spring model. The mesh for this model is made of discrete points or nodes assigned with masses and which are linked by springs, using triangular shapes [14], T-shapes [15] for planar models, or spread over cubic elements for volumetric models [10, 16]. In order to simulate the deformations of the modeled object, because of external forces and internal forces between the nodes, the displacement of each node is computed by applying Newton's second law, solving the resulting differential equations, usually using Euler of Runge-Kutta integration methods. The simplicity and low computational complexity of surface spring-mass models have proven their efficiency on planar deformable objects showing elastic properties. However, when dealing with volumetric objects, the surface mass-spring model often fails to maintain a constant volume. The deformations do not propagate very fast, limiting the approach to small elastic deformations. With volumetric mass-spring models, a slightly better volumetric behavior, although with still slow propagation of deformations, can be obtained at the cost of significantly higher computational costs. Linked volume models, which are less employed, can also be mentioned. Instead, the FEM provides a realistic physical model by relying on continuous mechanics. In order to compute the deformation of the object, the basic idea is to solve Navier-Stokes or Euler PDEs, which continuously describe the deformations of the object and consist in balancing internal forces, resulting from viscosity and elastic properties of the material, and external forces, such as gravity or collisions. Then, solving these PDEs can be addressed by the FEM. It relies on finding numerical solutions of the equations by first discretizing the deformable object into a mesh made of elements, usually tetrahedrons, and performing numerical explicit or implicit integration over this mesh. A significant limitation of FEM approaches such as [17, 18] is their computational complexity since they involve solving non-linear equations involving the displacements and the forces at the nodes of the elements. Approximations of these equations are thus inevitable for real-time computations and simulations. In the simplest case, a static linear relation between forces and displacement can be derived by means of the stiffness matrix, which last involves the elastic properties of the considered element by depending on two parameters, namely, the Young's modulus and the Poisson's ratio. The methods solving these equations can then be operated in real-time [19]. However, they are still restricted to 100 relatively small elastic deformations. This drawback has been circumvented with co-rotational approaches [20, 21], or by using more complex tensors to model the stiffness, such as the non-linear St. Venant-Kirchhoff strain tensor [22], handling some hyper-elastic deformations, but with a more considerable computational burden. Other approaches [23] have been able to cope with more general hyper-elastic materials, potentially in real-time. More recently, [24] proposed the MJED, an algorithm relying on a total Lagrangian FEM to handle, in real-time, materials showing hyper-elasticity, porosity, and viscosity behaviours.

2.2 *Mesh-Free Approaches*

We can introduce a different line of approaches which does not consider a geometrical mesh to model the object but instead a system of particles involving more or less realistic physical properties. With particle-based methods, which have been primarily developed in the computer graphics community [3], the object is represented by a set of 3D points storing attributes such as position, velocity, physical properties, and so on. In an initial description [25], referred to as coupled spatial particle system, the particles are dynamically evolving in a stochastic manner. Their attributes are potentially related to the behavior of other particles, with static or dynamic interactions, to model topological changes, for instance. This simple model has enabled us to simulate fuzzy and inelastic objects such as clouds, water, and so on. Another particle-based approach, referred to as SPH [26], employs a more continuous method by interpolating particular quantities defined at discrete locations on the whole space, smoothing properties between particles. This approach has been widely used to simulate fluid, based on Navier-Stokes equations, the various internal forces resulting from pressure, viscosity, or surface tension being computed with smoothing functions. With respect to mesh (or grid)-based techniques such as FEM, SPH has several advantages by simplifying the fluid equations since mass conservation is guaranteed. The involved internal forces are computed from weighted contributions from other particles instead of solving systems of equations for FEM. Due to its higher simplicity and flexibility, SPH particle systems enable to model highly viscous, viscoelastic, or even plastic materials, as well as topological changes such as splitting or merging events [27–30]. However, compared to mesh-based approaches, SPH approaches have problems when maintaining the incompressibility of the material, and they require a large number of particles (with respect to the number of vertices for the mesh) to perform simulations with equivalent resolution. Nevertheless, real-time simulations can be performed in the approaches proposed in [27–29]. A challenge for particle-based methods is to recover the surface of the object. It can be done explicitly, through the surface reconstruction between the particles [27, 28], or implicitly, based on level sets [30], or using both explicit and implicit representations [29]. The level sets implicitly define the surface by signed values for points in the volume. These values generally correspond to a distance (positive when inside the fluid, negative otherwise) to the explicit surface. Level sets are especially suited to handle severe topological changes.

2.3 Hybrid Approaches

Coupling both mesh and particle-based methods combines the advantages of both ones. An example is the particle level set method [31, 32] which introduces particles in the vicinity of the mesh's surface. Such a coupling enables the accurate simulation of highly viscoelastic and plastic objects, as well as fluids with topological changes.

3 FEM Elastic Model

Among the different physical modelling methods mentioned above, we have adopted a mesh-based FEM model for the perception system of the RoDyMan platform. Besides modelling potentially large elastic volumetric deformations, FEM models can act as reliable regularizers for registration on noisy vision data, which mesh-free methods might not guarantee. They can also now achieve real-time performances, and implementation is eased by the availability of open-source simulators embedding such models. These models assume the prior knowledge of a consistent mesh (which could be automatically reconstructed offline) and the material properties (through the Young's modulus and the Poisson's ratio) of the observed and manipulated object.

3.1 Finite Element Modelling

As introduced above, the FEM provides a realistic physical model by relying on continuum mechanics instead of finite differences for mass-spring systems, for instance. For a detailed description, the reader can refer to [33]. It consists of tessellating the deformable object into a mesh made of elements. Here, we rely on a volumetric linear FEM approach with tetrahedral elements (see Fig. 1), which are preferred over other topologies such as triangles or hexahedrons, due to their modeling accuracy

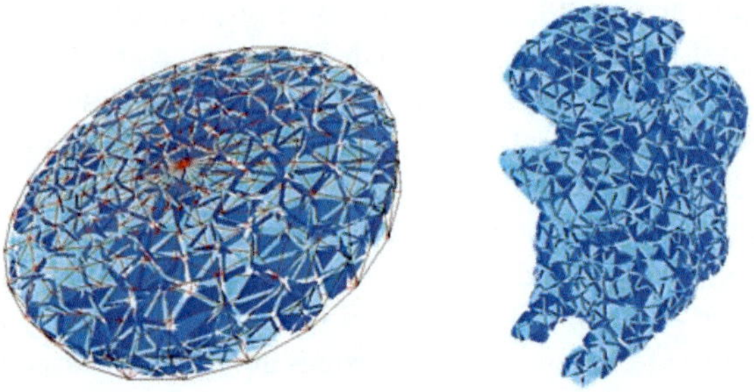

Fig. 1 Examples of tetrahedral meshes

(with respect to triangles, for instance), while being computationally efficient and convenient for meshing volumes with topologies of any complexity (with respect to hexahedrons).

The deformation field $\mathbf{u}_e$ over an element e is then approximated as a continuous interpolation of the displacement vectors $\hat{\mathbf{u}}_e$ of its four vertices, through the matrix $\mathbf{N}_e(\mathbf{x}) \in \mathbb{R}^{3\times 12}$, which contains the polynomial basis functions of the element

$$\mathbf{u}_e(\mathbf{x}) = \mathbf{N}_e(\mathbf{x})\hat{\mathbf{u}}_e, \quad (1)$$

where $\hat{\mathbf{u}}_e = \mathbf{x}_e - \mathbf{x}_{e,0}$.

3.2 *Linear Elasticity*

In order to model deformations and elasticity, we resort to the infinitesimal strain theory [33] to compute the Cauchy's linear strain tensor $\boldsymbol{\epsilon}_e$ within the tetrahedron. It can be linearly expressed with respect to $\hat{\mathbf{u}}_e$ as

$$\boldsymbol{\epsilon}_e = \mathbf{L}_e\hat{\mathbf{u}}_e, \quad (2)$$

with $\mathbf{L}_e \in \mathbb{R}^{6\times 12}$ a constant matrix. In order to relate the stress with the strain, we then rely on Hooke's law linear elasticity theory for a continuous isotropic material, which leads us to write the infinitesimal stress tensor $\boldsymbol{\sigma}_e$ in the element e as

$$\boldsymbol{\sigma}_e = \mathbf{C}_e\boldsymbol{\epsilon}_e, \quad (3)$$

where $\mathbf{C}_e \in \mathbb{R}^{6\times 6}$ symmetric matrix depending on two elastic parameters of the material, namely, the Young's modulus $E \in \mathbb{R}$ and the Poisson's ratio $\nu \in \mathbb{R}$. Folding (2) into (3) yields

$$\boldsymbol{\sigma}_e = \mathbf{C}_e\mathbf{L}_e\hat{\mathbf{u}}_e. \quad (4)$$

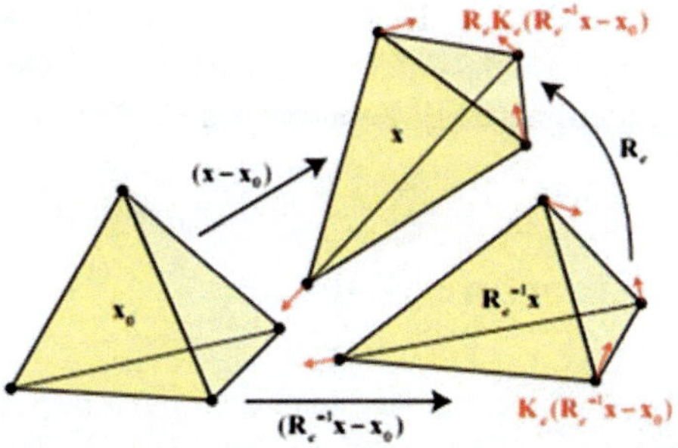

Fig. 2 The corotational approach

The strain energy in e can be computed as a function of $\boldsymbol{\sigma}_e$, $\boldsymbol{\epsilon}_e$, and the volume $V_e \in \mathbb{R}$ of the element. It can then be derived to determine the internal elastic forces $\mathbf{f}_e$ exerted on the four vertices of e. Based on (2) and (4), $\mathbf{f}_e$ can be linearly related to the displacements $\hat{\mathbf{u}}_e$ of the vertices as

$$\mathbf{f}_e = \mathbf{K}_e\hat{\mathbf{u}}_e = V_e\mathbf{L}_e^T\boldsymbol{\sigma}_e, \tag{5}$$

with $\mathbf{K}_e = V_e\mathbf{L}_e^T\mathbf{C}_e\mathbf{L}_e$, and $V_e \in \mathbb{R}$.

3.3 *The Corotational Approach*

The model using an infinitesimal approximation of the strain tensor, giving a constant $\mathbf{K}_e$ linearising the elastic forces, is insensitive to linear transformations. However, it is inaccurate when modeling large rotations of the elements causing, for instance, unexpected growth of volume. A workaround consists of the corotational approach proposed in [21, 34, 35], used for registration purposes in [36], which is a good compromise between the ability to model large elastic deformations and computational efficiency. The displacement of an element can be decomposed into a rigid rotation $\mathbf{R}_e$ and a pure deformation. As suggested in [35], we use a polar decomposition on the transformation matrix between the current and initial state of the vertices to extract $\mathbf{R}_e$. Then, the stiffness matrix $\mathbf{K}_e$ can be warped with respect to this rotation to accommodate rotation transformations, giving

$$\mathbf{f}_e = \mathbf{R}_e\mathbf{K}_e\hat{\mathbf{u}}_e^r = \mathbf{R}_e\mathbf{K}_e(\mathbf{R}_e^{-1}\mathbf{x}_e - \mathbf{x}_{e,0}), \tag{6}$$

where $\hat{\mathbf{u}}_e^r = \mathbf{R}_e^{-1}\mathbf{x}_e - \mathbf{x}_{e,0}$, with $\mathbf{R}_e^{-1}\mathbf{x}_e$ the back rotated deformed coordinates of the vertices of e to an unrotated frame. Notice that the forces $\mathbf{K}_e\hat{\mathbf{u}}_e^r$ are rotated to the current deformed element through $\mathbf{R}_e$. Figure 2 illustrates the principle of such a corotational approach.

In this way, the overall forces on the whole mesh can be summed to zero, while computational efficiency is ensured since $\mathbf{K}_e$ can be computed in advance, in contrast to non-linear FEM approaches.

4 Fracture Model

During deformable object manipulation, excessive stress may result in fractures of the object. Therefore, regardless of whether it might be a desirable effect, detecting and capturing such a topological event becomes essential for proper manipulation.

In the computer graphics community, various systems modeling fractures were proposed. The detection of fractures, or cracks, in 3D object simulation has been initially achieved using mass-spring systems [37] by removing springs whose lengths exceed a certain threshold. Based on the FEM, a physically consistent method using the linear elastic fracture theory [38] was proposed in [18] through the computation, for some given strains, of tensile forces undergone by the nodes. A fracture is then propagated from each detected fracturable node (or crack tip) by subdividing the neighboring tetrahedra according to a fracture plane. Similarly, a more computationally efficient solution was suggested in [21, 39] by resorting the fracture being propagated along element boundaries to element-wise stresses. Although improvements have been achieved from these widespread approaches, for both fracture detection and propagation [40], we propose here an approach based on [18]. One reason is the ability to cope appropriately with the provided noisy point cloud data continuously. Another reason is to preserve real-time performance.

4.1 Fracture Detection

Detecting fractures in the mesh is performed similarly to [18], by decomposing the internal forces exerted on the nodes into tensile and compressive forces, with the difference that [18] relies on a standard linear FEM method. For this purpose, the stress tensor undergone by a tetrahedron is separated into tensile and compressive components. Here, using a corotational approach, we compute $\boldsymbol{\sigma}_e$ as

$$\boldsymbol{\sigma}_e = \mathbf{C}_e \mathbf{L}_e (\mathbf{R}_e^{-1} \mathbf{x}_e - \mathbf{x}_{e,0}). \tag{7}$$

Then, the three eigenvalues, $\nu_i(\boldsymbol{\sigma}_e) \in \mathbb{R}$, and eigenvectors, $\mathbf{n}_i(\boldsymbol{\sigma}_e) \in \mathbb{R}^3$, associated to $\boldsymbol{\sigma}_e$ are computed. Positive eigenvalues correspond to tensile stresses, whereas negative eigenvalues correspond to compressive stresses. As in [18], tensile and compressive tensors can be then defined as

$$\boldsymbol{\sigma}_e^+ = \sum_{i=1}^{3} \max(0, \nu_i(\boldsymbol{\sigma}_e))\mathbf{M}(\mathbf{n}_i(\boldsymbol{\sigma}_e)), \tag{8}$$

$$\boldsymbol{\sigma}_e^- = \sum_{i=1}^{3} \min(0, \nu_i(\boldsymbol{\sigma}_e))\mathbf{M}(\mathbf{n}_i(\boldsymbol{\sigma}_e)), \tag{9}$$

with

$$\mathbf{M}(\mathbf{a}) = \begin{cases} \dfrac{\mathbf{a}\mathbf{a}^T}{\|\mathbf{a}\|} & \text{if} \quad \mathbf{a} \neq \mathbf{0}_3, \\ \mathbf{O}_3 & \text{if} \quad \mathbf{a} = \mathbf{0}_3, \end{cases} \tag{10}$$

for any vector $\mathbf{a} \in \mathbb{R}^3$.

Based on these tensile and compressive stresses and using (5), the internal forces exerted on the vertices of e can be decomposed as

$$\mathbf{f}_e^+ = V_e \mathbf{L}_e^T \boldsymbol{\sigma}_e^+, \tag{11}$$

$$\mathbf{f}_e^- = V_e \mathbf{L}_e^T \boldsymbol{\sigma}_e^-. \tag{12}$$

Considering a mesh with $n_X > 0$ vertices, it is possible to define the set X of all these vertices, $\mathbf{x}_j \in \mathbb{R}^3$, with $j = 1, \ldots, n_X$. Then, as in [18], the separation tensor, $\boldsymbol{\zeta}_j \in \mathbb{R}^6$ can be computed for each node $\mathbf{x}_j$ in the mesh using the sets $\left\{\mathbf{f}_{e,j}^+ \mid j \in e\right\}_e$ and $\left\{\mathbf{f}_{e,j}^- \mid j \in e\right\}_e$ of tensile and compressive forces, respectively, exerted by the elements $\{e \mid j \in e\}$ attached to the node

$$\boldsymbol{\zeta}_j^- = \frac{1}{2}(\sum_e \mathbf{M}(\mathbf{f}_{e,j}^+) - \mathbf{M}(\mathbf{f}_j^+) - \sum_e (\mathbf{f}_{e,j}^-) - \mathbf{M}(\mathbf{f}_j^-)), \tag{13}$$

with

$$\mathbf{f}_j^+ = \sum_e \mathbf{f}_{e,j}^+ \ \text{and} \mathbf{f}_j^- = \sum_e \mathbf{f}_{e,j}^-. \tag{14}$$

The tensor $\boldsymbol{\zeta}_j$ enables to evaluate deformation imbalance between tensile and compressive forces, while being invariant to imbalance resulting in rigid motions. A fracture at node $\mathbf{x}_j$ is then detected if the largest positive eigenvalue $\nu_j^+ \in \mathbb{R}$ of $\boldsymbol{\zeta}_j$ is above a certain threshold corresponding to the toughness of the material. The eigenvector $\mathbf{n}_j^+ \in \mathbb{R}^3$, corresponding to ν_j^+, defines the normal to the fracture plane at node $\mathbf{x}_j$.

4.2 Fracture Propagation and Remeshing

The mesh around the considered node shall be modified to propagate the fracture from the detected fracturable nodes in the direction of the fracture plane. The elements attached to the node and intersected by the fracture plane are subdivided according to the plane's orientation [18]. Other neighboring elements also need to be modified to maintain the consistency of the mesh. However, such a re-meshing process, which is proper and suitable for simulations purposes since it manages to maintain the fracture orientation, is computationally costly and tends to generate a continuously growing number of elements, which is not desirable to keep real-time performance. Besides, for registration concerns, by changing locally and drastically the resolution of the mesh, matching the re-meshed regions with the acquired point cloud (which has a constant and homogeneous resolution while being noisy) may lead to heterogeneous or unstable behaviors. Another approach, suggested in [21, 39, 41], consists of propagating the fracture along the boundaries of the neighbouring elements. Although it enables to keep the resolution constant, this method tends to produce jumbled fracture patterns and artefacts, which is also not desirable in our case when coping with the noisy point cloud data.

We have instead opted here for a solution preserving the mesh homogeneity and performing a neat fracture propagation (see Fig. 3). As in [18], the fracturable node is first replicated with the same positions. Then, within the set of elements attached to the original fracturable node, the nodes which are not intersected by the given fracture plane are determined. They are thus re-assigned to the original fracturable node or the replicated one, based on which side of the plane they are located. This procedure appears coarse, affects the volume of the model, and tends to degenerate the mesh. However, it remains acceptable if we consider objects undergoing simple fracture events, so that element losses in the mesh can be neglected.

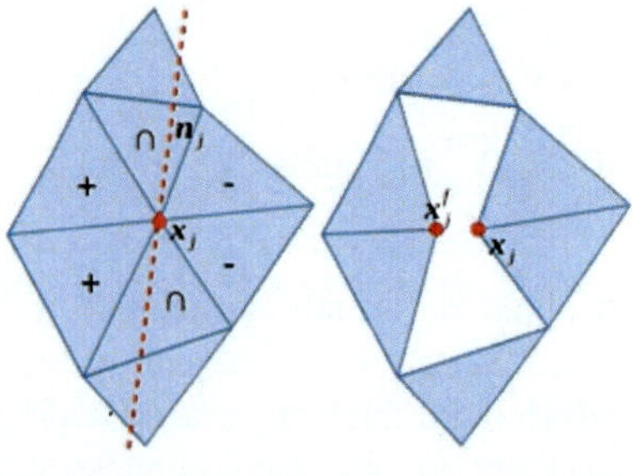

Fig. 3 Re-meshing procedure, here represented in the 2D case with triangles for simplicity. The fracturable node $\mathbf{x}_j$ is replicated, initially, with the same position providing a new vertex $\mathbf{x}_j^f$ for the mesh (it is here translated from $\mathbf{x}_j$ for clarity). The intersected elements are removed. The elements on the positive side of the fracture plane (determined by its normal $\mathbf{n}_j$) assigned to $\mathbf{x}_j$ and the others remain attached to $\mathbf{x}_j$

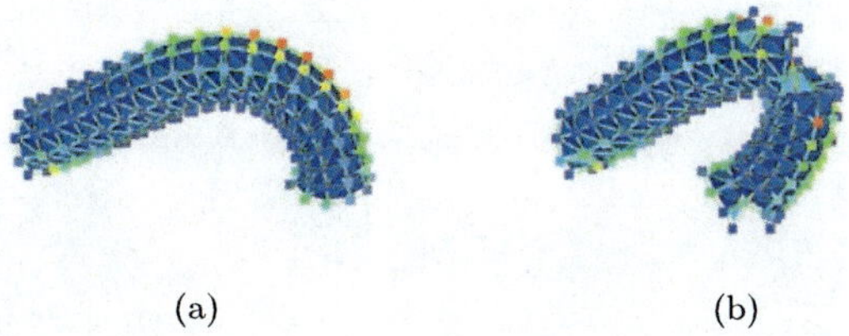

Fig. 4 Example of a fracture occurring on a cylindrical beam object

4.3 Comments on the Results

An example of a beam-like object being bent and fractured according to our fracture model is shown in Fig. 4. In Fig. 4a, we observe the fracture detection process with the value for each vertex of the fracture measure, i.e., the largest eigenvalue of the separation tensor, which is mapped to colours. In Fig. 4b, we can notice the fracture propagation step once fractures are detected.

5 Interaction Model

When manipulating multiple objects, objects interacting with their environments, or the manipulation tools, it becomes necessary to sense the potential contacts and the resulting deformations reliably. Here, we propose to use interaction models to handle these phenomenons.

Indeed, modelling the interactions between the considered objects constitutes a central issue to allow a joint registration of the deformations. We propose here a state-of-the-art model consisting of classical collision detection and collision response strategies. Collision being performed on surface meshes, we rigidly map each object's collision triangular surface mesh on their corresponding volumetric tetrahedral mesh.

5.1 Collision Detection

Collisions can be handled by employing different approaches to determine the intersections and the contacts between the interacting objects. Reviews of collision detection methods can be found in [42, 43]. Intersections are computed by performing successive distance tests based on a hierarchical set of bounding volumes around the polyhedral triangular collision meshes.

Here, we employ AABB hierarchies around the considered objects. First, as a broad phase, intersections between the coarsest AABBs are found. Then, as a narrow phase, we descend in the AABB hierarchies of the objects in a coarse-to-fine manner. Within the finest AABBs, a search for intersections between pairs of geometric primitives is performed. The primitives are triangles in our case. The computation of

Fig. 5 Illustration of the collision detection algorithm with the hierarchies of bounding volumes (left), and the computed pairs of collision points (middle, right), tested on the SOFA framework

intersections is based on a threshold on the distances between the primitives (triangle-to-triangle), and it returns corresponding pairs of collision (or contact) points on the colliding primitives, as illustrated in Fig. 5.

5.2 Collision Response

In order to model the contact forces between two colliding deformable objects i and j, we use a simple penalty-based method. For two detected collision points on the collision surfaces of the interacting objects, barycentric mappings on the tetrahedral meshes are carried out to determine two corresponding points $\mathbf{x}_i \in \mathbb{R}^3$ and $\mathbf{x}_j \in \mathbb{R}^3$, in both volumetric models. Then a penetration distance $\delta > 0$ can be defined as $\delta = d - D$, with $d = \|\mathbf{x}_i - \mathbf{x}_j\|$ and $D > 0$ a proximity threshold. As a collision normal $\mathbf{n}_{i,j}$, we choose $\mathbf{n}_{i,j} = \dfrac{\mathbf{x}_i - \mathbf{x}_j}{d}$.

A spring-like repulsive penalty force $\mathbf{f}^c_{j/i}(\mathbf{x}_i)$, exerted by $\mathbf{x}_j$ on $\mathbf{x}_i$, can then be computed as

$$\mathbf{f}^c_{j/i}(\mathbf{x}_i) = -k\delta\mathbf{n}_{i,j}, \tag{15}$$

with $k > 0$ a stiffness factor.

6 Discussion and Conclusion

This chapter introduced some background on physical simulation with FEM to regularize observed deformations in a perception system. Despite more complex and accurate models exist, relying on continuum mechanics and elasticity theory for deformation modelling offers some physical realism and the ability to deal with various magnitudes of deformations, with volumetric effects, which is not guaranteed by standard geometric and surface-based approaches. It requires only the knowledge of a couple of parameters (Young modulus and Poisson ratio) setting the mechanical properties of the involved object. However, the physical realism of the approach shall be mitigated since capturing the deformations in the considered perception system would consist in balancing internal forces from the FEM computations and external forces resulting from observations through the vision sensors. These latter forces have no relation with the object's real forces: the knowledge of the actual material properties does not have much significance. Deformation modelling acts as a regularizer for the observed deformations and shall not be considered as an accurate model for other purposes such as control or planning. The use of simple fracture and collision models also provides tools to handle topological changes and multiple interacting objects. The efficiency of the solvers and the implementation, through the SOFA platform, make these models suitable for real-time perception.

References

1. A. Petit, S. Cotin, V. Lippiello, B. Siciliano, Capturing deformations of interacting non-rigid objects using RGB-D data, in *2018 IEEE/RSJ International Conference on Intelligent Robots and Systems*, pp. 491–497, Madrid, E (2018)
2. A. Petit, V. Lippiello, B. Siciliano. Tracking fractures of deformable objects in real-time with an RGB-D sensor, in *2015 International Conference on 3D Vision*, pp. 632–639, Lyon, F (2015)
3. A. Nealen, M. Müller, R. Keiser, E. Boxerman, M. Carlson, Physically based deformable models in computer graphics. Comput. Gr. Forum **25**(4), 809–836 (2006)
4. J. Bender, M. Müller, M.A. Otaduy, M. Teschner, M. Macklin, A survey on position-based simulation methods in computer graphics. Comput. Gr. Forum **33**(6), 228–251 (2014)
5. C. Smith, Y. Karayiannidis, L. Nalpantidis, X. Gratal, P. Qi, D.V. Dimarogonas, D. Kragic, Dual arm manipulation: A survey. Robot. Auton. Syst. **60**(10), 1340–1353 (2012)
6. M. Tenorth, M. Beetz, A unified representation for reasoning about robot actions, processes, and their effects on objects, in *2012 IEEE/RSJ International Conference on Intelligent Robots and Systems*, pp. 1351–1358 (2012)
7. M. Bollini, S. Tellex, T. Thompson, N. Roy, D. Rus, Interpreting and executing recipes with a cooking robot, in *Experimental Robotics* (Springer, 2013), pp. 481–495
8. P. Kormushev, S. Calinon, D.G. Caldwell. Robot motor skill coordination with EM-based reinforcement learning, in *2010 IEEE/RSJ International Conference on Intelligent Robots and Systems*, pp. 3232–3237 (2010)
9. M.C. Gemici, A. Saxena, Learning haptic representation for manipulating deformable food objects, in *2014 IEEE/RSJ International Conference on Intelligent Robots and Systems*, pp. 638–645 (2014)
10. S.F.F. Gibson, B. Mirtich, *A survey of deformable modeling in computer graphics* (Technical report, Citeseer, 1997)
11. U. Meier, O. López, C. Monserrat, M.C. Juan, M. Alcaniz, Real-time deformable models for surgery simulation: A survey. Comput. Methods Prog. Biomed. **77**(3), 183–197 (2005)
12. D. Terzopoulos, J. Platt, A. Barr, K. Fleischer, Elastically deformable models. ACM Siggraph Comput. Gr. **21**(4), 205–214 (1987)
13. D. Terzopoulos, A. Witkin, M. Kass, Constraints on deformable models: Recovering 3D shape and nonrigid motion. Artif. Intell. **36**(1), 91–123 (1988)
14. H. Delingette, G. Subsol, S. Cotin, J. Pignon, Craniofacial surgery simulation testbed. Vis. Biomed. Comput. **1994**, 607–618 (1994)
15. E. Keeve, S. Girod, B. Girod, Craniofacial surgery simulation, in *Visualization in Biomedical Computing 1996* (Springer, 1996), pp. 541–546

16. Y. Chen, Q.-H. Zhu, A. Kaufman, S. Muraki, Physically-based animation of volumetric objects, in *Computer Animation '98*, pp. 154–160 (1998)
17. G. Debunne, M. Desbrun, M.-P. Cani, A.H. Barr. Dynamic real-time deformations using space & time adaptive sampling, in *28th Annual Conference on Computer Graphics and Interactive Techniques*, pp. 31–36 (2001)
18. J.F. O'Brien, J.K. Hodgins, Graphical modeling and animation of brittle fracture, in *26th Annual Conference on Computer Graphics and Interactive Techniques*, pp. 137–146 (1999)
19. S. Burion, F. Conti, A. Petrovskaya, C. Baur, O. Khatib, Identifying physical properties of deformable objects by using particle filters, in *2008 IEEE International Conference on Robotics and Automation*, pp. 1112–1117 (2008)
20. C.A. Felippa, B. Haugenm A unified formulation of small-strain corotational finite elements: I. Theory. Comput. Methods Appl. Mech. Eng. 194(21), 2285–2335 (2005)
21. M. Müller, M. Gross, Interactive virtual materials. In Graphics. Interface **2004**, 239–246 (2004)
22. G. Picinbono, H. Delingette, N. Ayache, Real-time large displacement elasticity for surgery simulation: Non-linear tensor-mass model, in *Medical Image Computing and Computer-Assisted Intervention*, pp. 643–652 (2000)
23. J. Teran, S. Blemker, V. Hing, R. Fedkiw. Finite volume methods for the simulation of skeletal muscle, in *2003 ACM SIGGRAPH/Eurographics Symposium on Computer Animation*, pp. 68–74 (2003)
24. S. Marchesseau, T. Heimann, S. Chatelin, R. Willinger, H. Delingette, Fast porous visco-hyperelastic soft tissue model for surgery simulation: Application to liver surgery. Prog. Biophys. Mol. Biol. **103**(2), 185–196 (2010)
25. W.T. Reeves, Particle systems-a technique for modeling a class of fuzzy objects. ACM SIGGRAPH Comput. Gr. **17**(3), 359–375 (1983)
26. J.J. Monaghan, Smoothed particle hydrodynamics. Ann. Rev. Astron. Astrophys. **30**(1), 543–574 (1992)
27. S. Clavet, P. Beaudoin, P. Poulin, Particle-based viscoelastic fluid simulation, in *2005 ACM SIGGRAPH/Eurographics Symposium on Computer Animation*, pp. 219–228 (2005)
28. M. Müller, D. Charypar, M. Gross, Particle-based fluid simulation for interactive applications, in *ACM Eurographics/SIGGRAPH Symposium on Computer Animation*, pp. 154–159 (2003)
29. M. Müller, R. Keiser, A. Nealen, M. Pauly, M. Gross, and M. Alexa. Point based animation of elastic, plastic and melting objects. In *2004 ACM SIGGRAPH/Eurographics Symposium on Computer Animation*, pages 141–151, 2004
30. S. Premžoe, T. Tasdizen, J. Bigler, A. Lefohn, R.T. Whitaker, Particle-based simulation of fluids. Computer Graphics Forum **22**(3), 401–410 (2003)
31. D. Enright, R. Fedkiw, J. Ferziger, I. Mitchell, A hybrid particle level set method for improved interface capturing. J. Comput. Phys. **183**(1), 83–116 (2002)
32. T.G. Goktekin, A.W. Bargteil, J.F. O'Brien, A method for animating viscoelastic fluids. ACM Transactions on Graphics **23**(3), 463–468 (2004)
33. R.D. Cook. *Finite Element Modeling for Stress Analysis*. Wiley, 1994
34. O. Etzmuß, M. Keckeisen, and W. Straßer. A fast finite element solution for cloth modelling. In *11th Pacific Conference on Computer Graphics and Applications*, pages 244–251, 2003
35. M. Nesme, Y. Payan, and F. Faure. Efficient, physically plausible finite elements. In *Eurographics*, 2005
36. N. Haouchine, J. Dequidt, I. Peterlik, E. Kerrien, M.-O. Berger, and S. Cotin. Image-guided simulation of heterogeneous tissue deformation for augmented reality during hepatic surgery. In *2013 IEEE International Symposium on Mixed and Augmented Reality*, pages 199–208, 2013
37. A. Norton, G. Turk, B. Bacon, J. Gerth, P. Sweeney, Animation of fracture by physical modeling. Vis. Comput. **7**(4), 210–219 (1991)
38. T.L. Anderson. *Fracture Mechanics: Fundamentals and Applications*. CRC Press, 2005
39. M. Müller, L. McMillan, J. Dorsey, and R. Jagnow. *Real-time Simulation of Deformation and Fracture of Stiff Materials*. Springer, 2001

40. L. Glondu, M. Marchal, G. Dumont, Real-time simulation of brittle fracture using modal analysis. IEEE Trans. Visual Comput. Graphics **19**(2), 201–209 (2013)
41. J. Smith, A. Witkin, D. Baraff, Fast and controllable simulation of the shattering of brittle objects. Computer Graphics Forum **20**(2), 81–91 (2001)
42. P. Jiménez, F. Thomas, C. Torras, 3D collision detection: A survey. Computers & Graphics **25**(2), 269–285 (2001)
43. M. Teschner, S. Kimmerle, B. Heidelberger, G. Zachmann, L. Raghupathi, A. Fuhrmann, M.-.P Cani, F. Faure, N. Magnenat-Thalmann, W. Strasser, and P. Volino. Collision detection for deformable objects. *Computer Graphics Forum*, 24:61–81, 2005

Non-rigid Tracking Using RGB-D Data

Antoine Petit, Vincenzo Lippiello, and Bruno Siciliano

Abstract In this chapter, the real-time non-rigid tracking system to continuously estimate the deformations of the manipulated objects is described, using visual and range data provided by an RGB-D sensor. Based on the models described in the previous chapter, the method enables to deal with various deformations (elastic and plastic), fractures, and contacts, while ensuring physical consistency, handling rigid motions, occlusions, and addressing these tasks in real-time. It relies on a prior visual segmentation of the object in the RGB image. The mesh is registered first in a rigid manner with a classical ICP algorithm between the visible surface of the mesh and the segmented point cloud. A non-rigid fitting phase is then performed by determining geometrical point-to-point correspondences with the point cloud, used to compute external forces exerted on the mesh. Deformations are computed by solving mechanical equations balancing these external forces with internal forces provided by the FEM models. A technique to estimate the elastic parameters of the object is proposed by minimizing a fitting error between the simulated deformations, actuated by the input operator force provided by a force sensor, and the deformations captured by the RGB-D camera. Conversely, estimating a contact force exerted on the object can be carried out using point cloud data by minimizing the deviation between the registered and the simulated deformations. The system has been evaluated on synthetic and real data, with various objects, deformation, and interaction scenarios, and by integrating it into manipulation experiments on the RoDyMan humanoid robotic platform. This chapter is based on the works presented in [1–3].

A. Petit (✉)
Therenva, 74F rue de Paris, 35000 Rennes, France
e-mail: antoine.a.petit@inria.fr

V. Lippiello · B. Siciliano
Department of Electrical Engineering and Information Technology, CREATE Consortium & University of Naples Federico II, PRISMA Lab, Via Claudio 21, 80125 Naples, Italy
e-mail: vincenzo.lippiello@unina.it

B. Siciliano
e-mail: bruno.siciliano@unina.it

B. Siciliano and F. Ruggiero (eds.), *Robot Dynamic Manipulation*,
Springer Tracts in Advanced Robotics 144,
https://doi.org/10.1007/978-3-030-93290-9_2

1 Brief Introduction

The problem of capturing the deformations of non-rigid objects is an active research field in computer vision that finds prospective applications in robotic manipulation, human/machine, tangible interactions, and motion capture.

For the perception system of the RoDyMan platform, this objective of sensing deformations led to handle several challenges such as modelling the considered manipulated objects and their material and aligning the designed models with the vision and range data provided by an RGB-D sensor.

As introduced in the previous chapter, a mechanical model based on the FEM was employed. The recent suitability of these models for real-time registration and tracking problems, as demonstrated by promising approaches [4–7], has corroborated our choice. Thus, we assume the prior knowledge of consistent meshes of the considered objects (which could be automatically reconstructed offline) and their material properties (through the Young's modulus and the Poisson's ratio). As a vision sensor, we rely on an off-the-shelf RGB-D camera.

The deformable registration and tracking method that we propose consists of fitting in real-time, frame-by-frame, the known mesh of the considered object with the point cloud data provided by an RGB-D sensor. The basic idea is to derive

Table 1 Main symbols used in this chapter

Definition	Symbol
Input image	$\mathcal{I}$
Number of pixels	$N > 0$
Binary pixel label	α_i
Set of the binary pixel pixel labels	$\alpha = \{\alpha_i, i = 1, \ldots, N\}$
Pixel of the image	$\mathbf{p}_i \in \mathbb{R}^2$
Input depth map	$\mathcal{D}$
Segmented image	$\mathcal{I}^s$
Segmented depth map	$\mathcal{D}^s$
Number of points in the cloud	$N_Y > 0$
3D point in the sensor frame	$\mathbf{y}_j \in \mathbb{R}^3$
Segmented point cloud	$\mathcal{Y} = \{\mathbf{y}_j, j = 1, \ldots, N_Y\}$
Number of vertices of the mesh	$N_X > 0$
Coordinates of a mesh's vertex	$\mathbf{x}_j \in \mathbb{R}^3$
Set of the vertices of the mesh	$\mathcal{X} = \{\mathbf{x}_j : j = 1, \ldots, N_X\}$
Set of the vertices of the visible surface	$\mathcal{X}_V$
Size of $\mathcal{X}_V$	$N_{\mathcal{X}_V} > 0$
Vector collecting all the vertices of the mesh	$\mathbf{x} \in \mathbb{R}^{3N_X}$
Young's modulus	$E \in \mathbb{R}$
Poisson's ratio	$\nu \in \mathbb{R}$

external forces exerted by the point cloud on the mesh and balance these forces with the internal forces computed from the FEM physical model by solving mechanical equations for the deformations. The FEM model thus acts as a regularizer over the computed deformations.

The external forces are computed from geometrical point-to-point correspondences between the point cloud and the mesh, relaxing the assumption of having a textured object or an object with a rough surface, for which 2D or 3D key points can be extracted and matched. We assume the mesh to be manually initialised correctly.

Based on this framework, other physical behaviours have been addressed besides pure elastic deformations, such as capturing fractures undergone by objects and interactions, which constitute an open research area in the field of 3D computer vision. For fractures, our approach involves the FEM physical model for fractures described in the previous chapter.

Several challenges must be overcome when considering multiple interacting objects to estimate the non-rigid objects' deformations jointly. The main challenge regards inter-occlusions between objects and deformations due to contacts. Tracking the different objects independently would result in inter-penetrations and physically unrealistic behaviours between objects due to inaccuracies and ambiguities in the visual or point cloud information. Instead, a simultaneous registration method relying on modelling the contacts between the objects would inherently handle these issues and reach a reliable behaviour.

As direct applications of this perception system for the RoDyMan project, we show its relevance when performing a simple planning task on the RoDyMan robotic platform (see Sect. 10). A method to estimate a point-wise force exerted on a deformable object is also described in Sect. 11.3 by using the external RGB-D sensor to confront the deformations measured by the registration technique with simulated deformations generated with respect to the force to be estimated. This method requires an accurate preliminary estimation of elasticity parameters of the material (Sect. 11), also based on the deviation between simulated deformations, generated from a measured force, and observed deformations, using RGB-D point clouds.

2 Related Work and Motivations

In this section, various works related to registration and tracking of non-rigid objects using vision sensors are reported, considering 3D/2D and 3D/3D modalities and the various estimation strategies to register isometric, elastic deformations, topological changes, and interactions, with regards to the motivations of the system designed for the RoDyMan project.

In the literature, the various approaches proposed to register deformable objects using vision and/or range data could be classified according to the underlying model of the considered object and its physical realism. Let us first clarify our scope and distinguish it from non-rigid reconstruction methods, for which at each frame provided by the vision/range sensor a single mesh is reconstructed as in [8, 9]. Instead, here the goal is to continuously estimate the rigid transformations and the deformations undergone by the object, modelled by a known mesh.

2.1 Registration Using Implicit Physical Modeling

Based on implicit physical models, approaches in [10–12] use a 1D parametric curve or 2D splines models (B-splines, radial basis functions) to track deformable objects in monocular images. This method relies on minimizing an energy function, involving an external energy term related to some image features, and an internal energy term regularizing curvature, bending, or twisting, compelling the model to vary smoothly. Adapting these techniques to register with 3D shapes or surfaces in monocular images is much more complex since 3D deformations can imply ambiguous 2D transformations, resulting in an underconstrained problem. A first attempt in [13], relying on 3D splines and inspired by the work in [11], densely processes gradient features to compute the data energy term. Less ambiguous feature-based approaches [14] have been preferred, and additional constraints are often added to solve ambiguities. With point cloud data, methods in [15, 16] employ an RGB-D sensor to register the acquired point cloud to a surface mesh by minimizing an error function accounting for geometric or direct depth and color errors and a stretching penalty function for the mesh. By means of a NURBS parametrization [15] or an optimized GPU implementation [16], real-time performance can be achieved. Although these two systems have shown promising and impressive results, they are still bounded to isometric or slightly elastic deformations, employing regularization functions proportional to squared distances between nodes of the mesh. In contrast, we wish to model elastic in a more physically realistic manner to handle volumetric effects and more significant strains. Another limitation of these methods is that they process mesh-to-input point cloud correspondences in their data error functions and are thus sensitive to missing data or unobserved areas of the considered object due to occlusions. We consider in this chapter also correspondences from the input point cloud to the mesh, through the use of a segmentation method, to restrict the input point cloud to the observed areas of the object, and based on these correspondences, the occluded or unobserved areas would not affect registration.

We can also mention template-free non-rigid reconstruction methods for which at each frame provided by the vision/range sensor, a single mesh of the sensed scene is reconstructed and tracked, as in [8, 9, 17, 18]. However, these dense and exhaustive methods, besides often failing fulfilling real-time constraints, are limited to isometric slightly elastic deformations.

2.2 Registration Using Explicit Physical Modeling

Instead, another formulation of the problem relies on explicit physics-based deformable models to perform registration. Concerning implicit methods, other sorts (such as non-linear elasticity) and magnitudes of deformations can be treated, inferring more consistently shape and/or volumetric regularization. Statistically, the solution can then be determined by setting internal and external forces equal or, equivalently, minimizing energy functions. Physics-based methods include discrete mass-spring-damper systems [4, 7, 19], or more straightforward approaches relying on the FEM, based on continuum mechanics. In [7], based on a mass-spring-damper system, 3D-3D correspondences, determined through a probabilistic inference, enable the computation of the external forces applied to the mesh. However, mass-spring systems are limited in elastic deformations' magnitude and do not guarantee specific mechanical properties, such as volume preservation, failing to handle volumetric effects. First attempts for registration employing the FEM for 3D surfaces in [20, 21] used linear elasticity FEM models. More recently, registration in monocular images is addressed in [22]by designing a stretching/shrinking energy using continuous mechanical constraints on 2D elements assuming linear elasticity and some 3D boundary conditions. A linear tetrahedral co-rotational FEM model, coping with more significant elastic deformations, is used in [6] through external forces being related to correspondences between tracked 3D feature points mapped to the 3D mesh employing a stereo camera system. This system has been then extended in [5] by involving a non-linear St-Venant Kirchoff FEM model to address hyperelastic behavior efficiently. At the same time, the registration was addressed through monocular keypoint-based external forces. This latter method proposes the most realistic physical elastic model within a real-time vision-based tracking system to the best of our knowledge. However, both approaches in [5] and [6] do not address issues such as capturing rigid motions and require textured materials.

2.3 Handling Topological Changes, Fractures and Cuts

Regarding the methods dealing with topological changes of a 3D deformable object, we can mention the 3D reconstruction methods in [23, 24] based on the *TransforMesh* system to address mesh evolution and topological changes. However, these are model-free methods, and they reconstruct a single mesh at each time step [24], or they are not suited for real-time applications [23]. More recently, some works [25, 26] addressed the issue of tracking cuts of deformable objects. They rely on the detection of changes in the configurations or structures of neighbouring visual features to predict a cut over a registered surface [26] or volumetric physics-based model [25], which is propagated on the considered meshes.

2.4 Multiple Objects Registration

Simultaneously estimating the 3D transformations undergone by multiple objects with vision data is already a well-studied problem for rigid or articulating objects. This consists of retrieving the 6D pose of all rigid objects or the various joint angles in articulated objects. Common approaches address the task by independently processing the different objects [27, 28]. Other works propose to register objects and capture interactions and cope with inter occlusions. In the case of articulated hand-hand, hand-object, and object-object interactions, methods are numerous [29–36]. In these works, the interactions are modelled through physics-based constraints, enabling to properly deal with collisions and occlusions, using single or multiple RGB or RGB-D cameras. However, to our knowledge, handling multiple deformable objects in interaction is quite a leftover problem. We can mention the system designed in [37], which introduces a collision detection process in its non-rigid reconstruction pipeline to deal with auto-collisions. However, this technique falls in the class of non-rigid surface reconstruction of the whole observed scene. Our method, instead, as a template tracking technique, would explicitly consider different known deformable interacting entities and in a volumetric manner.

2.5 Motivations and Contributions

Among the methods based on RGB-D or point cloud data having the closest goals, motivations, and constraints to ours, we can mention [6, 7, 15, 16]. Several contributions are proposed against them, such as handling various large deformations like elastic ones while ensuring physical consistency, handling rigid motions, occlusions, fractures, interactions, and addressing all these tasks with real-time or near real-time performance. For interactions, introducing the collision detection and response model into the registration framework has enabled to cope with contacts and with inter occlusions in the point cloud matching processes of the different considered objects. For this latter contribution, another improvement lies in the estimation of rigid transformations into the FEM resolution by integrating rigid ICP-based elastic forces. Another novelty has consisted of using the designed point cloud matching techniques and registration method for elasticity parameter and single contact force estimation and standard planning for manipulation on the RoDyMan platform.

2.6 Overview of the System

Our base tracking system for a single elastic object is represented in Fig. 1. As input data, we have the known 3D volumetric mesh of the object, a given RGB-D data. The procedure is sketched out as follows.

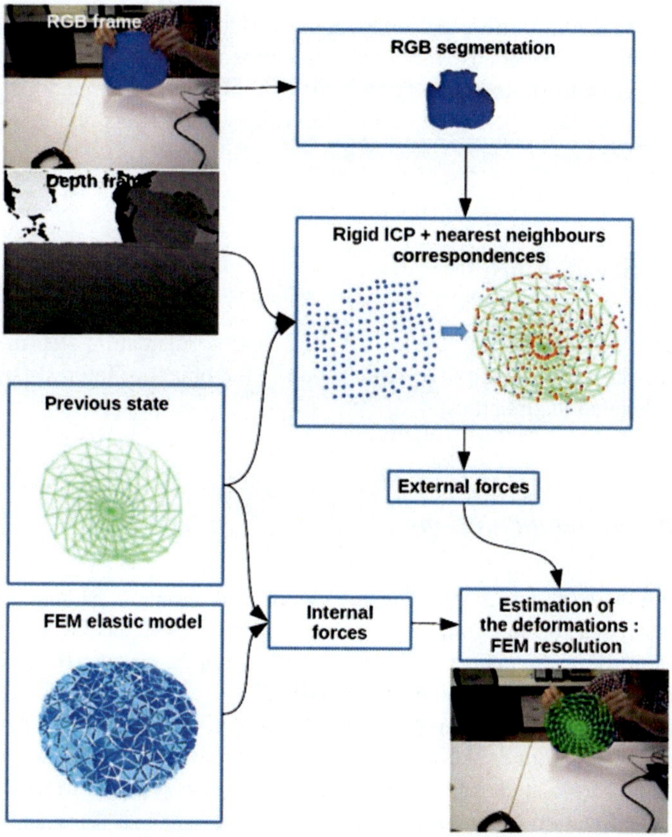

Fig. 1 Overview of our approach for deformable object tracking for a single object

1. Visual segmentation of the considered object, with a graph cut-based approach ensuring temporal coherence.
2. Using the resulted segmented point cloud, perform a rigid ICP to estimate a rigid transformation using closest point correspondences, from the point cloud to the mesh.
3. Using the resulting segmented point cloud, compute external linear elastic forces exerted on the vertices of the mesh from the point cloud to the mesh and conversely, based on closest point correspondences.
4. Numerical resolution of mechanical equations to compute the deformations, based on a tetrahedral linear co-rotational FEM model.

The different steps of this pipeline will be described in the following sections. The visual segmentation will be presented in Sect. 3, point cloud processing in Sect. 4, the rigid registration technique in Sect. 5, the non-rigid one in Sect. 6 and the resolution strategy will be introduced in Sect. 7.

Being the base for our different registration/tracking algorithms, these described methods will then be extended to the cases of fractures and interacting multiple object, as described respectively in Sects. 8 and 9. Finally, they will be applied for planning for robotic manipulation (Sect. 10), and for elasticity parameters and contact force estimation (Sect. 11).

3 Visual Segmentation

In this work, we advocate the use of a prior visual segmentation step in order to restrict each successively acquired point cloud to the object of interest (see Sect. 4 for a more detailed justification).

3.1 Grabcut Segmentation

We rely here on the efficient and widespread *GrabCut* method [38], based on graph cuts. In its original formulation, the *Grabcut* algorithm addresses the visual bilayer segmentation task as an energy minimization problem, based on statistical models of the foreground (the object) and the background.

For an input image $\mathcal{I}$, we denote within α, the set of the binary pixel pixel labels, $\alpha_i = 0$ for the background pixels and $\alpha_i = 1$ for the foreground. The labels' estimation, $\widehat{\alpha}$, can be formulated as the minimization of an energy-based Markov random field objective function, $E(\alpha)$, defined as

$$\widehat{\alpha} = arg \min_{\alpha} E(\alpha) \tag{1}$$

$$\text{with} \quad E(\alpha) = E_{data}(\alpha) + \gamma E_{smooth}(\alpha), \tag{2}$$

$$\text{and} \quad E_{data}(\alpha) = \sum_i U_i(\alpha_i), \tag{3}$$

where E_{data} is the data energy term, $U_i(\alpha_i)$ a unitary term accounting for the observation probability $p(\mathbf{p}_i \mid \alpha_i)$ for a pixel $\mathbf{p}_i$ to belong to the foreground or to the background, based on some image data (intensity, color, location, and so on) observed on the pixel, using the statistical models built for the background and the foreground. More formally, we have $U_i(\alpha_i) = -\log(p(\mathbf{p}_i \mid \alpha_i))$. Finally, E_{smooth} is the smoothness energy term whose goal is to favour smoothness or spatial coherence within the pixels.

In order to compute the optimal solution of the previous minimization problem and determine $\widehat{\alpha}$, a *graph cuts* minimization algorithm [39] is employed, providing the segmented frame $\mathcal{I}^s$.

Statistical models for the data energy function are GMMs based on color distributions, learned for both the foreground and background layers, which are initially determined by the user through a bounding box, manually defined around the foreground on the initial image. Besides, pixels outside this bounding box are definitely assigned to the background layer ($U_i(\alpha_i = 0) = \text{inf}$), whereas inside their label is unknown, so that energy minimization only has effects inside the bounding box.

3.2 Temporal Coherence and Real-Time Issues

Once the initial image is segmented through user interaction, the following frames are treated similarly, except that, instead of segmenting the whole frame, the area to segment effectively is updated frame-by-frame around the silhouette contour of the segmented area, providing temporal coherence in the segmentation process. More precisely, as shown in Fig. 2, the silhouette contour of the previous segmented foreground is extracted and the distance transform is computed over it, providing a signed distance map $\mathbf{d}$ to these contours (negative outside, positive inside). According to a fixed threshold, $d_t > 0$, on this distance map, we define a narrow strip around the contour. Such a narrow strip (grey area) has the property that $d_i < d_t$, with $d_i > 0$ is an element of $\mathbf{d}$, and in which the label of a pixel $\mathbf{p}_i$ is unknown, with $U_i(\alpha_i) = -\log(p(\mathbf{p}_i \mid \alpha_i))$. On the other hand, the inner side of the strip is assigned to the foreground (white area), if $d_i < -d_t$ and by setting $U_i(\alpha_i = 1) = inf$, otherwise it is assigned to the background (black area) if $d_i > d_t$ and by setting, in this case, $U_i(\alpha_i = 0) = inf$.

In this manner, temporal consistency is ensured since energy minimization is only effective within this strip, in the vicinity of the previous segmentation boundary, avoiding some outliers outside or inside, and reducing significant computations. Let us note that the GMMs are determined initially, making this segmentation process valid for sequences for which color distributions of both the foreground and the background are assumed relatively constant, which is a fair assumption for the robotic manipulation applications considered in this chapter.

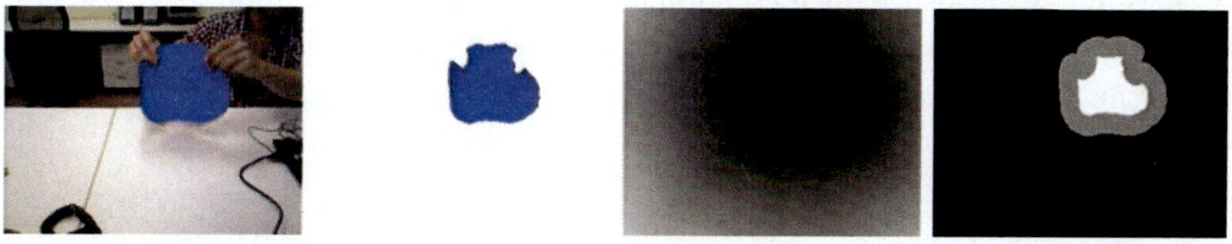

(a) Original frame (b) Segmentation (c) Distance map (d) Trimap

Fig. 2 Temporal consistency for segmentation. Segmentation will be effective on the strip (grey area on **d**) around the contour of the previous segmented frame (**b**), through the distance map to the contour (**c**)

4 Segmented and Sampled Point Cloud

We use the acquired RGB image sequence to visually segment the object of interest from its background and occlusions. Since we do not rely on some distinctive visual features, the point cloud provided by the depth sensor is indeed restricted to the considered object to avoid ambiguities in the matching process with the background or with occluding shapes, and be able to process correspondences from the input point cloud to the mesh. At frame k, using $\mathcal{I}^s$, a segmented depth map $\mathcal{D}^s$ is obtained by aligning and intersecting the original input depth map $\mathcal{D}$ with the segmented area in $\mathcal{I}^s$. Then, by back-projecting $\mathcal{D}^s$ in the sensor frame, the desired segmented point cloud $\mathcal{Y}$ is determined. For computational purposes, we limit the size of $\mathcal{Y}$ by firstly sampling $\mathcal{D}^s$ on a regular grid in the image plane.

5 Rigid Registration

The first step in our method is to register $\mathcal{Y}$ in terms of rigid translation and rotation transformations, initially considering the mesh of the object as rigid. Let us consider the set of vertices of the mesh $\mathcal{X}$, initially in their previous computed states, for the frame $k-1$. We suggest a classical rigid ICP algorithm [40] between $\mathcal{Y}$ and the vertices of the visible surface $\mathcal{X}_V$ of the mesh, transformed with respect to the previous RGB-D data. $\mathcal{X}_V$ is determined by performing a visibility test on the rendered depth map of the projected 3D mesh of the object, for the frame $k-1$. Through this procedure, which converges rapidly, fast rigid motions can be tracked and a fair initialisation for the non-rigid process can be obtained.

6 Point Cloud Matching for Non-rigid Registration

In order to register the segmented point cloud with the mesh in a non-rigid manner, we suggest an ICP-like procedure, given the sets $\mathcal{X}$ and $\mathcal{X}_V$ updated by the rigid transformation estimated as described above.

6.1 *Nearest Neighbor Correspondences*

By means of a k-dimensional tree search, nearest neighbour correspondences are determined both from the segmented point cloud to the visible surface of the mesh and from the visible surface of the mesh to the segmented point cloud. This step provides us with the sets of nearest neighbours $\mathcal{N}_{\mathcal{X}_V} = \{\mathbf{NN}_{\mathcal{Y}}(\mathbf{x}_j) : \mathbf{x}_j \in \mathcal{X}_V\}$ and

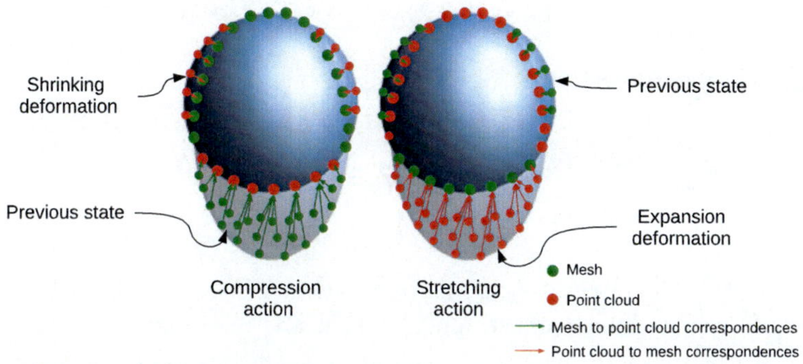

Fig. 3 Point cloud to mesh and mesh to point cloud correspondences. With shrinking deformations, mostly mesh to point cloud correspondences can attract the mesh towards the compressed point cloud, and conversely for expansion deformations

$\mathcal{N}_{\mathcal{Y}} = \{\mathbf{NN}_{\mathcal{X}}(\mathbf{y}_j) : j = 1, \ldots, N_Y\}$ in $\mathcal{Y}$ for $\mathcal{X}_V$, with the 1-NN function $\mathbf{NN}_{\mathcal{Y}}$, and in $\mathcal{X}_V$ for $\mathcal{Y}$, with the 1-NN function $\mathbf{NN}_{\mathcal{X}_V}$, respectively.

Both sets of correspondences are employed since relying on the sole geometrical proximity may lead to inconsistent matches using single point-to-point matches. Indeed, as illustrated in Fig. 3, from the segmented point cloud to the mesh, the correspondences enable to track expansion deformations under stretching forces, for which the observed segmented point cloud $\mathcal{Y}$ would spread over the visible surface of the mesh $\mathcal{X}_V$. The extended areas of $\mathcal{Y}$ with respect to $\mathcal{X}_V$ can be matched with the outer areas of $\mathcal{X}_V$ (red arrows on the right side of Fig. 3). These correspondences also enable to deal with occlusions and segmentation errors since the corresponding unobserved areas of the object would not affect the underlying areas of $\mathcal{X}_V$. Conversely, from $\mathcal{X}_V$ to $\mathcal{Y}$, the correspondences are instead more suited to track shrinking deformations under compression actions. The outer areas of $\mathcal{X}_V$ are coherently matched with the outer areas of the observed point cloud $\mathcal{Y}$ of the compressed object (green arrows on the left side of Fig. 3). As a drawback, unobserved areas (occlusions and segmentation errors) would affect the underlying areas $\mathcal{X}_V$ which would match with the closest areas of $\mathcal{Y}$.

As described hereafter, a trade-off must be found between these two sets of correspondences, whether the application deals with stretching or compression actions on the object and whether occlusions or segmentation errors are to be dealt with.

6.2 *Computation of External Forces*

Based on the two sets of mesh-to-point cloud and point cloud-to-mesh correspondences, given by $\mathcal{N}_{\mathcal{X}_V}$ and $\mathcal{N}_{\mathcal{Y}}$, an external elastic force $\mathbf{f}_{ext}(\mathbf{x}_j) \in \mathbb{R}^3$ can be computed for each $\mathbf{x}_j \in \mathcal{X}_V$ as follows

$$\mathbf{f}_{ext}(\mathbf{x}_j) = k_{ext}(\mathbf{x}_j - \mathbf{y}_j^f) \tag{4}$$

with $k_{ext} > 0$ the stiffness of these external spring elastic forces and

$$\mathbf{y}_j^f = \begin{cases} \lambda \mathbf{NN}_{\mathcal{Y}}(\mathbf{x}_j) + (1-\lambda)\frac{1}{N_K}\sum_{\mathbf{y}_j \in \mathcal{K}_j} \mathbf{y}_j & \text{if } N_K > 0, \\ \lambda \mathbf{NN}_{\mathcal{Y}}(\mathbf{x}_j) + (1-\lambda)\mathbf{x}_j & \text{if } N_K = 0, \end{cases} \tag{5}$$

where $\lambda \in \mathbb{R}$ is a fixed scalar, $\mathcal{K}_j = \{\mathbf{y}_i \in \mathcal{Y} : \mathbf{NN}_{\mathcal{X}_V}(\mathbf{y}_i) = \mathbf{x}_j\}$ is the set of points in $\mathcal{Y}$ whose nearest neighbours are $\mathbf{x}_j$, $\mathbf{y}_j^f$ acts as a virtual point attracting $\mathbf{x}_j$, and $N_K \geq 0$ is the size of $\mathcal{K}_j$. The component of $\mathbf{y}_j^f$ resulting from the multiple point cloud to mesh correspondences is scaled by the reciprocal of N_K to be homogeneous with the component resulting from the single mesh to point correspondence. The fixed scalar λ tunes the balance between the mesh-to-point cloud and point cloud-to-mesh correspondences, as a trade-off between the stretching or compression actions to be tracked, as suggested in Sect. 6.1. If $\mathcal{K}_j$ is empty, the missing point cloud-to-mesh correspondences are replaced by a self-contribution for the vertices $\mathbf{x}_j$, compelling it to remain at its current position. The stiffness k_{ext} can be set accordingly to the Young's modulus, using the equivalence between a mass spring model and a linear FEM model, as described in Sect. 7.2. However, modelling this way the external forces, using equivalent springs, can be coarse since vision dat affects the visible vertices only. Some outliers in the point cloud may result in aberrant correspondences and thus in aberrant forces exerted on some vertices. A simple solution has been to discard points in the cloud whose point-to-point distances with their nearest neighbors in the mesh are above a certain threshold with respect to the mean value and the standard deviation of the whole set of point-to-point distances. In this case, for the considered vertices $\mathbf{x}_j$, we have $\mathbf{f}_{ext}(\mathbf{x}_j) = 0$.

Finally, regarding points $\mathbf{x}_j$ in $\mathcal{X}$ which are not visible, we also set $\mathbf{f}_{ext}(\mathbf{x}_j) = 0$. The whole set of forces is finally concatenated in a vector $\mathbf{f}_{ext}$ of size N_X.

6.3 Weighting Forces Using Contours

A limitation of this method lies in tracking large elastic deformations due to stretching efforts, for instance. In this case, since correspondences are established based on 3D geometry, only vertices lying on the occluding contour of the mesh are attracted to the extended area in the point cloud. As a consequence, the forces attracting the contours are weak. We propose to emphasize them by weighting the vertices of the visible surface of the mesh given their distance to the occluding contour of the projected mesh. Based on the depth map $\mathcal{D}^M$ of the projected mesh, we compute the distance map of the occluding contour of the mesh. Then, the weight $w_j \in \mathbb{R}$ for the vertex $\mathbf{x}_j$ is set proportional to

$$w_j \propto e^{-\frac{d_j^M}{\sigma}}, \tag{6}$$

where $d_j^M > 0$ is the distance from $\mathbf{x}_j$ to the nearest contour of the projected mesh, $\sigma > 0$ is a parameter which is empirically set. Then, w_j is normalized so that we get an observation probability. Finally, the forces are computed as

$$\mathbf{f}_{ext}(\mathbf{x}_j) = w_j k_{ext}(\mathbf{x}_j - \mathbf{y}_j^f). \tag{7}$$

7 Solver

Estimating the deformations of the mesh consists in solving a dynamic system of ordinary differential equations, involving the internal and external forces, based on Lagrangian dynamics

$$\mathbf{M}\ddot{\mathbf{x}} + \mathbf{C}\dot{\mathbf{x}} + \mathbf{f} = \mathbf{f}_{ext} \tag{8}$$

$$\text{with} \mathbf{f} = \mathbf{K}\mathbf{x} + \mathbf{f}_0, \tag{9}$$

where $\mathbf{M} \in \mathbb{R}^{3N_X \times 3N_X}$ and $\mathbf{C} \in \mathbb{R}^{3N_X \times 3N_X}$ are the mass and damping matrices, respectively, $\mathbf{K} \in \mathbb{R}^{3N_X \times 3N_X}$ is the global stiffness matrix which sums the element-wise *rotated* stiffness matrices $\mathbf{K}_e = \mathbf{R}_e\mathbf{K}_e\mathbf{R}_e^{-1} \in \mathbb{R}^{3N_X \times 3N_X}$, written with respect to whole set of vertices, and $\mathbf{f}_0 = \mathbf{R}_e\mathbf{K}_e\mathbf{x}_0$ the corresponding global offset summing the element-wise ones, with $\mathbf{x}_0 \in \mathbb{R}^{3N_X}$ the position of the vertices before the deformation. An Euler implicit integration scheme is used to solve the system, along with a conjugate gradient method. Finally, $\mathcal{X}$ can then be updated using the resulting estimated positions $\mathbf{x}$ of the vertices of the mesh.

Algorithm 1 summarizes the whole process presented above.

7.1 Experimental Results

In order to evaluate the performance of our method and contributions, some experimental results are shown in this section, on some computer-generated data, and some real data. Various objects, deformations, and conditions are tested.

For the non-rigid registration phase, we have employed the SOFA simulator [41], which enables us to deal with various physical models and to evolve simulations in real-time.

Algorithm 1: Registration process

Data: Tetrahedral mesh of the object, RGB-D data
Result: Fitting the mesh with the RGB-D data
Initialization of the vertices $\mathcal{X}$ in the mesh;
Initialization of segmentation;
for *each new* $\mathcal{I}$ *and* $\mathcal{D}$ **do**
  Segmentation in $\mathcal{I}$, providing $\mathcal{D}^s$;
  Sampling and back-projection, giving $\mathcal{Y}$;
  Determine $\mathcal{X}_V$;
  Rigid ICP between $\mathcal{X}_V$ and $\mathcal{Y}$;
  Rigidly transform $\mathcal{X}$;
  Non-rigid process:
  3D correspondences from $\mathcal{Y}$ to $\mathcal{X}_V$;
  3D correspondences from $\mathcal{X}_V$ to $\mathcal{Y}$;
  Computation of external forces $\mathbf{f}_{ext}$;
  In SOFA simulator:
  Computation internal forces $\mathbf{f}$;
  Computation of the new positions $\mathbf{x}$:
  by solving $\mathbf{M}\ddot{\mathbf{x}} + \mathbf{C}\dot{\mathbf{x}} + \mathbf{f} = \mathbf{f}_{ext}$ (Euler implicit + conjugate gradient);
end

7.2 *Results for Tracking on Synthetic Data*

Relying on the SOFA simulator, we have first generated a sequence involving the deformations of a cylindrical elastic object, modeled by the FEM corotational approach. The sequence involves the elastic deformations of a cylindrical elastic object modelled by the FEM corotational approach. It has a Young's modulus of $E = 800$ Pa and Poisson's ratio of $\nu = 0.3$. Based on a manually designed cylindrical surface triangular mesh with radius/height dimensions of 0.11×0.02 m, the volumetric tetrahedral mesh was generated using the 3D Delaunay triangulation through the CGAL library.[1] A disordered placement of the nodes has been adopted here on this simple cylindrical geometric topology, not to lose generality to an arbitrary object topology. A 3D Delaunay triangulation would be preferred as a mesh generation routine. The resulting mesh is made of 1369 elements and 497 vertices and featured in Fig. 4a.

In order to simulate deformations and generate the considered sequence, an elastic stretching force is applied in the $-z$ direction of $\mathcal{W}$ (see Fig. 5), on one point on the border of the object (point 1) and few other points being fixed on the opposite border (points 4,5,6). Two compression forces are applied along y and $-Y$ (points 2,3) directions in $\mathcal{W}$. The applied forces result in a fast elongation deformation of the object, with a maximum elongation above 50%, and in a critical bending deformation along y-axis, as it can be seen in Fig. 7. For the tracking phase, segmentation aspects are not considered in these experiments. We only process the visible vertices of the rendered object in the sequence, and as ground truth, the positions of the whole set of

[1] http://www.cgal.org.

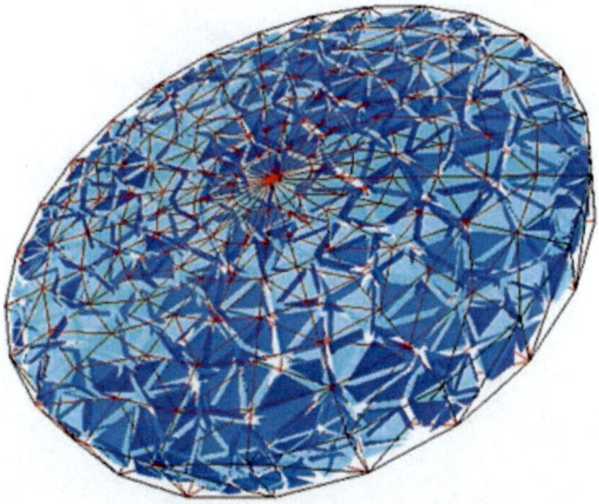

Fig. 4 Processed mesh

points are stored for evaluation. Results can be visually observed in Fig. 5, featuring the original target (red) and the tracking 3D mesh (blue), and in Fig. 6 the 3D errors between the vertices of the registered mesh and the corresponding points in the point cloud are plotted.

The following models and methods have been compared:

- volumetric mass-spring model;
- standard FEM model;
- corotational FEM model;
- corotational FEM model along with contour weighting.

On these results, we set the Young's modulus and the Poisson's raio for the FEM approaches equal to the values assigned for the simulated object ($E = 800$ Pa, $\nu = 0.3$). We set $\lambda = 0.5$, balancing point cloud to mesh and mesh to point cloud forces. Also, we set $k_{ext} = 10\,\text{Nm}^{-1}$ for the stiffness of the external forces, which appeared to provide satisfying results. We consider a volumetric mass-spring model on the tetrahedrons of the generated mesh by attaching springs to connected vertices.

Determining the spring parameters of a mass-spring model equivalent to the FEM model in terms of elasticity behavior is not trivial, especially when dealing with unordered and irregular tetrahedrons, as pointed out in [42, 43]. Using the analytical expression proposed in [43] to derive the stiffness $k_m(i, j)$ of an equivalent spring between connected vertices $\mathbf{x}_i$ and $\mathbf{x}_j s$ of the tetrahedral mesh

$$k_m(i, j) = \sum_e \frac{2\sqrt{2}}{25} \left(\cdot V_e \frac{12}{\sqrt{2}} \right)^{\frac{1}{3}} E, \tag{10}$$

with the sum over the elements e adjacent to the edge (i, j). The resulting stiffnesses are too large to register the investigated deformations. The performance of a mass-spring model with spring stiffnesses scaled by 0.05 (defined as k_{tuned}), which appeared to fit best on this sequence, is thus also shown.

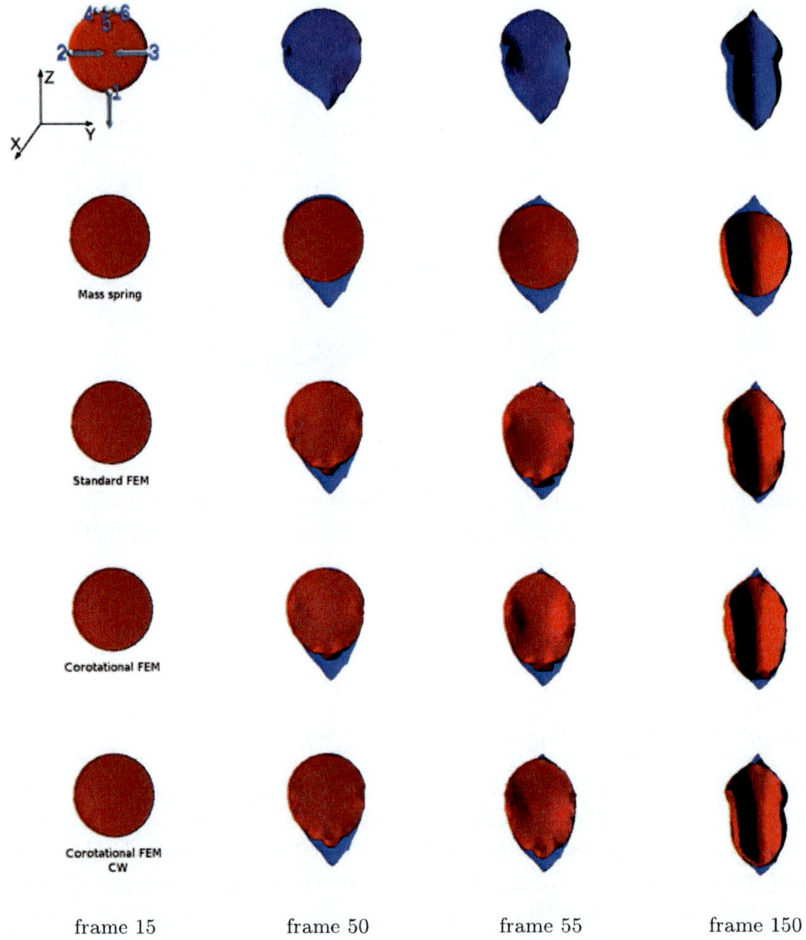

Fig. 5 Results of the deformable tracking process. In the first row, it is featured the ground truth. In the second row, the tracking with the equivalent mass spring system is displayed. In the third row, the linear FEM is depicted. In the fourth row, the corotational FEM is shown. Finally, in the fifth row, the contour weighting is added

We observe that the equivalent mass-spring model outperformed by the FEM approaches and the mass-spring model with manually tuned stiffnesses. This latter performs worse than the FEM models on the first frames when triggering the deformations (frame 50) due to the stretching deformations, which are severe over this phase. However, the standard linear FEM tends then to fail to track the large bending deformations which occur until steady-state, enhancing the sensitivity of this approach regarding rotation transformations within elements, as it can also be observed in Fig. 7. On the whole sequence, the benefit of using a co-rotated approach

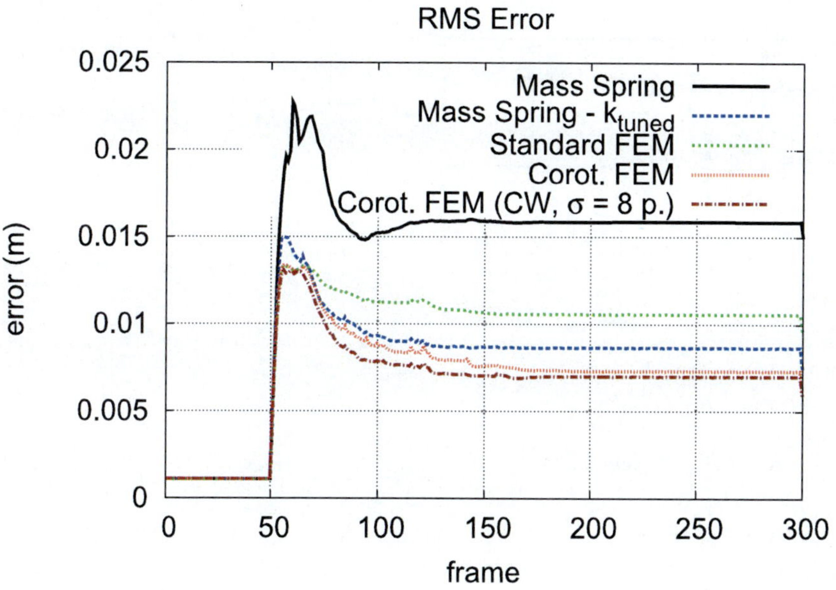

Fig. 6 Errors of the deformable tracking process with the different tested approaches, the mass-spring system with the equivalent stiffness, the mass-spring system with tuned stiffness (k_{tuned}), the standard FEM, the corotational FEM, and the corotational FEM with contour weighting (with $\sigma = 8$ pixels)

Fig. 7 Deformations on the generated sequence (**a**), on the tracked mesh with the standard FEM (**b**), and with the corotational FEM (**c**)

and the advantage of using the contour weighting function (with $\sigma = 8$ pixels) to track significant bending and stretching deformations can thus be particularly stressed out.

In order to evaluate the sensitivity with respect to tuning parameters, in Fig. 8 the errors for different external stiffnesses k_{ext} and for different values of λ are represented. For the external stiffness, the value of k_{ext} acts as a gain. It tunes the balance between tracking large stretching deformations, especially on the silhouette borders, and regularization effects. Its influence is especially observable on the large initial deformations. With λ a good compromise is found with $\lambda = 0.7$, showing the relevance of point cloud to mesh correspondences in this case for which stretching actions prevail.

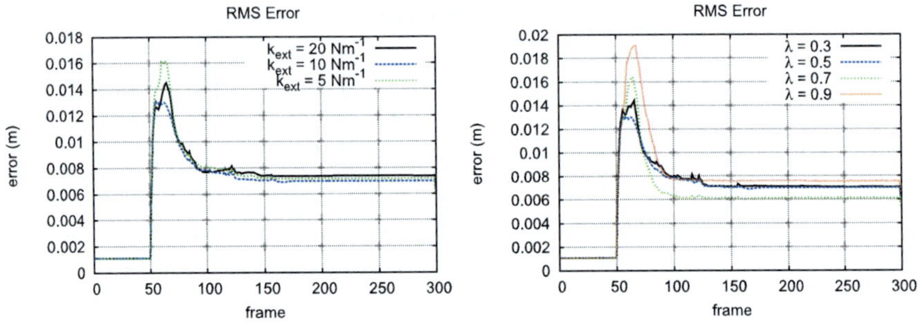

Fig. 8 Errors with respect to k_{ext} (**a**) and λ (**b**), tuning the balance between mesh to point cloud and point cloud to mesh forces

7.3 Results on Real Data

In order to carry out experiments on real data, the point cloud of the investigated scene is acquired from a calibrated RGB-D camera Asus Xtion, 320×240 RGB. Depth images were processed using a standard laptop with an NVIDIA GeForce 720M graphic card, along with a 2.4GHz Intel Core i7 CPU. Here, the segmentation process is involved in the loop and, since fast real-time performance is required, it relies on a CUDA implementation. The results presented here deal with a textureless and smooth elastic object made with silicon.

Concerning the "pizza-like" object, the idea was to test motions and deformations similar to those involved in the pizza-making process. The involved mesh is generated in a similar manner to Sect. 7.2, with radius/height dimensions of 0.12×0.01 m, and consisting in 574 vertices and 1675 elements. Note that the point cloud's resolution is chosen through sampling to coincide approximately with the resolution of the visible mesh.

On the first featured sequence, the object undergoes large rigid motions and various isometric and elastic deformations. The Young's modulus and the Poisson's ratio of the material are unknown and good compromises have been found with $E = 300$ Pa and $\nu = 0.3$, by setting $k_{ext} = 1$ Nm^{-1}.

Qualitative results are presented in Fig. 9, comparing our method using the corotational FEM approach with other models. In the first row, input RGB images are shown. The second row features the corresponding segmented frame, while the third row shows the 3D mesh tracking the object with the mass-spring model. The fourth shows the standard FEM model, while the fifth shows the proposed approach (with $\sigma = 30$ pixels).

We can visually notice the ability of the proposed method to segment the visible part of the object correctly. Besides, it tracks rigid motions, in contrast to the mass-spring, to accurately register stretching deformations. Finally, in contrast to the standard linear FEM, it is robust to folding/bending deformations thanks to the

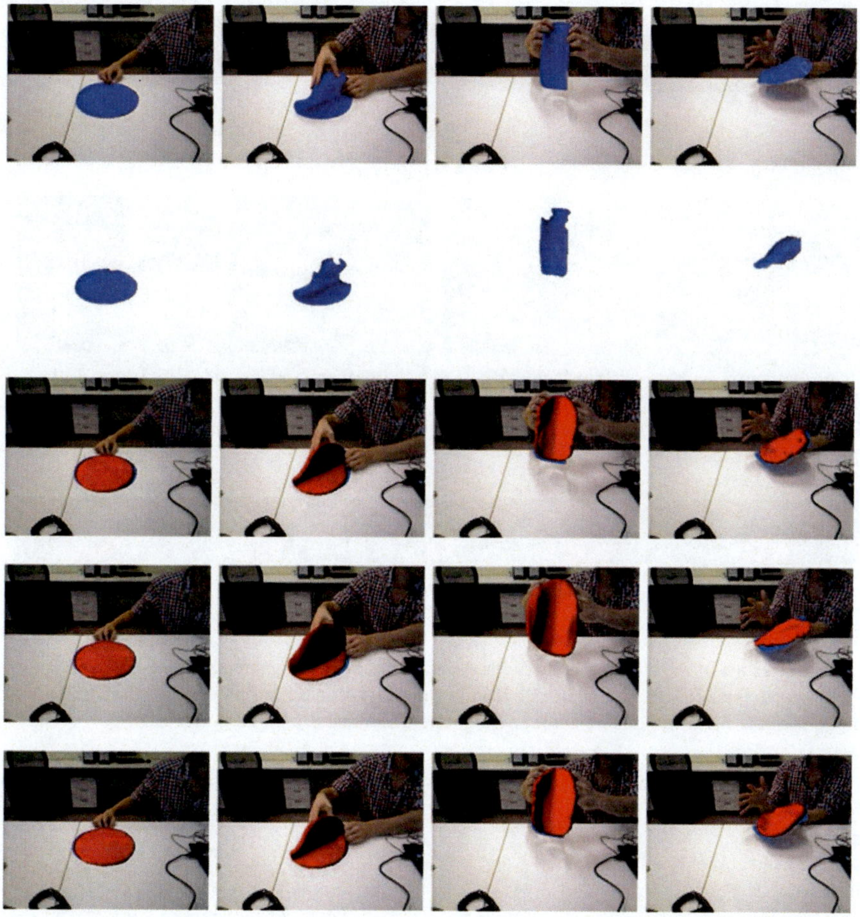

Fig. 9 Results of the tracking process for the pizza-like object, with the input images (first row), the segmented frames (second row), and the registered mesh reprojected in the input image, for: the mass-spring model (third row), the standard FEM model (fourth row), and the corotational model with the contour weighting technique (fifth row)

corotational approach. Robustness to occlusions due to the hands manipulating the object (third column in Fig. 9) or segmentation errors can also be observed.

The second sequence has been worked out to stress the robustness of the proposed method under large elastic deformations due to stretching actions. The results for the proposed method are shown in Fig. 10, while they are compared to different models during stretching in Fig. 11. Since the object is essentially stretched along the image plane, involving few bending deformations along the optical axis, we propose an evaluation criterion to compute the mean 3D distance between the contour of the projected mesh in the image and the contour of the segmented silhouette of the object. This is achieved by searching from the 3D vertices, corresponding to the

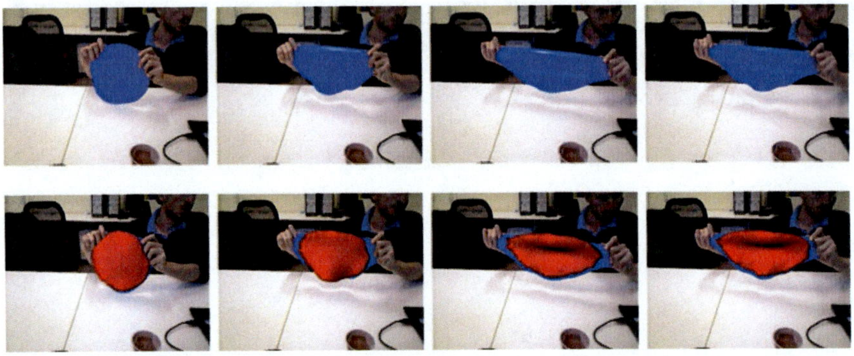

(e) frame 1 (f) frame 20 (g) frame 50 (h) frame 150

Fig. 10 Results of the tracking process for the silicon pizza, with the input images (first row), and the tracking results in the second row

Fig. 11 Results of the tracking process for the silicon pizza during stretching for the mass spring model (**a**), the standard FEM model, with (**b**) and without (**c**) contour weighting, with the corotational model, with (**d**) and without the contour weighting technique (**e**), on frame 150

Fig. 12 Contour fitting error for the different tested approaches (mass spring, standard and corotational FEM)

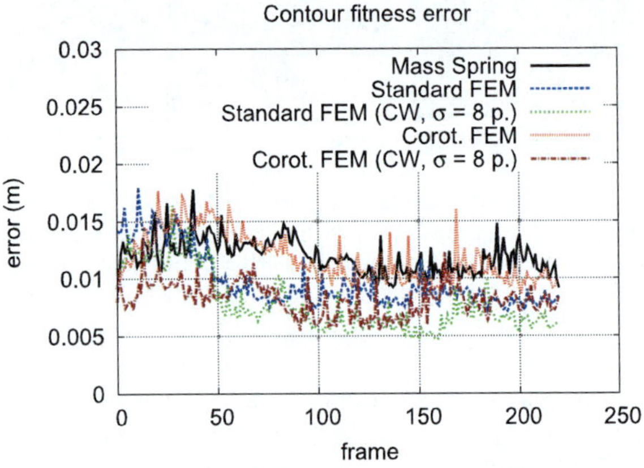

extracted 2D contour of the mesh, for the nearest 3D points, corresponding to the 2D contour of the segmented silhouette. This way, the benefit of using the corotational approach along with the contour weighting function can also be stressed out (see Fig. 12).

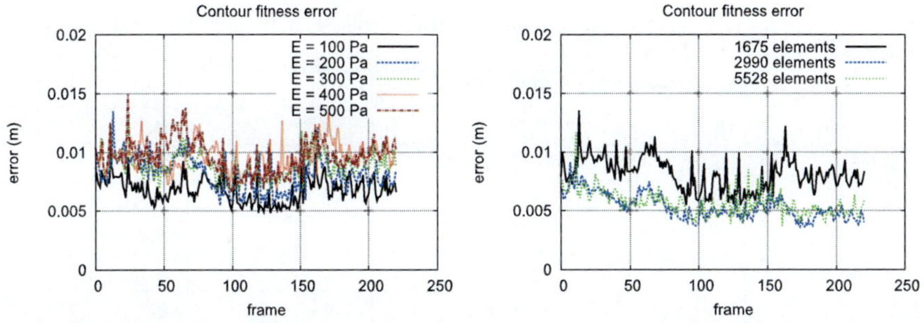

Fig. 13 Contour fitting error, for different stiffnesses (**a**) and resolutions (**b**)

Fig. 14 Results of the tracking process with the proposed approach, for $E = 100$ Pa (**a**) and $E = 500$ Pa (**b**) with the standard resolution, and for resolutions of 2990 (**c**) and 5528 (**d**) elements, with $E = 100$ Pa

Concerning sensitivity to material and tuning parameters, different stiffnesses of the material have been tested, $E = 300$ Pa appearing, as stated before, to be a good compromise between the ability to register large deformations, while providing sufficient regularization and robustness to noise, which can result in spurious registrations, and self-collisions, especially around the borders of the mesh, as seen in Fig. 14a, b with the contour fitting errors. Different resolutions of the mesh have been tried adjusting the resolution of the point cloud accordingly through the manually set sample step. It can be observed in Fig. 13b that higher resolutions manage to fit better with the contour, and also, as observed in Fig. 14c, d to capture higher frequency details, up to a certain point for which the benefit becomes minor.

7.3.1 Computational Costs

Regarding computational aspects, the computation times of the various phases of the algorithm, for the different methods compared in this chapter, are shown in Table 2. *Visibility* corresponds to the process of determining the visible vertices of the rendered mesh and, in the case of using the contour weighting mode, extracting the vertices lying on the contour. *Ext. forces* is the step involving the determination of the closest points between the mesh and the point cloud and the computation of the subsequent external forces exerted on the mesh. *Resolution* consists instead of the resolution of the Lagrangian mechanical equations to compute the deformations. The presented figures are the averages of the execution times per frame (in milliseconds)

Table 2 Execution times, in milliseconds, for the different phases of the approach, and the various models and methods employed in this paper

	Mass Spring	Stand. FEM	Corot.	Corot. - CW
Segmentation	10.7	10.5	10.7	10.7
Rigid ICP	3.0	2.5	2.7	2.6
Visibility	8.1	8.2	7.6	7.4
Ext. forces	3.4	3.5	3.5	4.0
Resolution	2.8	3.3	4.0	4.1
Total	28.0	27.9	28.6	28.8

for the sequence presented in Fig. 9. As noticed, the suggested method (corotational model with the contour weighting mode) runs on the sequence at around 35 frame per second. We can also observe that the computational costs for the resolution phase being relatively small within the whole process, overall execution times are relatively independent of the selected model.

8 Fractures

8.1 A Pure Physics-Based Approach

For fractures, once deformations are computed, topological changes due to fractures, as modelled in the previous chapter, are handled. For each vertex $\mathbf{x}_j \in \mathcal{X}$, we detect if it is a fracture node, based on Sect. 1.4.1. Fracturable nodes undergo the remeshing procedure presented in Sect. 1.4.2. Let us note that, for remeshing, an element attached to different fracturable nodes is treated only once by the first investigated node. At the end of the procedure, a new mesh $\mathcal{X}$ is obtained and it will be directly employed for the next frame.

8.2 Experimental Results

In order to evaluate the performance of our method and contributions, some experimental results are shown in this section, in a qualitative manner on some real data.

8.2.1 Implementation and Experimental Set-Up

For the non-rigid registration phase, we have employed the SOFA simulator [41], which enables us to deal with various physical models and to evolve simulations

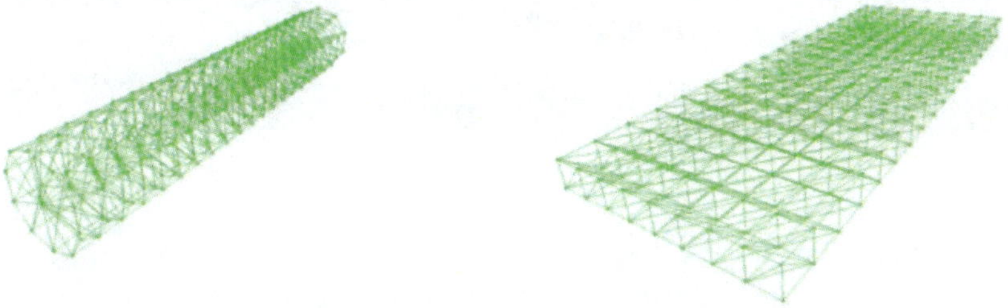

Fig. 15 Meshes of both the cylindrical bar (left) and plate (right) objects

in real-time. In order to carry out experiments on real data, the point cloud of the investigated scene is acquired from a calibrated RGB-D camera Asus Xtion, 320×240 RGB and depth images being processed. A standard laptop with an NVIDIA GeForce 720M graphic card has been used, along with a 2.4 GHz Intel Core i7 CPU. The segmentation process relies on a CUDA implementation. The results presented here deal with elastic bar and plate objects made of modeling clay. For the cylindrical bar object, the involved mesh, depicted in Fig. 15a, has a circumferential/radial/height resolution of $10 \times 20 \times 2$, resulting in 220 vertices and 570 tetrahedral elements. The material is here poorly elastic, the Young's modulus being empirically set to $E = 2.0$ and the Poisson's ratio to $\nu = 0.01$. The plate object (see Fig. 15b) instead has a length/width/height resolution of $18 \times 9 times 2$, resulting in 324 vertices and 816 elements. For material properties, we set $E = 2.5$ and $\nu = 0.01$, in this case. Three fracture cases are demonstrated here: fracture due to opening by excessive bending for the bar object and fractures due to shearing and opening by stretching for the plate object, tearing the object. Let us note that in the first case the parameter $0 \leq \lambda \leq 1$, which tunes the balance between mesh-to-point cloud and point cloud-to-mesh forces (see [44]), is set to $\lambda = 0.5$. In the second case, $\lambda = 0.1$, so that mesh's vertices lying on the fracture crack of the object will be driven towards the nearest observed point on the border of the crack. Let us note that in [44], the registration process is iteratively repeated (three iterations were performed in the experiments). Here, only one iteration is executed to maintain real-time performance since fracture detection remains costly.

8.2.2 Results

As seen in Fig. 16, opening though bending deformations can be tracked quite correctly for the bar object. In this way, internal tensile forces computed on the nodes remain physically consistent, making the fracture model valid, as shown in the third column. For the plate object, tearing it through out-of-plane shearing (see Fig. 17) can also be tracked, despite some artifacts that can be noticed on some unfractured

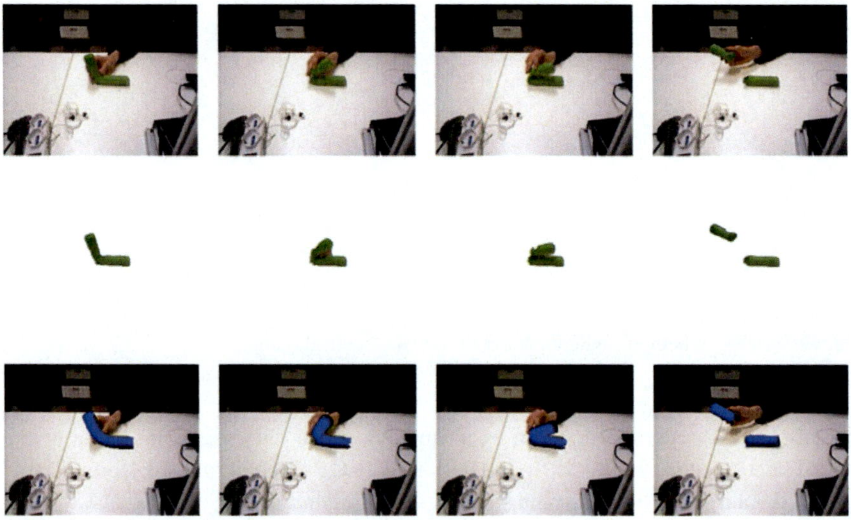

Fig. 16 Results of the tracking process for the cylindrical bar object, with the input images (first row), the segmented frames (second row), and the registered mesh reprojected in the input image (third row)

regions. However, when opening the plate object by stretching it, as shown in Fig. 18, only vertices lying on the occluding contour of the mesh are attracted to the expanded areas in the point cloud since correspondences are established based on 3D geometry, without any discriminative descriptors. As a consequence, forces attracting the contours and stretching the mesh are weak. Therefore, in this case, fracture appears with an unavoidable delay when both fractured parts split away. However, the fracture crack is relatively coherent, and both parts are finally consistently recovered.

Regarding computational aspects, in Table 3 are shown the mean computation times of the various phases of the algorithm for the sequence presented in Fig. 17. *Ext. forces* is the step involving the determination of the closest points between the mesh and the point cloud and the computation of the subsequent external forces exerted on the mesh. *Resolution* corresponds to the computation of the internal forces and in the resolution of the Lagrangian mechanical equations to compute the deformations based on the computed external and internal forces. *Fracture detection* consists of computing the separation tensors to detect fractures on the nodes of the mesh. Finally, *Remeshing* shows the time dedicated to remesh the mesh around the fracturable nodes. As noticed, the suggested method runs on this sequence at around 27 frame per second.

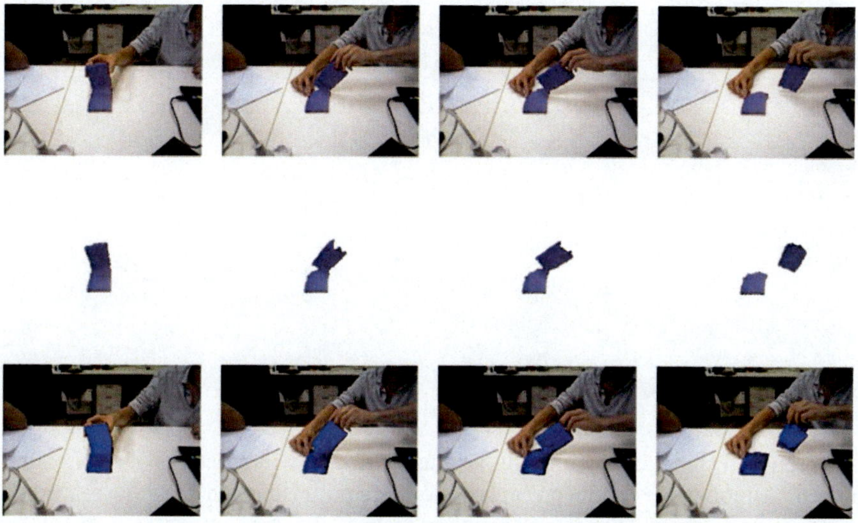

Fig. 17 Results of the tracking process for the plate object, torn through out-of-plane shearing, with the input images (first row), the segmented frames (second row), and the registered mesh reprojected in the input image (third row)

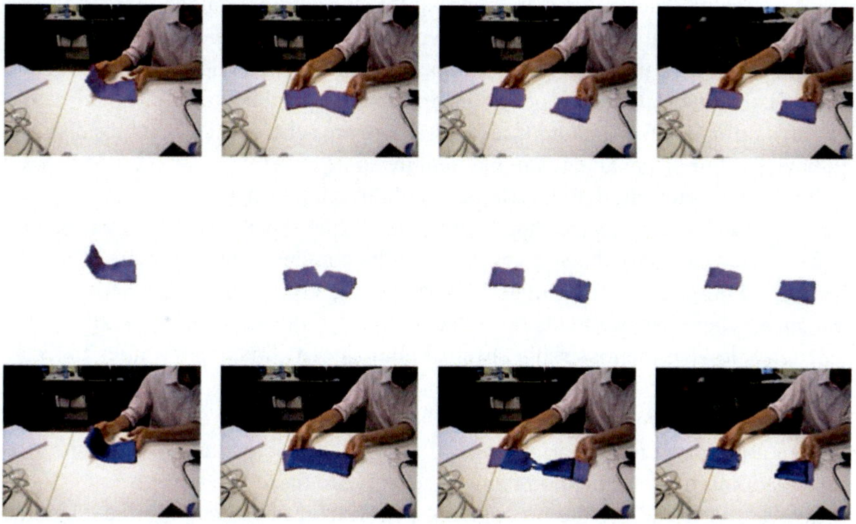

Fig. 18 Results of the tracking process for the plate object, torn through opening, with the input images (first row), the segmented frames (second row), and the registered mesh reprojected in the input image (third row)

Table 3 Mean execution times, in milliseconds, for the different phases of the approach

Phases	Mean execution times (ms)
Segmentation	10.1
Ext. forces	3.1
Resolution	2.3
Int. forces and fracture detection	2.3
Remeshing	0.6
Total	37.5

8.3 Comments on the Results

The proposed approach is a first simple attempt to deal with fractures of deformable objects. Although the results presented above are promising and show the relevance of employing a physically rigorous method to detect fracture, several issues shall be discussed.

Stiffness of the materials. Experiments are shown on two different simple objects with particular material properties (high stiffness). With more elastic materials, the mesh's estimated deformations and internal forces would be rougher, resulting in a more variable separation tensor and so potentially spurious detections of fractures.

Tuning material parameters. A cumbersome process can lie in determining the object's coherent material properties, which consist in the Young's modulus, the Poisson's ratio, and the fracture threshold. There are set empirically in this chapter, and having a fair physical realism may require some effort.

Tracking stretching deformations. As observed, our method faces some problems when tracking opening tearing fractures due to the expansion of the area covered by the object since only the vertices lying on the contour of the mesh are attracted. The method proposed in [44] suggests to weight the visible surface's vertices of the mesh, given their distance to the occluding contour of the mesh. However, this technique tends to alter the material's physical homogeneity, making fractures to likely occur around the contour of the mesh.

Sensitivity to occlusions. Since in the segmentation no distinction is made between a fracture crack and unobserved areas of the object due to occlusions or segmentation errors, our method remains sensitive to these latter events.

Remeshing. Handling the propagation of a fracture by removing the elements attached to a fracturable node and intersected by the fracture plane leads to progressive degeneration of the mesh if several fracture events occur.

9 Multiple Objects

The frame-by-frame registration problem, described in Sect. 2.6 can be extended to multiple objects that can interact and collide.

As illustrated in Fig. 19, the solution consists in integrating the suggested collision detection and response models (Sects. 1.5 and 1.5.2) into the rigid and non-rigid registration framework based on successive RGB-D data.

Thus, point cloud segmentation, matching, rigid and non-rigid registration techniques are adapted as described below.

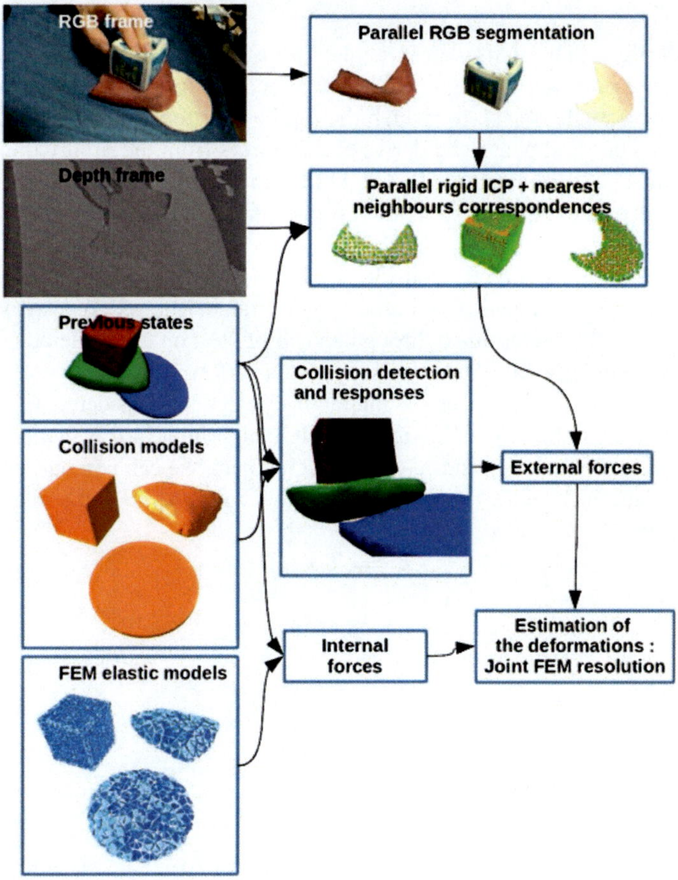

Fig. 19 Overview of our approach for deformable object tracking for multiple interacting objects

9.1 Preliminary Parallel Visual Segmentation

The visual segmentation step presented section3.1, based on a graph cut minimisation over foreground/background colour models, is carried out in a parallel manner on each considered object $\mathcal{O}_i$. This step provides us with a 3D target point cloud Y_i for each $\mathcal{O}_i$. We limit the size of each Y_i by sampling it on a regular grid in the image plane.

9.2 Parallel Rigid Pose Estimation

Each mesh $\mathcal{X}_i$, initially in its state computed for the previous frame, is then rigidly registered to each observed segmented point cloud $\mathcal{Y}_i$. In contrast to the approach proposed for a single object, for which the rigidly transformed meshes are used to initialise the non-rigid phase, we derive elastic forces from the rigidly transformed meshes. The idea is to integrate these rigid transformations in a smooth and simultaneous manner into the FEM simulation framework and thus to properly cope with collisions.

We actually do not directly consider $\mathcal{X}_i$ but its vertices which appears visible from the previous estimated state in the camera frame. The set $\mathcal{X}_{V,i}$, associated to the object $\mathcal{O}_i$, is thus determined through a visibility test on the rendered mesh with respect to its previous state. This visibility test is performed on the whole rendered scene so that occlusions between the interacting object are considered.

As introduced in Sect. 5, a classical rigid ICP algorithm [40] is then carried out for each object, between $\mathcal{Y}_i$ and $\mathcal{X}_{V,i}$. For each object, we can compute the rigidly transformed mesh $\mathcal{X}_i^t = \{\mathbf{x}_{i,j}^t : j = 1, \ldots, N_{X_{i,t}}\}$, with $N_{X_{i,t}} > 0$ the dimension of $\mathcal{X}_i^t$, as follows

$$\mathbf{x}_{i,j}^t = \mathbf{T}_i \mathbf{x}_{i,j}, \tag{11}$$

with $\mathbf{T}_i \in SE(3)$ the rigid transformation estimated with the ICP.

We then derive elastic forces $\mathbf{f}_{ext,i}^r$ between both sets $\mathcal{X}_i^t$ and $\mathcal{X}_i$ as

$$\mathbf{f}_{ext,i}^r(\mathbf{x}_{i,j}) = k_{i,ext}^r(\mathbf{x}_{i,j}^t - \mathbf{x}_{i,j}), \tag{12}$$

with $k_{ext,i}^r > 0$ a stiffness parameter.

9.3 Parallel Point Cloud Matching for Non-rigid Registration

Following the single object approaches, we derive the external forces exerted by the observed segmented point clouds Y_i on the different corresponding sets $\mathcal{X}_{V,i}$. We use sets of external forces $\mathbf{f}_{ext,i}^d \in \mathbb{R}^3$ related to geometrical information as introduced in Sect. 6.2. For each object $\mathcal{O}_i$, based on the two sets of mesh-to-point cloud and point cloud-to-mesh correspondences, we can compute an external elastic force $\mathbf{f}^d_{ext,i}$ exerted on each $\mathbf{x}_{i,j}$ in $\mathcal{X}_{V,i}$ as follows

$$\mathbf{f}^d_{ext,i}(\mathbf{x}_{i,j}) = k^d_{ext,i}(\mathbf{x}_{i,j} - \mathbf{y}^f_{i,j}), \tag{13}$$

where, as described in [44], $\mathbf{y}^f_{i,j} \in \mathbb{R}^3$ is a linear combination of points in the point clouds which are matched to $\mathbf{x}_{i,j}$, either from mesh-to-point cloud and from point cloud-to-mesh correspondence sets, while $k^d_{ext,i} > 0$ is the stiffness of these external elastic forces, which is here common to each object.

9.4 Resolution

The objective of the resolution process is to balance the sets of external forces capturing rigid and non-rigid transformations of the objects and the interactions between themselves, with the internal forces based on the different deformation FEM models presented above. The estimation consists in solving for each object the dynamic system of linear ODEs involving the internal and the different external forces, based on Lagrangian dynamics

$$\begin{aligned} \mathbf{M}_i\ddot{\mathbf{x}}_i + \mathbf{C}_i\dot{\mathbf{x}}_i + \mathbf{f}_i &= \mathbf{f}^d_{ext,i} + \mathbf{f}^r_{ext,i} + \mathbf{f}^c_i, \\ \text{with} \quad \mathbf{f}_i &= \mathbf{K}_i\mathbf{x}_i + \mathbf{f}_{0,i}, \end{aligned} \tag{14}$$

where $\mathbf{x}_i \in \mathbb{R}^{3N_{X_i}}$ is the vector containing the positions of the vertices of the object $\mathcal{O}_i$, $\mathbf{M}_i \in \mathbb{R}^{3N_{X,i} \times 3N_{X,i}}$ and $\mathbf{C}_i \in \mathbb{R}^{3N_{X,i} \times 3N_{X,i}}$ are the mass and damping matrices, respectively, $\mathbf{K}_i \in \mathbb{R}^{3N_{X,i} \times 3N_{X,i}}$ is the global stiffness matrix which sums the element-wise *rotated* stiffness matrices $\mathbf{K}^r_e = \mathbf{R}_e\mathbf{K}_e\mathbf{R}^{-1}_e \in \mathbb{R}^{3N_{X,i} \times 3N_{X,i}}$, written with respect to whole set of vertices, and $\mathbf{f}_{0,i}$ the corresponding global offset summing the element-wise ones $\mathbf{R}_e\mathbf{K}_e\mathbf{x}_{e,0}$. The terms $\mathbf{f}^d_{ext,i} \in \mathbb{R}^{3N_{X,i}}$, $\mathbf{f}^r_{ext,i} \in \mathbb{R}^{3N_{X,i}}$ and $\mathbf{f}^c_i \in \mathbb{R}^{3N_{X,i}}$ are vectors containing the external forces, the ones resulting from the interaction model, and the collision responses between object $\mathcal{O}_i$ and the other objects, respectively. The Euler implicit integration scheme is used to solve the system with respect to each $\mathbf{x}_i$, along with a conjugate gradient method.

9.5 Experimental Results

Based on the multiple interacting objects tracking algorithm (see Sect. 9), the results presented here report different scenarios involving objects of different shapes, potentially textureless, and with different interactions.

Fig. 20 Tetrahedral meshes for the brain, squirrel and crocodile, plus the visual and collision models for the crocodile toy

The first sequence shows a stuffed toy and an elastic rubber toy brain colliding. The second is a silicon elastic disk wrapping a silicon crocodile puppet filled with plastic balls. In the third sequence, the disk, a silicon object, and a foam cube are stacked and squeezed together.

The point clouds of the investigated scene are acquired using different structured light commodity RGB-D cameras. An Asus Xtion, 320×240 RGB and depth images, is employed for the first sequence. An Intel RealSense SR300 (960×540 images) is employed for the second and third test sequences.

In order to create the geometric and deformation models of some of the considered objects (the squirrel, the crocodile, the toy brain, the light red silicon object), surface meshes of the undeformed objects were first reconstructed offline. We have used the RGB-D based dense 3D reconstruction techniques KinectFusion [45] and InfiniTAM [46] which is implemented for the RealSense SR300. The large resolution and short range of the RealSense SR300, with respect to the Asus Xtion, enables capturing and modeling finer grain geometries, such as the one of the crocodile. Finally, some re-meshing and smoothing tasks have been carried out with a modelling engine to get fair, closed, and clean surface meshes of the objects. The corresponding volumetric tetrahedral meshes have been generated by achieving 3D Delaunay triangulations within the surface meshes, thanks to the CGAL library. As an approximation, we assume the isotropy of the materials in order to apply the deformation model described before. Some of the considered tetrahedral meshes are featured in Fig. 20, as well as the visual and collision surface models for the crocodile. For computational efficiency, the collision mesh is made coarser than the visual mesh.

The software implementation relies on the SOFA simulator [41], which provides the FEM elastic and the collision models. Qualitative results are shown in Figs. 21 and 22 for the different scenarios, as in the provided video. Note that initialization

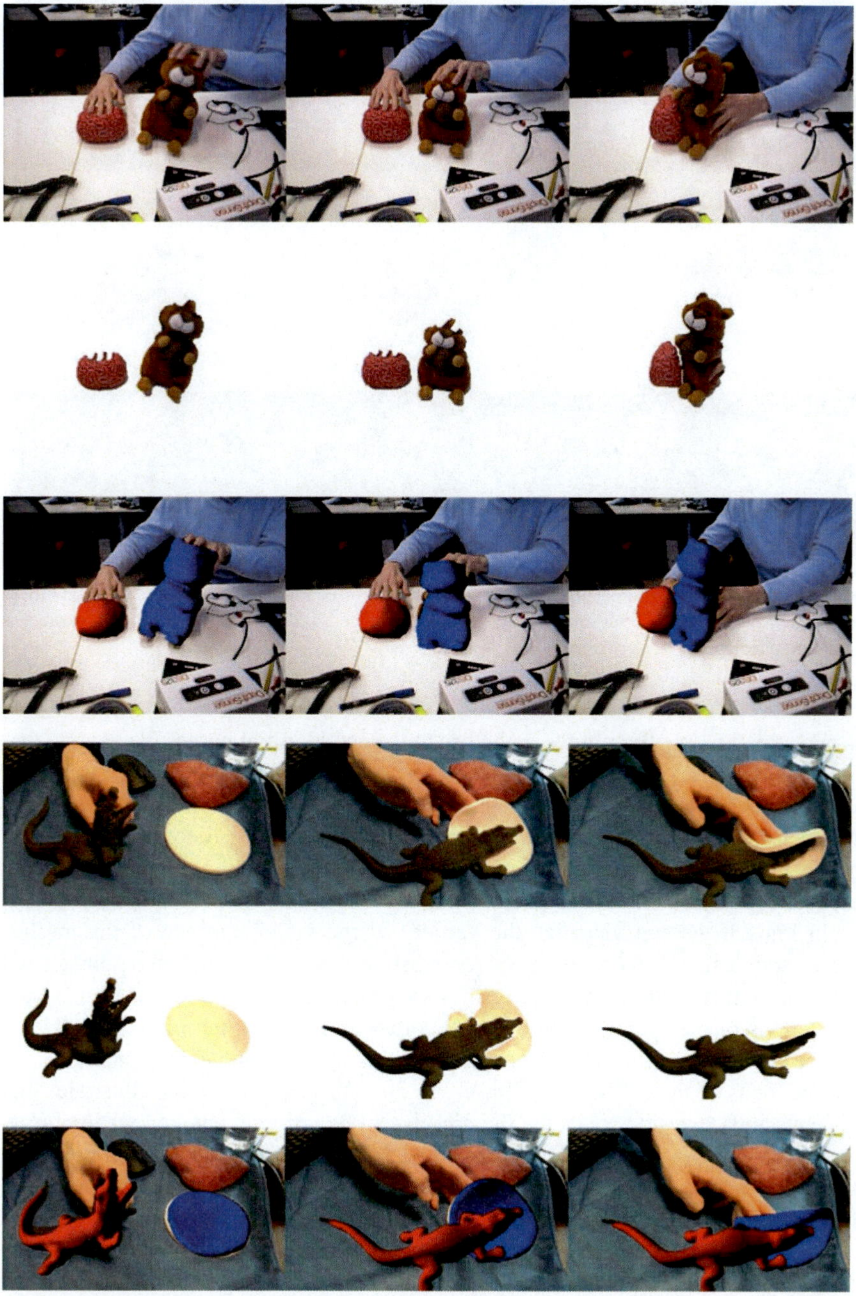

Fig. 21 Results of the proposed registration method for two interacting objects. For each block, the sequence of rows is: input images, the segmented images, and the registered mesh re-projected in the input image

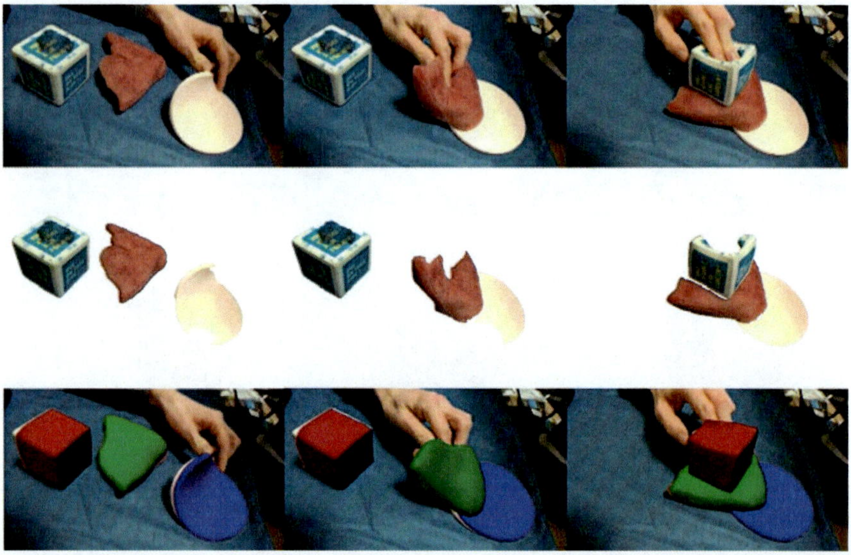

Fig. 22 Results of the proposed registration method for three interacting objects. For each block, the sequence of rows is: input images, the segmented images, and the registered mesh re-projected in the input image

is performed manually, rigidly, and close to the actual initial poses. We assume that the objects are not in contact initially.

The case of the silicon disk wrapping the crocodile toy is examined. The collision between both objects enables to constraint sufficiently the silicon disk so that it gets registered coherently, even with severe bending deformations, occlusions by the crocodile, and errors in the segmentation.

In Fig. 23, we can visualize the vertices of the visible surface of the meshes (green squares), which are processed for registration with the segmented point clouds (brown squares). Different samplings for the point clouds are used according to the resolutions of the meshes. The detected collisions between the interacting objects can be seen on the third row with segments between the found contact points.

The focus on the collision regions in Fig. 24 with wireframe views illustrates that contacts and occlusions between the objects can be properly handled when objects are squeezed, wrapped, or stacked together. Independently tracking them leads to inter-penetrations. In Fig. 25 are shown some failure situations when forces resulting from the rigid ICP registration are not integrated with the resolution.

As for computational issues, segmentation being addressed in a parallel manner, and collision requiring extra computations when objects are getting into proximity, the method does not achieve fast performance: 5–10 frame per second are reached, depending on the number of objects and on the resolutions of the collision surface meshes.

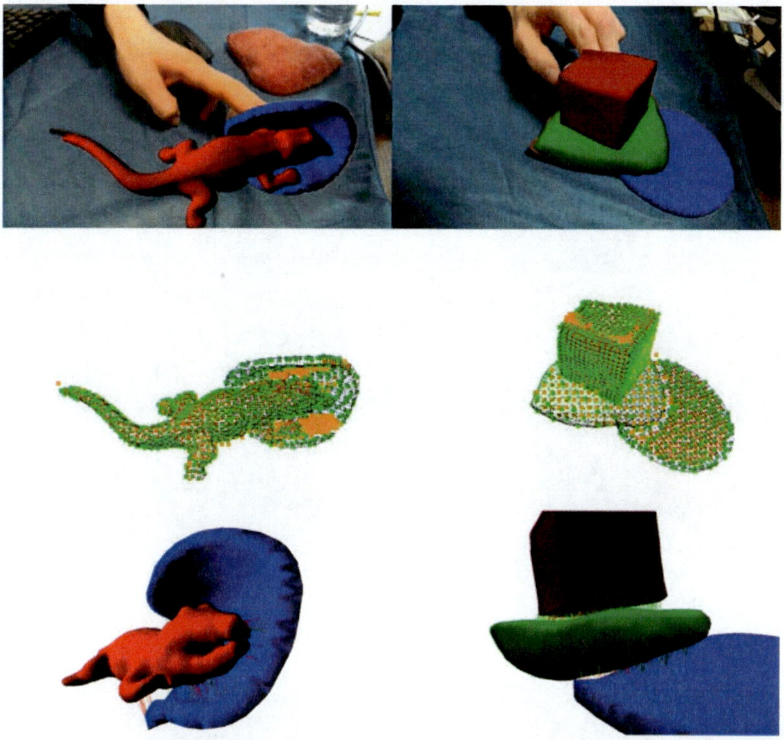

Fig. 23 Visualization of the registered point clouds (brown squares) for the meshes (greens squares as vertices on the visible surfaces) of examples with two and three interacting objects. The displayed segments on the third row show the detected collisions

9.6 Comments on the Results

The experiments presented above show some promising results for the method. However, a few limitations of the method shall be discussed. The system requires some tuning for both the FEM models of the objects and the collision detection and response models to obtain stability and proper behaviour. Some instabilities or ripples can indeed appear when the involved object undergoes fast rigid motions or when too many interactions are involved i.e., when two many objects of complex shapes and collision models are involved). In these cases, the addition of numerous and potentially significant and opposite external forces coming from the point cloud correspondences, the rigid motion estimation, and the collision responses can make the Lagrangian dynamics equations hardly to be solved using the classical implemented conjugate gradient solver. Hence, some tuning can be required to achieve a successful resolution of the whole system. Besides, contact detection distances need to be set relatively high in some cases (see Fig. 23, third row).

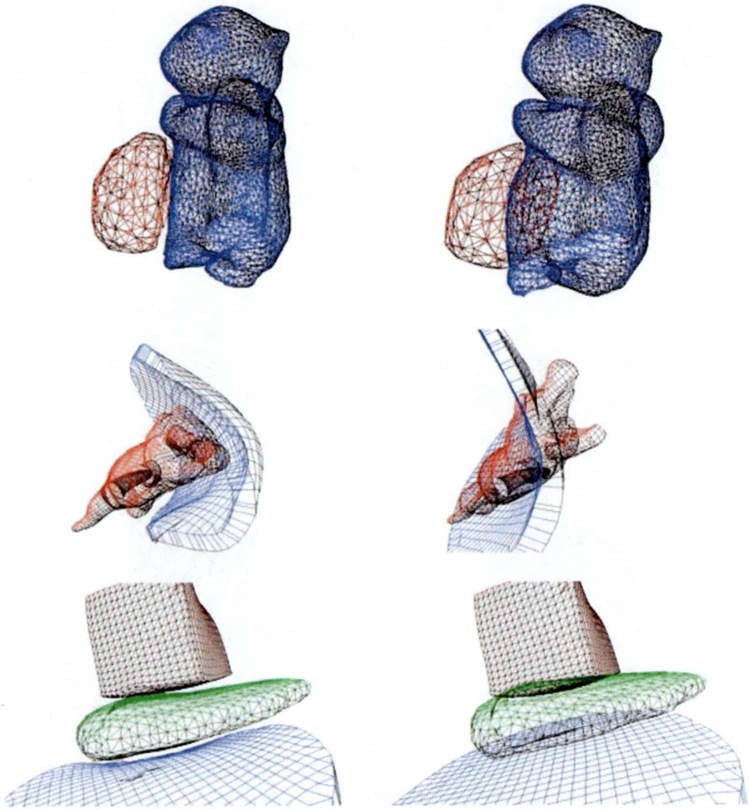

Fig. 24 For the different examples, comparison between registration based on a collision detection model (left column), and without such a model, resulting in large inter-penetrations (right column)

Fig. 25 For the different examples, failure cases when the proposed rigid registration approach is not included in the resolution

The collision detection and response models are also very classical, and a more elegant approach could be undertaken guaranteeing a more physical response by resorting to Signorini's model and by solving the contacts as a linear complementary constraint problem [47]. Including friction through Coulomb's law could also provide some improvements.

Although the system has proven to be fast when considering a single object, multiple object computations linearly increase with the number of objects and even more when contacts occur. One crucial burden in the computations lies in the segmentation process, which is handled independently for each considered object. A simultaneous segmentation method, in a pyramidal manner, could be designed to obtain better efficiency.

10 Application to Robotic Manipulation

In the scope of the RoDyMan project, we present in this section some experiments integrating the tracking method described and validated above into a robotic manipulation task involving a deformable object, that is, the silicon pizza dough. Based on the proposed registration method, the scenario consists of a kinematic control in the operational space to follow a trajectory aiming at manipulating the pizza from one hand to the other on the RoDyMan robotic platform.

10.1 Trajectory Planning and Control

For each hand, the idea is to follow a path that can be decomposed into three paths. The first one is a circular path with constant orientation to make the hand supporting the pizza dough approaching the other one. The second path is also circular, while this time performing a rotation of 180° around the x-axis of $\mathcal{W}$ and a rotation of 50° around the z-axis of $\mathcal{W}$ until the pizza gets overturned. The third is again a circular path that wants to move back the hand to its initial position. In a feed-forward manner, the final position of the second path is computed based on the data provided by the registration algorithm initially, which enables to retrieve the position of the vertex in the mesh corresponding to the object's CoM in its rest shape (see the point C_P in Fig. 26). In detail, it is given by the hand's CoM, C_R, on which the pizza shall be laid, plus a translation along the z-axis of $\mathcal{W}$, and an offset, as seen in Fig. 26. This offset is approximated as the length of the path between C_R and C_P and passing through P_C. The CLIK-based controller controls the joints of the robot's torso and head so that the optical axis of the RGB-D sensor intersects throughout the manipulation task.

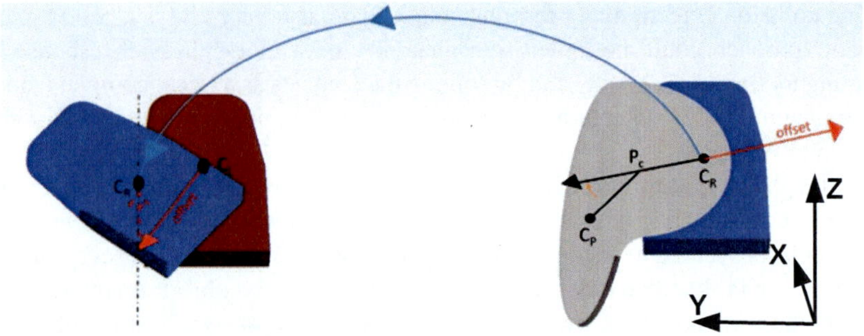

Fig. 26 Trajectory planning for the second circular path

10.2 Experimental Set-Up

The images and the point clouds of the investigated scene are provided by the Asus Xtion RGB-D sensor mounted on the head (320 $times$240 RGB and depth images). The computer hardware and software have the same characteristics as given in Sect. 7.3.

10.3 Comments on the Results

Two passes, way and back, are performed to bring the silicon pizza dough from the right hand to the left and conversely. The resulting trajectories in the $y-z$-plane of the CoM of both hands can be observed in Fig. 28a. Starting from the label dot 1, three phases are addressed. At first, a circular path brings the right hand close to the left (label dot 2). Then, the corrected circular path flips the pizza (label dot 3). Finally, a third circular path brings the hand back to its initial position. The reverse trajectory is then achieved by the left hand.

For the registration phase of the pizza dough during the manipulation phase, the results are presented in Fig. 27. The input RGB images are shown in the first row. The corresponding segmented frames are placed in the second row. Te 3D mesh tracking the object is finally depicted in the third row. The trajectory of C_P is shown in Fig. 28b. Note that the position of C_P is smoothed using a constant velocity Kalman filter.

When the pizza gets flipped, the occlusions caused by the hands prevail too much: the deformation registration process is thus stopped to avoid spurious deformations on the mesh that may not be recovered afterwards when the pizza gets reasonably visible again. Hence, only the rigid ICP process is maintained during these phases, from step 3 to step 4 and step 7 to 8.

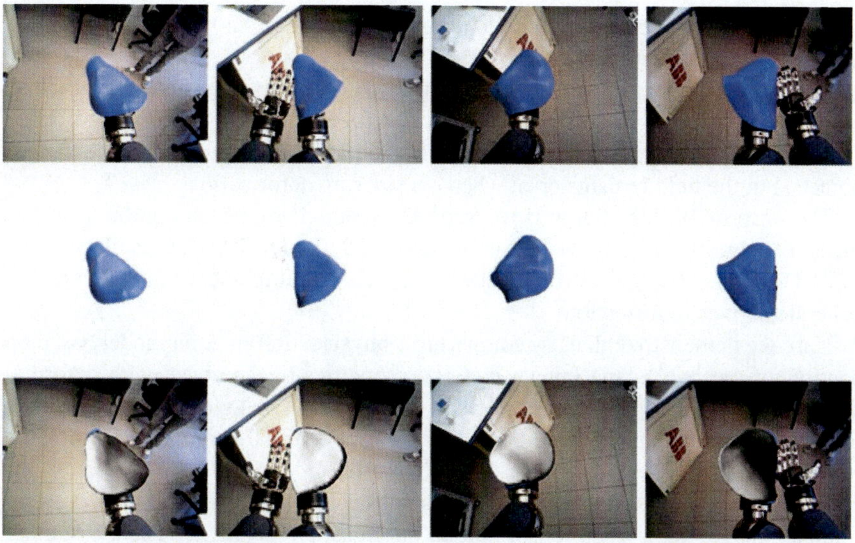

Fig. 27 Results of the tracking process for the silicon pizza, with the input images (first and fourth rows), the segmented frames (second row), and the registered mesh re-projected in the input image with the proposed method

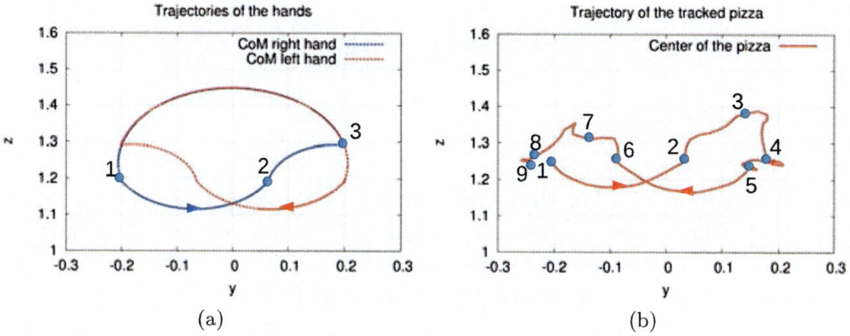

Fig. 28 Trajectories in $y - z$-plane of both hands' COMs (**a**), and trajectory of the vertices corresponding to the object's CoM at rest (**b**)

11 Application to Elasticity Parameter Estimation and Contact Force Estimation

The measure of the contact forces can be a crucial requirement in various applications such as capturing and synthesising human manipulation tasks or controlling robotic hands. Classical techniques use mechatronics sensors, such as force transducers mounted on the manipulator, the manipulated object, or human hands through gloves. This may represent a severe problem in applications like minimally invasive surgical robots, where sensors cannot be easily installed on surgical instruments due to sterilisation or electrification.

Capturing interactions in manipulation based on computer vision has aroused much interest recently and may represent a convenient, minimally invasive, and cheap sensing set-up. Some efforts have been focused on sensing interactions between rigid objects, but the field remains open when considering deformations.

The purpose of this section is to propose a methodology enabling the measurements of contact forces between the operator and a deformable object through an RGB-D camera. The task is very challenging since a multiplicity of interactions can generate a given deformation.

Here we demonstrate that, by comparing a physical deformation model based on continuum mechanics and known material properties of the object with deformations measured through vision, it is possible to retrieve a single point-wise contact force exerted by an operator (a human hand, a robot end-effector, and so on) on the considered object.

Material properties, represented by elasticity parameters, are estimated in a preliminary step using a force sensor and an optimisation technique. Then, assuming that the object lies on a flat surface and that the tool/object interaction consists of a known single contact point, a similar optimisation technique is used to infer the value of the 3D force exerted on that point by fitting the simulated deformations with those estimated by the vision system. This approach is based on the techniques suggested in [3].

11.1 Related Work

Force sensing for robotics or virtual reality applications is typically based on transducers [48]. These devices can be placed either on the object to be manipulated or on the operator, embedded on skins or gloves [49–51]. Otherwise, force is estimated through transduction mechanisms mounted on the robot joints.

While most of these technologies rely on pressure sensing, other recent approaches use visual information as a cue to measure force. For instance, the changes in the appearance of the fingertip are measured through photodetectors or an external camera, and they are processed to estimate contact forces using statistical models, as in [52–57]. These technologies are limited to measure the normal force and cannot simultaneously consider shear or slip. A promising approach [58], also based on visual observations, relies on the GelSight sensor [59]. In this case, the deformations, measured by marker-based visual tracking techniques on an embedded elastomer medium, are interpreted as known responses to the external load exerted on the sensor.

All the above force-sensing methods require precise calibration and providing the interacting tool/hand and/or the manipulated object with cumbersome and expensive equipment, limiting the range of motion. Often they are tailored to particular objects, making difficult the generalisation, whereas we wish to design a generic data-driven system.

Force sensing based on external sensing devices represents an appealing alternative. Vision sensors appear as the most simple, cheap, convenient technology to propose. In the literature, several works have suggested using an external vision system to capture the interactions and sense contact forces in the case of object manipulation by a human or any manipulation tool. In the motion capture field, this approach has been demonstrated by using marker-based or markerless vision tracking to capture and synthesise hand/object interaction [32, 60, 61], providing a kinematic analysis of the interactions through discriminative or generative processes and introducing some physics-based constraints to deal with occlusions or collision detection. Some other approaches [62, 63] propose to bridge the gap between the kinematics provided by the motion capture systems and the contact forces by linking physical constraints to visual observations.

The techniques proposed in [34, 64] go further by employing rigid body and contact dynamics to fully model interactions and to link these models with kinematics provided by external vision systems. In this way, the contact forces between a human and the ground [64] or between a hand and a manipulated object [34] can be estimated in a physically realistic manner. To the best of our knowledge, these works are the only cases proposed in the literature where external visual tracking is used as a cue for contact force estimation in rigid and articulated body interactions.

What we propose in this chapter is to use an external vision system to infer the contact forces deriving from the manipulation of deformable objects. This is achieved by confronting the object's physical properties and its deformation model with the output of the vision system. Our approach is close to the idea in [58] of relating measured deformations to forces. However, the method presented in [58], however, relies on an embedded sensing device and addresses force estimation through empirical relationships between deformations and forces. In our approach, as a significant contribution, we propose to use realistic physical modelling of a deformable object to infer interaction forces exerted on it by processing external vision data. We focus here on a static case, for which the general idea is to estimate a point-wise contact force for which the resulting deformations best fit the deformations measured by the vision system. We assume the contact point to be known and the deformable object to be isotropic. Relying on a physical model implies knowing the mechanical properties of the object. Here, we employ the FEM to model the object and its elasticity described by the Young's modulus and the Poisson's ratio. As a second contribution of this work, we suggest estimating these two parameters. This is handled during a preliminary step by exploiting the vision data and a force sensor, which is also used to validate our vision-based force estimation framework.

11.2 Elasticity Parameter Estimation

Our system consists of estimating the elasticity parameters of the object using the point cloud data (provided by an RGB-D sensor and observing the deformations) and a force sensor. For the estimation of these parameters, we follow the data-driven approaches described in [65, 66]. These approaches minimise a fitting error between the simulated deformations, based on the designed deformation and interaction models presented above and generated by the input operator force acquired from the sensor, and the deformation captured by the RGB-D sensor. These two methods also employ finite elements for the deformation model. The work [66] goes further by proposing a framework that sequentially tracks the shape and estimates both material and dynamic parameters (i.e., the damping) through the dynamic deformation model and solely based on a complete vision capture set-up with various RGB-D sensors around the scene. Here, we reason statically for modeling, and we limit this preliminary process as an estimation process of the Young's modulus and the Poisson's ratio of the material, employing a single RGB-D sensor and the force sensor on a robotic operator, with a set-up similar to the one proposed in [65]. The deformations are indeed generated by applying an effort (a compression, in our case) on the deformable object, and we observe the deformations with the vision sensor once static equilibrium is reached. These deformations can be simulated, starting from the same initial rest shape, provided the deformation and interaction model presented above, the elasticity parameters, and the input measured contact force. As a data-driven approach, our problem is thus addressed by minimising the deviation between these simulated deformations and the observed ones with respect to the elasticity parameters. This deviation is defined by a fitting function accounting for the sum of squared distances between the measured real displacements, observed on the acquired object point cloud, and the simulated deformations. Such a function is defined as

$$e(E, \nu) = \mathrm{dist}(\mathrm{sim}(E, \nu, \mathbf{f}^{op}, \mathbf{x}_{i_c}), Y), \tag{15}$$

where $\mathbf{f}^{op} = \left[f_x^{op}\ f_y^{op}\ f_z^{op}.\right]^T \in \mathbb{R}^3$ is the measured force exerted by the operator on the object, $\mathbf{x}_{i_c} \in \mathbb{R}^3$ is the i-th contact point, Y is the acquired point cloud, dist$(\cdot)$ is the distance function, and sim$(\cdot)$ is the function simulating deformations. The point cloud Y is segmented on the considered object by running the segmentation phase described in Sect. 3 to obtain $\mathcal{Y}$.

For a relevant significant error function, we design it by employing the matching technique presented in Sect. 6 between the segmented point cloud and the visible part of mesh of the virtual object and conversely. Based on the two sets of mesh-to-point cloud and point cloud-to-mesh correspondences, the error function is calculated as:

$$e(E, \nu) = \frac{1}{N_{\mathcal{X}_V}} \sum_{i=0}^{n_{X_V}} (\mathbf{x}_i - \mathbf{NN}_{\mathcal{Y}}(\mathbf{x}_i))^2 + \tag{16}$$

$$\frac{1}{n_Y} \sum_{j=0}^{N_Y} (\mathbf{y}_j - \mathbf{NN}_{\mathcal{X}_V}(\mathbf{y}_j))^2. \tag{17}$$

The optimization problem for (E, ν) is nonlinear, and the evaluation of the objective function is expensive. At the same time, its gradients are non-trivial to compute, making gradient-based optimization methods prohibitive. We thus employ the gradient-free Nelder-Mead method, which is an extension of the downhill simplex method to the nonlinear case. We point out that, during the Nelder-Mead processs, for each evaluation of the objective function, the mesh is initially reset to its rest shape before applying the virtual contact force $\mathbf{f}^o p$ on the known vertex of the mesh, given the elasticity parameters (E, ν). For each evaluated value of (E, ν), a simulation is then started until a static equilibrium is reached. From this static equilibrium, the matching process to derive equation (16) is handled.

11.3 Contact Force Estimation

By inverting the method for parameter estimation, the problem of recovering the contact force exerted on the manipulated object relies on the interaction model between the object, the deforming operator, and the ground, presented in Sect. 1.5. Similarly, we then employ a fitting process, this time with respect to the force, between simulated deformations given the interaction model and the deformations observed using the vision. Since the material parameters are now determined, we can use the deformation model and the registration technique described before as the visual observation, providing a regularized and complete observation of the deformations.

Here, we consider the static case to estimate the interaction force, for which the static equilibrium state of the deformed material due to the exerted contact force is reached. Our system consists in determining the force for which the resulting simulated deformations best fit the mesh deformed by the vision data. Formally, we minimize the least square error e between the deformations sensed through vision, $\mathbf{x}_i^{vision} \in \mathbb{R}^3$, and the deformations simulated based on the interaction model involving, $\mathbf{x}_i^{sim}(\mathbf{f}^{op}) \in \mathbb{R}^3$, with respect to $\mathbf{f}^{op}$

$$e(\mathbf{f}^{op}) = \sum_i (\mathbf{x}_i^{vision} - \mathbf{x}_i^{sim}(\mathbf{f}^{op}))^2 \tag{18}$$

Given the full interaction model presented above, the derivation of the error (18) is non-trivial, and it requires the inversion of the model, making iterative methods (i.e., Newton's one) tricky to handle. Instead, as for the estimation of the material parameters, since we aim at performing a quite global process without any strong guess on this force, we suggest a Nelder-Mead optimization framework to minimize e with respect to $\mathbf{f}^{op}$. In practice, to compute e for a given force $\mathbf{f}^{op}$ in the Nelder-Mead process, we start from the state of the mesh deformed through the vision registration. We then substitute all the external forces due to the point cloud data by $\mathbf{f}^{op}$ on the known vertex in contact with the operator. A simulation is evolved based on this force and the interaction model presented in Sect. 1.5. After a few iterations in the simulation process, e can be computed and thus measures the ability of $\mathbf{f}^{op}$ to reproduce the actions of the forces provided by vision on the object.

11.4 Experimental Results

The results presented here involve a deformable object, a stuffed toy undergoing a compression deformation effort applied by a tool fixed on the end-effector of a Kuka LWR arm, equipped with a force sensor at the wrist (see Fig. 29). The point clouds of the investigated scenes are acquired using a calibrated RGB-D camera Asus Xtion, with 320×240 RGB and depth image resolution. To estimate the elasticity parameters and the applied contact compression force, we process the data of a single RGB-D camera taken at static equilibrium.

A surface mesh of the undeformed object was reconstructed offline to build the deformation model of the stuffed toy by using an RGB-D based dense 3D recon-

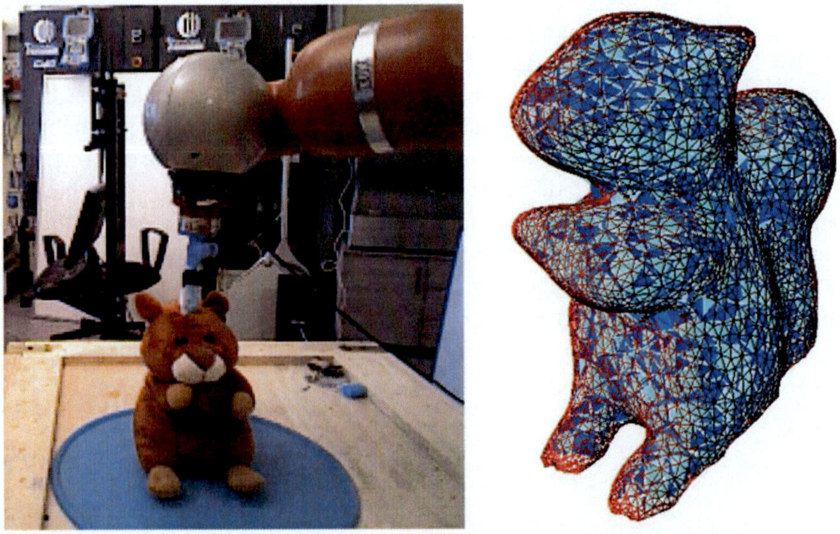

Fig. 29 On the left, the experimental set-up with the tool mounted on the robotic arm, equipped with a force sensor, to compress the object. On the right, the surface triangular (in red) and the volumetric tetrahedral mesh (in blue) of the stuffed toy

struction technique [45] and flying around the object with the Xtion sensor. Then, we manually segment the part of the scene featuring the object. Finally, some re-meshing and smoothing procedures are carried out with a modeling engine to get a fair, closed, and clean surface mesh of the object. The volumetric tetrahedral mesh was generated by carrying out a 3D Delaunay triangulation on the surface mesh with the CGAL library. As a compromise between modeling accuracy and real-time constraints, we generated a volumetric mesh with 951 vertices and 5015 tetrahedral elements (see Fig. 29). As an approximation, we assume the isotropy of the material of the stuffed toy to apply the suitable deformation model. We employed the SOFA simulator for modeling purposes, which enables us to deal with various physical models and evolve simulations in real-time. In terms of hardware, a standard laptop with an NVIDIA GeForce 720M graphic card has been used, along with a 2.4 GHz Intel Core i7 CPU.

11.4.1 Elastic Parameters Estimation

For the estimation of the elastic parameters, we first measure the contact force exerted by the tool mounted on the robotic arm to compress the object, that is $\mathbf{f}_{mech}^{op} = \left[0.17\ 1.125\ 4.006\right]^T$ N. Due to the particular shape of the considered object, the application of this pointwise contact force in simulation may result in the loss of the static equilibrium. For this reason, we constrain the system by fixing some vertices' position on the lower part of the shape, close to the contact area with the flat surface. In this way, the object may not get bent excessively or turned over, and its base remains quite rigidly attached to the flat contact surface.

Following the basic implementation of the Nelder-Mead algorithm, since the parameter space for (E, ν) is dimension two, three candidate samples are sorted after each iteration of the optimization while performing the reflection, expansion, contraction, and shrinking steps. This arrangement provides the best, a good, and the worst candidate. We also integrate the specific boundaries for both E and ν in the process, in the sense that inequalities $E > 0$ and $0 < \nu < 0.5$ should be preserved during the different steps. If an inequality is violated, E or ν is reset slightly below or above.

We tested our parameters estimation technique with two different initial configurations. In Figs. 31 and 32 we can observe the trajectories of the three sorted candidates for these two configurations, along E and ν, throughout the iterations of the Nelder-Mead process. The figures also show the fitting errors for the three candidates after each iteration. In the first case, the initial values are pretty far from the actual estimated one, stressing out the robustness of the estimation for rough initial guesses. At the same time, in the second configuration, the process starts closer to the solution. For both configurations convergence is achieved respectively towards $(E, \nu) = (4268.65\ \text{Pa}, 0.412031)$ and $(E, \nu) = (4328.12\ \text{Pa}, 0.415625)$. As it can be noticed on the plot of the fitting error in Fig. 30, the non-convexity can be espe-

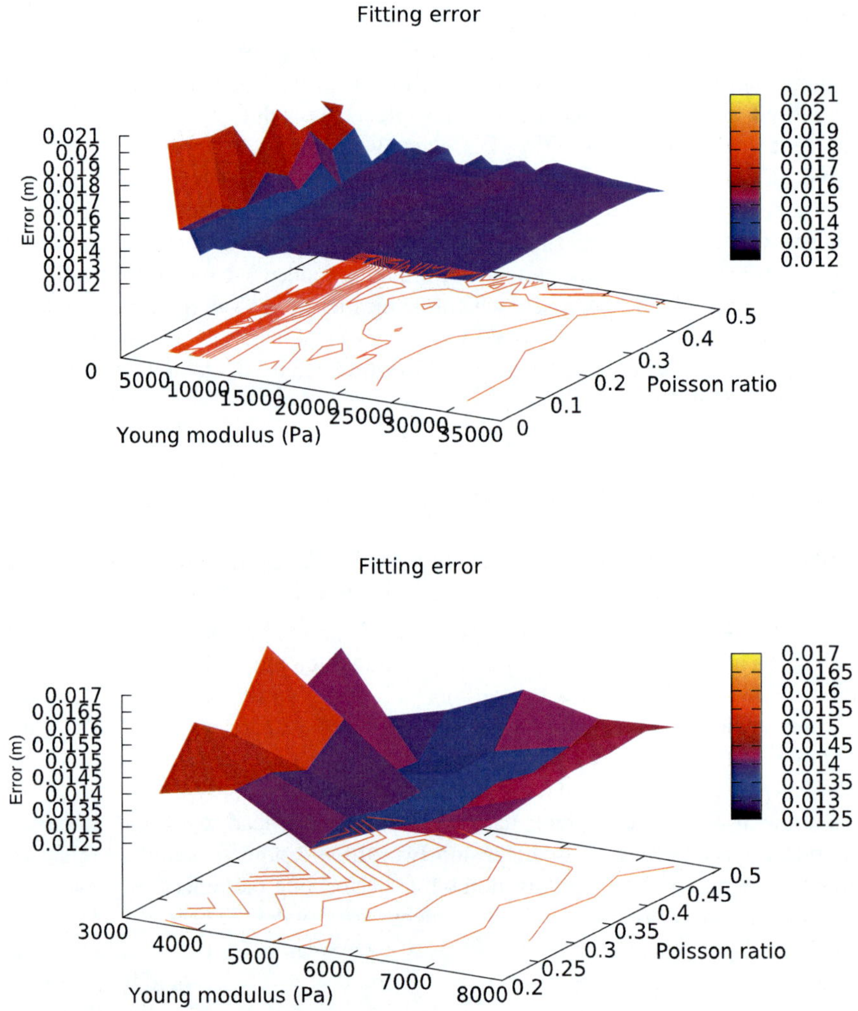

Fig. 30 The top figure shows the fitting error with respect to the elasticity parameters. The bottom one depicts a closer view around the global minimum

cially observable for the ν parameter, along which the error is relatively flat, resulting in some local minima.

11.4.2 Contact Force Estimation

Intending to test the operator contact force estimation based on the vision tracking system, we first proceed by setting the material parameters of the deformation models

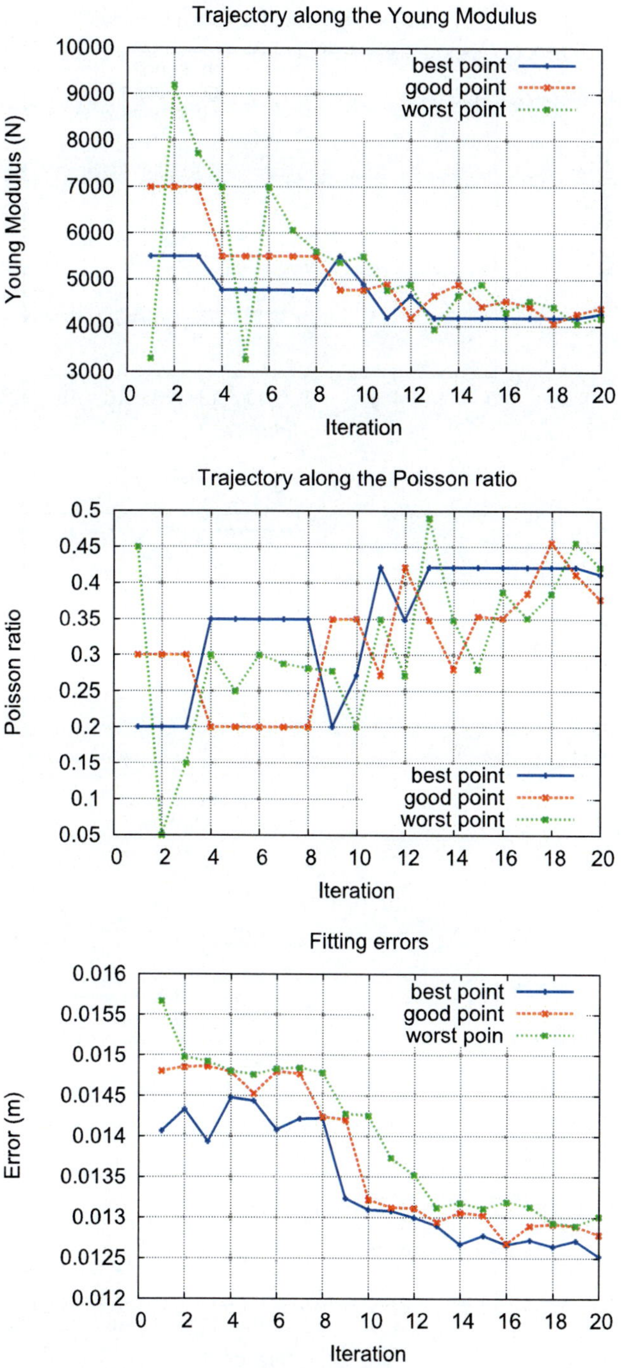

Fig. 31 Nelder-Mead process for elasticity parameters estimation for the first initial configuration

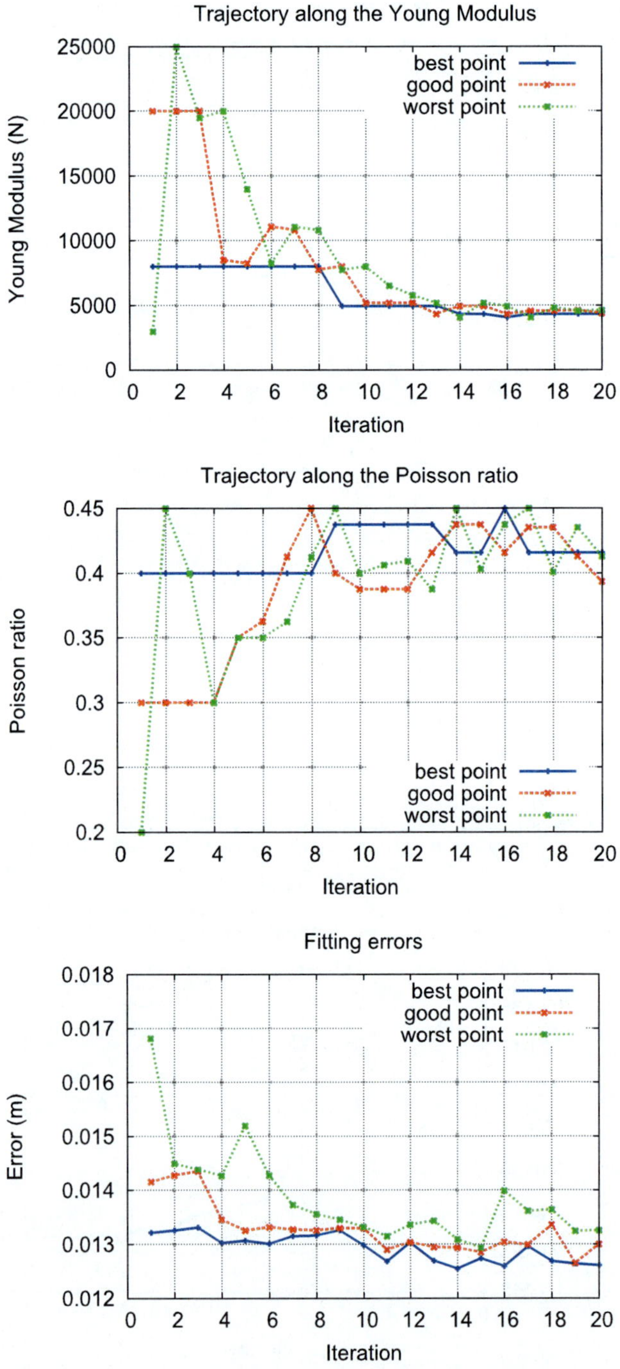

Fig. 32 Nelder-Mead process for elasticity parameters estimation for the second initial configuration

(a) (b) (c)

Fig. 33 Registration process, with: **a** Preliminary segmentation, **b** Fitting result between the mesh and the segmented point cloud, **c** Registered mesh reprojected in the image

used in the vision system. Here, we use $\mathrm{d}(E, \nu) = (4268.65\ \mathrm{Pa}, 0.412031))$. The result of the registration process can be observed in Fig. 33.

For each evaluation of the error function, the registered mesh is relaxed from the forces exerted by the vision while applying the point-wise contact force to evaluate the known vertex. Fixing vertices as boundary conditions to constraint the simulation is not necessary since we measure here the ability of this force to keep the static equilibrium already reached by the action of the vision forces. Since the parameter space is three-dimensional, four samples are sorted after each iteration, namely the best, the worst, and two intermediate samples. Two different initial configurations are tested here. The former starts quite far, without any particular guess on the intensity and direction of the force. The latter is close to the actual value of the force given by the sensor $\mathbf{f}^{op}_{mech} = \left[0.17\ 1.125\ 4.006\right]^T$ N. Figures 34 and 35 show in both cases the trajectories of the four candidates for the estimate of the contact force $\mathbf{f}^{op}$ and the corresponding fitting errors. In both cases, the algorithm converges respectively towards a force of $\mathbf{f}^{op} = \left[0.618\ 0.0687929\ 3.54801\right]^T$ N and $\mathbf{f}^{op} = \left[-0.462414\ 0.247626\ 3.71292\right]^T$ N which are relatively close to the value sensed by the force sensor mounted on the robot, thus validating our whole model. Convergence is of course reached much faster in the second case.

11.5 Comments on the Results

The proposed framework consists of (i) estimating the material parameters based on a known exerted force to develop a deformation model, and (ii) estimating the force based on the known deformation model and on a registration technique that allows measuring deformations. The results presented in this chapter are promising, but several issues shall be discussed.

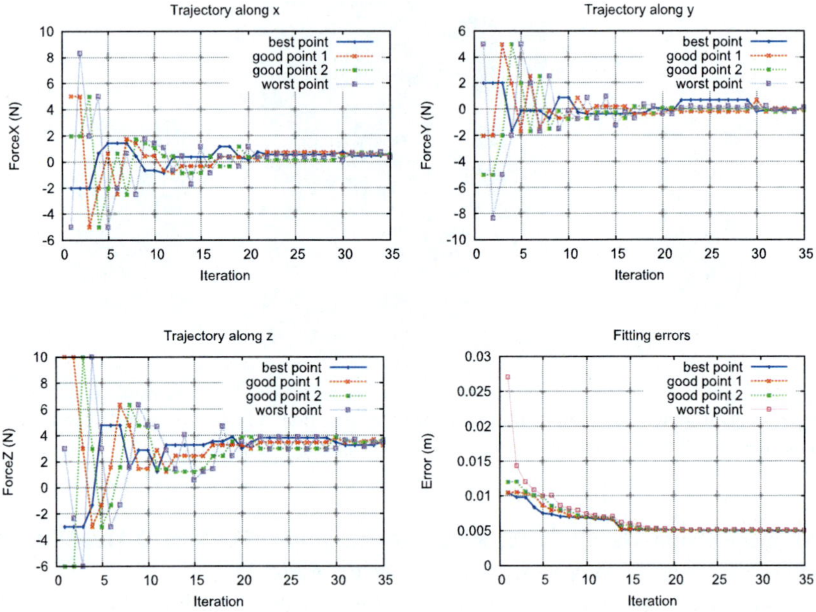

Fig. 34 Nelder-Mead process for force estimation, for the first initial configuration

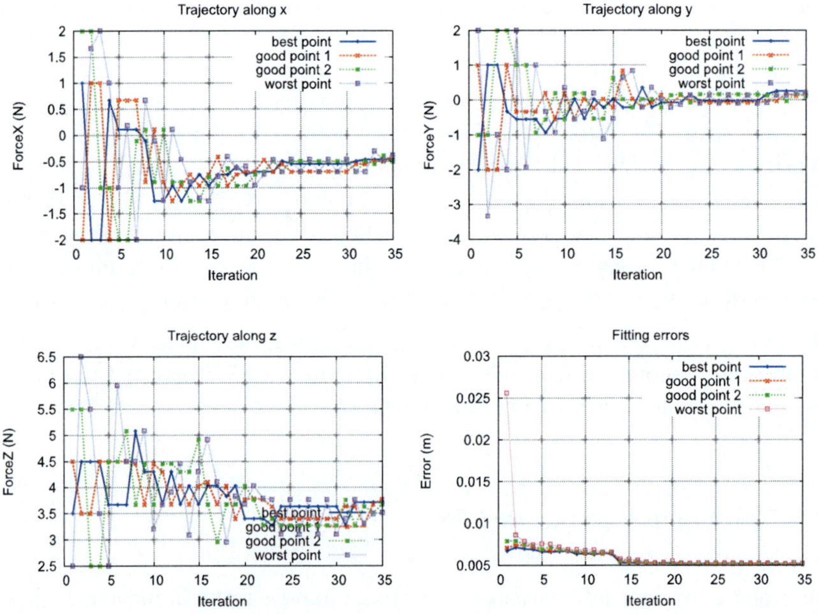

Fig. 35 Nelder-Mead process for force estimation, for the second initial configuration

Our imaging set-up is based on a single Asus Xtion RGB-D sensor, providing relatively low resolution, partial and noisy point cloud data around the object. It results in non-convex shapes for the fitting error function in the elasticity estimation process, especially concerning ν, or in registration errors. A more sophisticated set-up to capture deformations, such as the one proposed in [66], with a set of RGB-D sensors at different viewpoints, would give more accurate results for the estimation of both the mechanical parameters and the contact force.

Besides, we assumed the contact point between the object and the operator to be known, as well as the contact between the object and the underlying flat surface. Further development of our approach would be, following the method described in Sect. 9, the design of a vision system able to capture the interaction between the object and its interacting environment: the manipulation tool, the table, and so on, enabling the detection of contact points, some proper priors for the segmentation and registration of the different entities.

In the estimation of the force exerted by the operator, we use a gradient-free Nelder-Mead optimization method. This has the advantage of being relatively easy to implement and robust to a coarse initialization. It is, however, relatively slow to run. Indeed, for force estimation, each evaluation of the error function requires at least five successive simulations to obtain a reliable error for deformations of the registration process. An iteration in the Nelder-Mead algorithm requires, in this case, seven evaluations, around 350 ms, given that one simulation takes around 10 ms, making the process quite far from being real-time if considering a stream a successive RGB-D data. A possible improvement would be to investigate an efficient inversion of the whole interaction model and some local optimization techniques such as quadratic programming, as proposed in [67].

Finally, our system is designed for the static case, for which deformations have reached a static equilibrium. The presented method might be adapted to a dynamic case by benefiting from a measure, through vision, of the kinematics of the object or interacting entities. Hence, based on Lagrangian dynamics, the system could track online both the deformations and the force.

12 Discussion and Conclusion

Real-time perception is a crucial issue for robotic nonprehensile dynamic manipulation tasks, especially when considering deformable objects such as tissues, organs, clothes, or food, which has to be manipulated in a very challenging way, as in the case of a pizza being stretched and tossed by a humanoid pizza chef robot.

In this chapter, we have described a series of contributions to register and track deformable objects undergoing fast rigid motions, large deformations, and behaviours such as fractures and interactions with other objects or with the environment, based on the deformations models introduced in the previous chapter.

The recent development of physics-based modelling methods for deformable elastic objects for registration purposes and the availability of real-time implementations led us to choose such an approach to track a textureless and smooth object subjected to various large deformations with an RGB-D sensor. The main idea is to continuously fit the volumetric deformation models onto the successive 3D point clouds provided by the RGB-D camera. At the same time, the FEM modelling acts as a regularizer for the registration.

For the primary case of a single object, the use of the corotational FEM model, an efficient segmentation method, and a classical point cloud registration techniques have made our system a promising real-time tracking method able to handle various deformations and motions while being robust to segmentation errors or occlusions.

Regarding fractures, our approach, as the first attempt to deal with such topological changes in a computer vision and registration problem, consists of integrating into the FEM deformation model a model for fracture events. Fracturable nodes are firstly detected in the considered mesh by decomposing the internal forces into tensile and compressive components. In a second step, we perform a simple local re-meshing of the elements around the fracture nodes. The registration process with the RGB-D data is the same as for a single object to compute the deformations. The investigation of fracture events over the mesh and the eventual re-meshing is purely physics-based. Experimental results on two different objects and three sorts of fractures show the relevance of this simple physical-based method. However, this can be improved by handling more properly the re-meshing phase by dealing with more significant elastic and plastic deformations and handling occlusions by distinguishing them with fracture cracks.

We have also designed a system able to register multiple interacting deformable elastic objects simultaneously. This system also relies on the corotational FEM elastic models to accurately and efficiently model elasticity and an interaction model consisting of collision detection and response models to cope with contacts between objects. Based on RGB-D input data, point cloud matching and registration processes are achieved on segmented data in a parallel manner, enabling to deal with occlusions. Rigid motions, deformations, and contacts of all the considered objects are then estimated in a common solver, providing a realistic behaviour and coherently constraining the registration of the involved entities. The method has been tested on simple scenarios with elastic objects but could be extended to other physical models to articulate objects, such as for robotic manipulation applications.

A first step towards robotic manipulation of such objects has been achieved by performing a simple manipulation task on the RoDyMan platform.

Finally, another application has focused on designing a system able to recover a single contact force exerted on a deformable object by resorting to an external vision and registration system and fitting simulated deformations with the observed ones. This is achieved by taking advantage of the physical FEM modelling and employing a basic interaction model with the manipulator and the environment, proposing a simple manipulation scenario. The deformation model requires two mechanical parameters, Young's modulus and Poisson's ratio, which we initially estimate by fitting simulated deformations with the point cloud given by an RGB-D sensor. The entirely determined elastic model then feeds the registration system and the simulator, which are matched to the contact force. We address these two optimization problems using gradient-free Nelder-Mead methods. Some promising results have been obtained on a simple case of a single applied compression force on a known contact point at static equilibrium.

References

1. A. Petit, S. Cotin, V. Lippiello, B. Siciliano. Capturing deformations of interacting non-rigid objects using RGB-D data, in *2018 IEEE/RSJ International Conference on Intelligent Robots and Systems*, pp. 491–497, Madrid, E (2018)
2. A. Petit, V. Lippiello, G.A. Fontanelli, B. Siciliano, Tracking elastic deformable objects with an RGB-D sensor for a pizza chef robot. Robot. Auton. Syst. **88**, 187–201 (2017)
3. A. Petit, V. Lippiello, B. Siciliano, Tracking fractures of deformable objects in real-time with an RGB-D sensor, in *2015 International Conference on 3D Vision*, pp. 632–639, Lyon, F (2015)
4. C. Elbrechter, R. Haschke, H. Ritter, Bi-manual robotic paper manipulation based on real-time marker tracking and physical modelling, in *2011 IEEE/RSJ International Conference on Intelligent Robots and Systems*, pp. 1427–1432 (2011)
5. N. Haouchine, J. Dequidt, M.-O. Berger, S. Cotin, Monocular 3D reconstruction and augmentation of elastic surfaces with self-occlusion handling. IEEE Trans. Vis. Comput. Gr. **21**(12), 1363–1376 (2015)
6. N. Haouchine, J. Dequidt, I. Peterlik, E. Kerrien, M.-O. Berger, S. Cotin, Image-guided simulation of heterogeneous tissue deformation for augmented reality during hepatic surgery, in *2013 IEEE International Symposium on Mixed and Augmented Reality*, pp. 199–208 (2013)
7. J. Schulman, A. Lee, J. Ho, P. Abbeel, Tracking deformable objects with point clouds, in *2013 IEEE International Conference on Robotics and Automation*, pp. 1130–1137 (2013)
8. A. Bartoli, V. Gay-Bellile, U. Castellani, J. Peyras, S. Olsen, P. Sayd, Coarse-to-fine low-rank structure-from-motion in *2008 IEEE Conference on Computer Vision and Pattern Recognition*, pp. 1–8 (2008)
9. A. Weiss, D. Hirshberg, M.J. Black, Home 3D body scans from noisy image and range data, in *2011 IEEE International Conference on Computer Vision*, pp. 1951–1958 (2011)
10. A. Bartoli, A. Zisserman, Direct estimation of non-rigid registrations, in *British Machine Vision Conference*, pp. 899–908 (2004)
11. M. Kass, A. Witkin, D. Terzopoulos, Snakes: active contour models. Int. J. Comput. Vis. **1**(4), 321–331 (1988)
12. J. Pilet, V. Lepetit, P. Fua, Fast non-rigid surface detection, registration and realistic augmentation. Int. J. Comput. Vis. **76**(2), 109–122 (2007)
13. D. Terzopoulos, A. Witkin, M. Kass, Constraints on deformable models: recovering 3D shape and nonrigid motion. Artif. Intell. **36**(1), 91–123 (1988)
14. M. Salzmann, J. Pilet, S. Ilic, P. Fua, Surface deformation models for nonrigid 3D shape recovery. IEEE Trans. Pattern Anal. Mach. Intell. **29**(8), 1481–1487 (2007)
15. A. Jordt, R. Koch, Direct model-based tracking of 3D object deformations in depth and color video. Int. J. Comput. Vis. **102**, 1–17 (2013)
16. M. Zollhöfer, M. Nießner, S. Izadi, C. Rehmann, C. Zach, M. Fisher, C. Wu, A. Fitzgibbon, C. Loop, C. Theobalt, M. Stamminger, Real-time non-rigid reconstruction using an RGB-D camera. ACM Trans. Gr. **33**(4), 1–12 (2014)
17. B. Allain, J.-S. Franco, E. Boyer, An efficient volumetric framework for shape tracking, in *2015 IEEE Conference on Computer Vision and Pattern Recognition*, pp. 268–276 (2015)
18. R.A. Newcombe, D. Fox, S.M. Seitz, Dynamic fusion: reconstruction and tracking of non-rigid scenes in real-time, in *2015 IEEE Conference on Computer Vision and Pattern Recognition*, pp. 343–352 (2015)

19. V. Lippiello, F. Ruggiero, B. Siciliano, Floating visual grasp of unknown objects using an elastic reconstruction surface, in *Robotics Research: The Fourteenth International Symposium, in Springer Tracts in Advanced Robotics 70* ed. by C. Pradalier, R. Siegwart, G. Hirzinger (Springer, 2011), pp. 329–344
20. L.D. Cohen, I. Cohen, Deformable models for 3-D medical images using finite elements and balloons, in *1992 IEEE Conference on Computer Vision and Pattern Recognition*, pp. 592–598 (1992)
21. T. McInerney, D. Terzopoulos, A finite element model for 3D shape reconstruction and nonrigid motion tracking, in *1993 IEEE International Conference on Computer Vision*, pp. 518–523 (1993)
22. A. Malti, R. Hartley, A. Bartoli, J.-H. Kim. Monocular template-based 3D reconstruction of extensible surfaces with local linear elasticity, in *2013 IEEE Conference on Computer Vision and Pattern Recognition*, pp. 1522–1529 (2013)
23. K. Varanasi, A. Zaharescu, E. Boyer, R. Horaud, Temporal surface tracking using mesh evolution, in *European Conference on Computer Vision*, pp. 30–43 (2008)
24. A. Zaharescu, E. Boyer, R. Horaud, Topology-adaptive mesh deformation for surface evolution, morphing, and multiview reconstruction. IEEE Trans. Pattern Anal. Mach. Intell. **33**(4), 823–837 (2011)
25. C.J. Paulus, N. Haouchine, S.-H. Kong, R.V. Soares, D. Cazier, S. Cotin, Handling topological changes during elastic registration. Int. J. Comput. Assist. Radiol. Surg. **12**(3), 461–470 (2017)
26. A. Tsoli, A.A. Argyros, Tracking deformable surfaces that undergo topological changes using an RGB-D camera, in *2016 Fourth International Conference on 3D Vision*, pp. 333–341 (2016)
27. I. Badami, J. Stückler, S. Behnke, Depth-enhanced Hough Forests for object-class detection and continuous pose estimation, in *Workshop on Semantic Perception, Mapping and Exploration* (2013)
28. A. Tejani, D. Tang, R. Kouskouridas, T.-K. Kim, Latent-Class Hough Forests for 3D object detection and pose estimation, in *European Conference on Computer Vision*, pp. 462–477 (2014)
29. H. Hamer, K. Schindler, E. Koller-Meier, L. Van Gool, Tracking a hand manipulating an object, in *2009 IEEE International Conference On Computer Vision*, pp. 1475–1482 (2009)
30. K. Kim, V. Lepetit, W. Woo, Keyframe-based modeling and tracking of multiple 3d objects, in *2010 9th IEEE International Symposium on Mixed and Augmented Reality*, pp. 193–198 (2010)
31. N. Kyriazis, A. Argyros, Scalable 3D tracking of multiple interacting objects, in *2014 IEEE Conference on Computer Vision and Pattern Recognition*, pp. 3430–3437 (2014)
32. I. Oikonomidis, N. Kyriazis, A.A. Argyros, Full DoF tracking of a hand interacting with an object by modeling occlusions and physical constraints, in *2011 IEEE International Conference on Computer Vision*, pp. 2088–2095 (2011)
33. I. Oikonomidis, N. Kyriazis, A.A. Argyros, Tracking the articulated motion of two strongly interacting hands, in *2012 IEEE Conference on Computer Vision and Pattern Recognition*, pp. 1862–1869 (2012)
34. T.-H. Pham, A. Kheddar, A. Qammaz, A.A. Argyros, Towards force sensing from vision: Observing hand-object interactions to infer manipulation forces, in *2015 IEEE Conference on Computer Vision and Pattern Recognition*, pp. 2810–2819 (2015)
35. M. Salzmann, R. Urtasun, Physically-based motion models for 3D tracking: A convex formulation, in *2011 IEEE International Conference on Computer Vision*, pp. 2064–2071 (2011)
36. S. Sridhar, F. Mueller, M. Zollhöfer, D. Casas, A. Oulasvirta, C. Theobalt, Real-time joint tracking of a hand manipulating an object from RGB-D input, in *European Conference on Computer Vision*, pp. 294–310 (2016)
37. K. Guo, F. Xu, T. Yu, X. Liu, Q. Dai, Y. Liu, Real-time geometry, albedo, and motion reconstruction using a single RGB-D camera. ACM Trans. Gr. **36**(4), 32:1–32:13 (2017)
38. C. Rother, V. Kolmogorov, A. Blake, Grabcut: interactive foreground extraction using iterated graph cuts. ACM Trans. Gr. **23**, 309–314 (2004)

39. Y. Boykov, O. Veksler, R. Zabih, Fast approximate energy minimization via graph cuts. IEEE Trans. Pattern Anal. Mach. Intell. **23**(11), 1222–1239 (2001)
40. Y. Chen, G. Medioni, Object modelling by registration of multiple range images. Image Vis. Comput. **10**(3), 145–155 (1992)
41. F. Faure, C. Duriez, H. Delingette, J. Allard, B. Gilles, S. Marchesseau, H. Talbot, H. Courtecuisse, G. Bousquet, I. Peterlik, S. Cotin, Sofa: A multi-model framework for interactive physical simulation, in *Soft Tissue Biomechanical Modeling for Computer Assisted Surgery*, pp. 283–321 (2012)
42. A.V. Gelder, Approximate simulation of elastic membranes by triangulated spring meshes. J. Gr. Tools **3**(2), 21–41 (1998)
43. B.A. Lloyd, G. Székely, M. Harders, Identification of spring parameters for deformable object simulation. IEEE Trans. Vis. Comput. Gr. **13**(5), 1081–1094 (2007)
44. A. Petit, V. Lippiello, B. Siciliano, Real-time tracking of 3D elastic objects with an RGB-D sensor, in *2015 IEEE International Conference on Intelligent Robots and Systems*, pp. 3914–3921 (2015)
45. R.A. Newcombe, A.J. Davison, S. Izadi, P. Kohli, O. Hilliges, J. Shotton, D. Molyneaux, S. Hodges, D. Kim, A. Fitzgibbon. Kinect Fusion: real-time dense surface mapping and tracking, in *10th IEEE International Symposium on Mixed and Augmented Reality*, pp. 127–136 (2011)
46. O. Kähler, V.A. Prisacariu, D.W. Murray, Real-time large-scale dense 3d reconstruction with loop closure, in *European Conference on Computer Vision*, pp. 500–516 (2016)
47. J. Allard, F. Faure, H. Courtecuisse, F. Falipou, C. Duriez, P.G. Kry, Volume contact constraints at arbitrary resolution, in *ACM SIGGRAPH 2010 Papers*, pp. 1–10 (2010)
48. B. Siciliano, O. Khatib, *Springer Handbook of Robotics*, 2nd edn. (Springer Science & Business Media, 2016)
49. R.S. Dahiya, G. Metta, M. Valle, G. Sandini, Tactile sensing-from humans to humanoids. IEEE Trans. Rob. **26**(1), 1–20 (2010)
50. N. Wettels, J.A. Fishel, Z. Su, C.H. Lin, G.E. Loeb, Multi-modal synergistic tactile sensing, in *Tactile Sensing in Humanoids-Tactile Sensors and Beyond Workshop, 9th IEEE-RAS International Conference on Humanoid Robots* (2009)
51. H. Yousef, M. Boukallel, K. Althoefer, Tactile sensing for dexterous in-hand manipulation in robotics-A review. Sens. Actuat. A **167**(2), 171–187 (2011)
52. T.R. Grieve, J.M. Hollerbach, S.A. Mascaro, Force prediction by fingernail imaging using active appearance models, in *2013 World Haptics Conference*, pp. 181–186 (2013)
53. D. Hristu, N. Ferrier, R.W. Brockett, The performance of a deformable-membrane tactile sensor: Basic results on geometrically-defined tasks, in *2000 IEEE International Conference on Robotics and Automation*, pp. 508–513 (2000)
54. S.A. Mascaro, H.H. Asada, Photoplethysmograph fingernail sensors for measuring finger forces without haptic obstruction. IEEE Trans. Robot. Autom. **17**(5), 698–708 (2001)
55. Y. Sun, J.M. Hollerbach, S.A. Mascaro, Predicting fingertip forces by imaging coloration changes in the fingernail and surrounding skin. IEEE Trans. Biomed. Eng. **55**(10), 2363–2371 (2008)
56. Y. Sun, J.M. Hollerbach, S.A. Mascaro, Estimation of fingertip force direction with computer vision. IEEE Trans. Rob. **25**(6), 1356–1369 (2009)
57. S. Urban, J. Bayer, C. Osendorfer, G. Westling, B.B. Edin, P. Van Der Smagt, Computing grip force and torque from finger nail images using gaussian processes, in *2013 IEEE/RSJ International Conference on Intelligent Robots and Systems*, pp. 4034–4039 (2013)
58. W. Yuan, R: Li, M.A. Srinivasan, and E.H. Adelson. Measurement of shear and slip with a GelSight tactile sensor, in *2015 IEEE International Conference on Robotics and Automation*, pp. 304–311 (2015)
59. K. Sato, K. Kamiyama, N. Kawakami, S. Tachi, Finger-shaped gelforce: Sensor for measuring surface traction fields for robotic hand. IEEE Trans. Haptics **3**(1), 37–47 (2010)
60. L. Ballan, A. Taneja, J. Gall, L. Van Gool, and M. Pollefeys. Motion capture of hands in action using discriminative salient points, in *European Conference on Computer Vision*, pp. 640–653 (2012)

61. N. Kyriazis, A. Argyros, Physically plausible 3D scene tracking: the single actor hypothesis. In *2013 IEEE Conference on Computer Vision and Pattern Recognition*, pp. 9–16 (2013)
62. Y. Wang, J. Min, J. Zhang, Y. Liu, F. Xu, Q. Dai, J. Chai, Video-based hand manipulation capture through composite motion control. ACM Trans. Gr. **32**(4), 43 (2013)
63. W. Zhao, J. Zhang, J. Min, J. Chai, Robust realtime physics-based motion control for human grasping. ACM Trans. Gr. **32**(6), 1–12 (2013)
64. M.A. Brubaker, L. Sigal, D.J. Fleet, Estimating contact dynamics, in *2009 IEEE International Conference on Computer Vision*, pp. 2389–2396 (2009)
65. B. Frank, R. Schmedding, C. Stachniss, M. Teschner, W. Burgard, Learning the elasticity parameters of deformable objects with a manipulation robot, in *2010 IEEE/RSJ International Conference on Intelligent Robots and Systems*, pp. 1877–1883 (2010)
66. B. Wang, L. Wu, K. Yin, U. Ascher, L. Liu, H. Huang, Deformation capture and modeling of soft objects. ACM Trans. Gr. **34**(4), 94:1–94:12 (2015)
67. F. Largilliere, V. Verona, E. Coevoet, M. Sanz-Lopez, J. Dequidt, C. Duriez, Real-time control of soft-robots using asynchronous finite element modeling, in *2015 IEEE International Conference on Robotics and Automation*, pp. 2550–2555 (2015)

Smoothed Particle Hydrodynamics-Based Viscous Deformable Object Modelling

Jung-Tae Kim, Fabio Ruggiero, Vincenzo Lippiello, and Bruno Siciliano

Abstract Materials like fluids are long since important research objects of continuum mechanics as well as of computer graphics. Smoothed particle hydrodynamics(SPH) is one of the representation methods employed for continuous materials. Its simplicity in implementation and its realistic representation are drastically improved during the last decades. More recently, highly viscous fluids like honey, jam, and bread dough based on the SPH formulation have gained attention with impressive results. In this chapter, a novel implicit viscosity method is proposed. The internal viscosity forces are recursively calculated from the difference of the nearby velocities of the particles until they are small enough to be neglected. The proposed approach has longer time-steps compared with existing explicit viscosity methods, resulting in shorter computation time. Besides, the proposed method uses a physical viscosity coefficient, not an artificial one like in existing implicit viscosity methods, which helps predict the viscous behavior of continuous materials more accurately. The obtained results show that the computational time for the proposed approach is predictable, while the accuracy in modelling the viscosity behaviour is similar or higher than existing methods.

J.-T. Kim (✉)
MIKNMEK Inc., CheongAm-ro 87, Nam-gu, Pohang-si, Gyeongsangbuk-do 37673, Republic of Korea

F. Ruggiero · V. Lippiello · B. Siciliano
Department of Electrical Engineering and Information Technology, PRISMA Lab, CREATE Consortium & University of Naples Federico II, Via Claudio 21, 80125 Naples, Italy
e-mail: fabio.ruggiero@unina.it

V. Lippiello
e-mail: vincenzo.lippiello@unina.it

B. Siciliano
e-mail: bruno.siciliano@unina.it

B. Siciliano and F. Ruggiero (eds.), *Robot Dynamic Manipulation*, Springer Tracts in Advanced Robotics 144,
https://doi.org/10.1007/978-3-030-93290-9_3

1 Brief Introduction

Continuous materials (e.g., solid objects, fluids) were widely and deeply investigated a long time after Augustin-Louis Cauchy firstly formulated them in the 19th century. Various methods for modelling continuous materials are categorized as Lagrangian versus Eulerian approach, mesh-based versus mesh-free approach, or a hybrid combination. Below, a list of well-established methods is revised (Table 1).

The finite difference method(FDM) is an Eulerian grid method constructing regular grids [1], while the finite volume method(FVM) generates sub-domains referred to as cells [2]. The finite element method(FEM) is instead a grid method requiring mesh generation for their particle elements [3, 4]. Within the GSMs, gradient smoothing operations based on relevant gradient smoothing domains are employed to approximate derivatives [5]. The SPH consists of Lagrangian particles carrying the convection properties like mass, pressure, and velocity [6, 7]. Among the hybrid methods, it is worth mentioning the FLIP approach [8, 9], which is represented with Lagrangian particles while the projection step is performed on an Eulerian grid, and the MPM [10, 11], which handles collision and fracture of Lagrangian particles with the use of an Eulerian grid. The PBFM [12] applies geometric constraints of the PBDM [13] to enforcing constant density in the SPH framework. The integration of the PIC approach [14] and the FLIP one was proposed in [15] for animating granular materials such as the sand. The combination of the SPH's incompressibility solver with the position-based velocity correction for volume preservation of viscoelastic fluids was presented in [16].

Each of the methods listed above has its benefits and issues compared to the others. Recently, thanks to the drastically increasing computation power of computers and

Table 1 Main symbols used in this chapter

Definition	Symbol
Material or substantial derivative	$\frac{\mathrm{D}}{\mathrm{D}t}$
Fluid density	$\rho > 0$
Space dimension	$r = \{2, 3\}$
Flow vector velocity field	$\mathbf{v} \in \mathbb{R}^r$
Hydrostatic pressure	$p > 0$
Viscous stress	$\mathbf{T} \in \mathbb{R}^{r \times r}$
Body force due to gravity, surface tension, or friction	$\mathbf{f}^{body} \in \mathbb{R}^r$
ith fluid sample point	$\mathbf{x}_i \in \mathbb{R}^r$
Dynamic viscosity coefficient	$\mu > 0$
Neighbourhood of the i-the particle	$\mathcal{N}_i$
Mass of the ith particle	$m_i > 0$
Reynold number	$Re > 0$

the development of novel algorithms, the mesh-free Lagrangian modelling methods, like the SPH, are becoming more practical [6]. The intuitive and straightforward formulation of the SPH approach is suitable to describe highly deformable objects with complex surfaces. For this reason, such a method fits well the continuous material handled by this chapter.

The SPH method was firstly designed in [17–19] for analyzing astrophysical problems. Naturally, it emphasizes convection properties like mass rather than geometric volume conservation, precisely the benefits and drawbacks of a Lagrangian approach compared to an Eulerian approach. While SPH methods were applied to highly dense continuous materials like water, addressing the incompressibility of the materials would be a significant improvement. The first SPH method to simulate less compressible continuous materials, the so-called XSPH variant for matter diffusion, was proposed in [7, 20, 21] by including an artificial viscosity approach, the equation of state, and boundary conditions. Another contribution in applying the SPH method to incompressible materials can be found in [22]. An SPH technique based on a projection method to model incompressible flows was proposed in [23] by projecting a velocity onto a divergence-free subspace using pressure correction. An ISPH approach, in which the incompressibility of the material is satisfied through a pressure Poisson's equation, was proposed in [24]. A WCSPH method was proposed in [25] by using Tait's equation, resulting in fast computation and less density fluctuation. The success of the PCISPH method proposed in [26, 27] raised the popularity of implicit methods over explicit ones. The IISPH method in [28] computes density deviation not based on the position but based on the velocity, resulting in a robust time-integration scheme. The DFSPH approach computes impulse force to maintain the initial constant density and the divergence-free velocity field [29].

This chapter deals with highly viscous continuous materials like honey, jam, and bread dough. Based on the SPH formulation, several viscosity methods exist modelling incompressible behaviour of continuous materials. A conventional SPH method to model the incompressible flows with a low Reynolds number was proposed in [30]. A double density relaxation procedure to enforce incompressibility and particle anti-clustering was used in [31]. The work in [32] integrated the pressure force and the viscosity force in [33] with an additional elastic force derived from a modification of Hooke's law.

The Cross's model [34] for variable viscosity under shear stress in non-Newtonian fluids was employed in [24]. Later, a similar approach was used in [35, 36], resulting in accurate modelling of viscous jet buckling. An implicit viscosity integration method was presented in [37]: it was good at generating rotational viscous fluid behaviours like coiling or buckling with long time-steps. At the same time, a different implicit viscosity formulation was proposed in [38]. An impulse force-based implicit viscosity method, which is a similar approach to a previous incompressibility solver [29], was introduced in [39]. The DC-PBD solver proposed in [40] handles both position-based and velocity-based constraints efficiently so that large-scaled and highly viscous fluids are animated with high speed. A further implicit viscosity solver, which is controllable by physical viscosity coefficients, was proposed in [41]. Besides these SPH-based methods, other approaches model continuous viscous materials as well. Namely, a variant of the MAC algorithm was proposed in [42] to simulate wax, while an implicit Eulerian method for simulating free-surface viscous fluids was proposed in [43].

Recently, the vorticity diffusion of continuous material related to the turbulence or the eddy near solid boundaries in high Reynolds number flows has gained attention within the research community. For instance, the implicit viscosity formulation in [38] was improved in [44] to include vorticity diffusion. The DVH was proposed in [45] to generate a highly accurate vorticity field in the 2D space. However, this chapter deals with high viscous fluids with a related flow characterized by a low Reynolds number and a slow velocity. Therefore, the vorticity diffusion is not addressed here.

The method introduced in this chapter takes inspiration from the implicit viscosity method addressed in [37–39]. Hence, the proposed method has long time-steps as the other implicit viscosity methods do, resulting in a shorter computation time to simulate the continuous material. In addition, the following novelties are introduced by this chapter: (i) the proposed method uses a physical viscosity coefficient, instead of an artificial one as commonly done in the literature, and thus it is possible to predict the viscous behaviour of the continuous materials accurately; (ii) while most of the existing implicit viscosity methods use optimization methods (e.g., conjugate gradient or precomputed Jacobian to find the proper velocities of the particles), the proposed approach can control the viscosity accuracy and predict the computation time, as it will be verified in the proposed experiments.

The outline of the chapter is as follows. The mathematical background for Navier-Stokes equations used to represent the continuous materials and the SPH formulation, which is a mesh-free Lagrangian method implementing the Navier-Stokes equations, are briefly revised in the next section. The proposed approach is presented in Sect. 3 along with a brief description of existing viscosity approaches. At the same time, Sect. 4 gives additional ways to improve the introduced SPH implementation to be more accurate and fast in simulating the continuous materials. Experiments are described in Sect. 5 where the performance of the proposed algorithm is competitive with the conventional methods. Conclusions are provided within Sect. 6.

2 Theoretical Background about Navier-Stokes's Theorem and SPH

The basic concept of Navier-Stokes equations is briefly introduced in this section. The SPH method is revised in a nutshell as well.

2.1 Navier-Stokes' Theorem for Continuous Materials

The two equations regarding the conservation of mass and momentum for continuous materials are considered [46–48]. The former is also known as the mass continuity equation

$$\frac{\mathrm{D}\rho}{\mathrm{D}t} = -\rho\nabla \cdot \mathbf{v}. \tag{1}$$

The latter equation is about the momentum conservation

$$\rho\frac{\mathrm{D}\mathbf{v}}{\mathrm{D}t} = -\nabla \cdot p\mathbf{I}_r + \nabla \cdot \mathbf{T} + \rho\,\mathbf{f}^{body}. \tag{2}$$

For any material property $\mathbf{A}_i$ of the fluid sample point $\mathbf{x}_i$ with proper dimension, in the Eulerian approach, the material derivative is defined as $\frac{\mathrm{D}\mathbf{A}}{\mathrm{D}t} = \frac{\partial\mathbf{A}}{\partial t} + (\mathbf{v} \cdot \nabla)\mathbf{A}$, in which $\frac{\partial}{\partial t}$ is the time derivative at a fixed Eulerian sample point $\mathbf{x}_i$ and $(\mathbf{v} \cdot \nabla)\mathbf{A}$ is the so-called advection term. In the Lagrangian approach , he material derivative is defined as $\frac{\mathrm{D}\mathbf{A}}{\mathrm{D}t} = \frac{d\mathbf{A}}{dt}$, in which $\frac{d}{dt}$ is the time derivative at an advected Lagrangian sample point $\mathbf{x}_i$ [49, 50].

2.2 SPH Formulation

As briefly mentioned in Sect. 1, the SPH formulation is a mesh-free Lagrangian approach initially developed for astrophysical problems in [17, 18]. They treated the stars distributed sparsely in the universe as particles with mass and other properties, and they researched about interval physical properties among neighbour stars. Lately, the SPH formulation was applied to compressible fluid problems in continuum mechanics, in which each particle represents a collection of nearby atoms or molecules.

According to the general SPH formulation [7, 20, 50], a physical quantity $\mathbf{A}$ of any point $\mathbf{x} \in \mathbb{R}^r$ of the continuous material can be calculated by

$$\mathbf{A}(\mathbf{x}) = \int_{-\infty}^{\infty} \mathbf{A}(\mathbf{x}')\, W(\mathbf{x} - \mathbf{x}', h)\, d\mathbf{x}', \tag{3}$$

where $W : \mathbb{R}^r \to \mathbb{R}^{\geq 0}$ is an interpolating kernel, and h is the radius of the kernel domain. The value for the outer of the kernel domain is zero. Such a kernel satisfies the following two properties

$$\int_{-\infty}^{\infty} W(\mathbf{x} - \mathbf{x}', h)\, d\mathbf{x}' = 1, \qquad \lim_{h\to 0} W(\mathbf{x} - \mathbf{x}', h) = \delta(\mathbf{x} - \mathbf{x}'), \tag{4}$$

where $\delta : \mathbb{R}^n \rightarrow \mathbb{R}^{\geq 0}$ is the Dirac delta function, whose output is equal to zero everywhere except for $\mathbf{x} = \mathbf{x}'$, and whose integral over the domain is equal to one. Moreover, decreasing monotonically with distance may be an additional useful property for a kernel function W.

The discrete approximation of (3) is given by

$$\mathbf{A}_i = \sum_j \frac{m_j}{\rho_j} \mathbf{A}_j \, W(\mathbf{x}_i - \mathbf{x}_j, h), \tag{5}$$

where $m_j > 0$ and $\rho_j > 0$ are the mass and the density of the j-neighbour particle, respectively. The gradient $\nabla \mathbf{A}$ for Eq. (5) can be defined using the second golden rule of SPH, which is to rewrite formulae with the density placed inside operators [20, 38, 51], as

$$\nabla \mathbf{A}_i = \frac{1}{\rho_i} \sum_j m_j (\mathbf{A}_j - \mathbf{A}_i) \nabla W(\mathbf{x}_i - \mathbf{x}_j, h). \tag{6}$$

Similarly, the Laplacian $\nabla^2 \mathbf{A}$ for equation (5) can be defined as

$$\nabla^2 \mathbf{A}_i = \sum_j \frac{m_j}{\rho_j} (\mathbf{A}_j - \mathbf{A}_i) \nabla^2 \, W(\mathbf{x}_i - \mathbf{x}_j, h), \tag{7}$$

as proposed in [33, 52] by applying the second golden rule of SPH twice for incompressible continuous materials.

From now in this chapter, W_{ij} will be used as short notation for $W(\mathbf{x}_i - \mathbf{x}_j, h)$.

3 Viscosity Property and Various Viscosity Methods for SPH

The Navier-Stokes equation (2) for incompressible fluids with a Lagrangian approach can be written as follows for the ith element

$$\frac{\mathrm{D}\,\mathbf{v}_i}{\mathrm{D}t} = -\frac{1}{\rho_i} \nabla \cdot p_i \, \mathbf{I}_r + \frac{1}{\rho_i} \nabla \cdot \mathbf{T}_i + \mathbf{f}_i^{body}, \tag{8}$$

where it is possible to recognize the net motion force, $\mathbf{f}_i^{motion} = \frac{\mathrm{D}\mathbf{v}}{\mathrm{D}t}$, equal to the sum of the pressure force, $\mathbf{f}_i^{pressure} = -\frac{1}{\rho_i} \nabla \cdot p \, \mathbf{I}_r$, occurred by pressure difference, the viscosity force, $\mathbf{f}^{vis} = \frac{1}{\rho_i} \nabla \cdot \mathbf{T}_i$, occurred by shear stress, and the body force, $\mathbf{f}_i^{body}$, acting on a continuous fluid and given by gravity, inertial accelerations, elastostatic accelerations, and so on. Incompressibility of continuous materials is guaranteed by solvers like IISPH [28] or DFSPH [29], which generates the proper pressure force $\mathbf{f}_i^{pressure}$ to keep constant the density, that is, $\rho_i = \rho_0$, or at least $\frac{\mathrm{D}\rho_i}{\mathrm{D}t} = 0$. The SPH is purely based on Lagrangian approach, hence the material derivative for a velocity $\mathbf{v}_i$ is calculated as $\frac{\mathrm{D}\mathbf{v}_i}{\mathrm{D}t} = \frac{d\mathbf{v}_i}{dt}$ [49].

Concerning incompressible fluids, the viscosity force $\mathbf{f}^{vis}$ is defined as

$$\begin{aligned} \mathbf{f}_i^{vis} &= \frac{1}{\rho_i}\nabla \cdot \mathbf{T}_i = \frac{1}{\rho_i}\nabla \cdot \left(2\,\mu \frac{\nabla \mathbf{v}_i + \nabla \mathbf{v}_i^T}{2}\right) \\ &= \frac{1}{\rho_i}\Big(\mu \underbrace{\nabla \cdot \nabla \mathbf{v}_i}_{\nabla^2 \mathbf{v}_i} + \mu \underbrace{\nabla \cdot (\nabla \mathbf{v}_i)^T}_{\nabla(\nabla \cdot \mathbf{v}_i)=0}\Big) = \frac{1}{\rho_i}\mu \nabla^2 \mathbf{v}_i, \end{aligned} \tag{9}$$

where $\nabla^2 \mathbf{v}_i$ is the Laplacian of the velocity.

Regarding the SPH formulation, a double discretization of the velocity generates error accumulation, and it is prone to be too much sensitive with respect to the velocity of the particles and the kernel function. Hence, the conventional explicit viscosity methods [21, 24, 30] use alternative formulations by employing the first derivative kernel function only, while the existing implicit viscosity methods [37–39] try not to calculate the viscosity force explicitly. The formulation proposed in [33] is employed in this chapter to define the viscosity force $\mathbf{f}_i^{vis}$ as

$$\mathbf{f}_i^{vis} = \frac{1}{\rho_i}\mu \nabla^2 \mathbf{v}_i = \frac{1}{\rho_i}\mu \sum_j \frac{m_j}{\rho_j}(\mathbf{v}_j - \mathbf{v}_i)\, \nabla^2 W_{ij}. \tag{10}$$

The goal of most of the viscosity methods is to find out the velocity $\mathbf{v}_i$ satisfying the Navier-Stokes equation (8). However, it is not easy to find out such a velocity because the viscosity force $\mathbf{f}_i^{vis}$ is relative to both the velocity of a particle and the velocities of its neighbours, as evident from (10).

To easy explain the presented methodology, all the particles in the continuous material are assumed to be initially stable, i.e., $\mathbf{v}_i = \mathbf{v}_0\ \forall i$ where $\mathbf{v}_0 \in \mathbb{R}^r$ is a constant velocity, for instance zero. Let $\mathbf{f}_i^{ext} = \mathbf{a}_i \in \mathbb{R}^r$ be any other force applied to the ith particle, assimilated to an acceleration $\mathbf{a}$, except its viscosity force $\mathbf{f}_i^{vis}$. Then, consider that only a particle i of the continuous material is subject to an external force $\mathbf{f}_i^{ext}$, while the other particles are not $\mathbf{f}_{j\neq i}^{ext} = \mathbf{0}_r$. When the ith particle is accelerated by the external force $\mathbf{f}_i^{ext}$, its closest particles drag it to be less accelerated. Such a dragging force is equal to the viscosity force $\mathbf{f}_i^{vis}$ for the ith particle as

$$\mathbf{f}_i^{vis} = \sum_j \mathbf{f}_{i \leftarrow j}^{vis}, \tag{11}$$

where $i \leftarrow j$ means the effect of a particle j toward a particle i. Therefore, folding equation (10) into equation (8), with the assumption $\mathbf{v}_i = \mathbf{v}_0\ \forall k$, yields

$$\frac{\Delta \mathbf{v}_i}{\Delta t} \simeq \frac{1}{\rho_i}\mu \sum_j \frac{m_j}{\rho_j}(\mathbf{v}_j - (\mathbf{v}_i + \Delta \mathbf{v}_i))\, \nabla^2 W_{ij} + \mathbf{a}_i = -\frac{1}{\rho_i}\mu\, \Delta \mathbf{v}_i \sum_j \frac{m_j}{\rho_j}\, \nabla^2 W_{ij} + \mathbf{a}_i, \tag{12}$$

where the equality holds if the change of velocity, $\Delta\mathbf{v}$, is constant during a given time-step $\Delta t > 0$. Then, the increase of velocity for the particle i, $\Delta\mathbf{v}_i$, caused by the external force $\mathbf{f}_i^{ext}$ can be obtained from

$$\Delta\mathbf{v}_i \simeq (\Delta t \cdot \mathbf{a}_i)/(1 + \Delta t \cdot \frac{\mu}{\rho_i} \sum_j \frac{m_j}{\rho_j} \nabla^2 W_{ij}). \tag{13}$$

Moreover, the viscosity force occurred by a neighbourhood particle j can be retrieved from $\mathbf{f}_{i\leftarrow j}^{vis} = -\frac{1}{\rho_i}\mu\,\Delta\mathbf{v}_i \frac{m_j}{\rho_j}\nabla^2 W_{ij}$. According to the third Newton's law, $m_i\mathbf{f}_{i\leftarrow j}^{vis} + m_j\mathbf{f}_{j\leftarrow i}^{vis} = \mathbf{0}_r$, the dragging force from a particle j to the particle i also affects the particle j itself such that

$$\mathbf{f}_{j\leftarrow i}^{vis} = \frac{1}{\rho_j}\mu \frac{m_i}{\rho_i}((\mathbf{v}_i + \Delta\mathbf{v}_i) - \mathbf{v}_j)\nabla^2 W_{ij}. \tag{14}$$

Then, the force $\mathbf{f}_{j\leftarrow i}^{vis}$ can be regarded as a new external force $\mathbf{f}_j^{ext'} \in \mathbb{R}^r$ for the particle j, that is, $\mathbf{f}_j^{ext'} = \mathbf{f}_{j\leftarrow i}^{vis}$.

The described procedure is resumed within the schematic diagram in Fig. 1. Assume that five particles exist, each of which is denoted with subscript $j1$, $j2$, $j3$, $j4$, and i, serially and stably placed in a row. The particle i is affected by four neighbourhood particles $j1$, $j2$, $j3$, and $j4$ in the SPH formulation. The particle i is subject to a given external force $\mathbf{f}_i^{ext}$ vertically applied in the upward direction. Since the surrounding particles are still stable, the initial viscosity force is given by $\mathbf{f}_i^{vis} = \sum_j \mathbf{f}_{i\leftarrow j}^{vis}$, where $j = \{j1, j2, j3, j4\}$. Then, the summation of the provided external force $\mathbf{f}_i^{ext}$ and the viscosity force $\mathbf{f}_i^{vis}$ is the net motion force $\mathbf{f}_i^{motion}$ of the ith particle. Subsequently, the change of velocity $\Delta\mathbf{v}_i$ caused by the motion force $\mathbf{f}_i^{motion}$ creates the viscosity forces, $\mathbf{f}_{j1\leftarrow i}^{vis}$, $\mathbf{f}_{j2\leftarrow i}^{vis}$, $\mathbf{f}_{j3\leftarrow i}^{vis}$, and $\mathbf{f}_{j4\leftarrow i}^{vis}$. Each of these viscosity forces can be regarded as new external forces, $\mathbf{f}_{j1}^{ext'}$, $\mathbf{f}_{j2}^{ext'}$, $\mathbf{f}_{j3}^{ext'}$, and $\mathbf{f}_{j4}^{ext'}$, for the neighbourhood particles.

Now, consider the case where multiple particles in a continuous material are subject to external forces $\mathbf{f}_{j\neq i}^{ext}$, simultaneously. Following the previous Eqs. (13)–(14), a particle i might be subject to multiple viscosity forces $\sum_j \mathbf{f}_{i\leftarrow j}^{vis}$ from its neighbourhoods, so that the new external force $\mathbf{f}_i^{ext'}$ can be extended to

$$\begin{aligned} \mathbf{f}_i^{ext'} = \sum_j \mathbf{f}_{i\leftarrow j}^{vis} &= \frac{1}{\rho_i}\mu \sum_j \frac{m_j}{\rho_j}\Big((\mathbf{v}_j + \Delta\mathbf{v}_j) - (\mathbf{v}_i)\Big)\nabla^2 W_{ij} \\ &= \frac{1}{\rho_i}\mu \sum_j \frac{m_j}{\rho_j}(\Delta\mathbf{v}_j)\,\nabla^2 W_{ij}. \end{aligned} \tag{15}$$

Furthermore, removing the assumption that $\mathbf{v}_i = \mathbf{v}_0\ \forall i$ yields

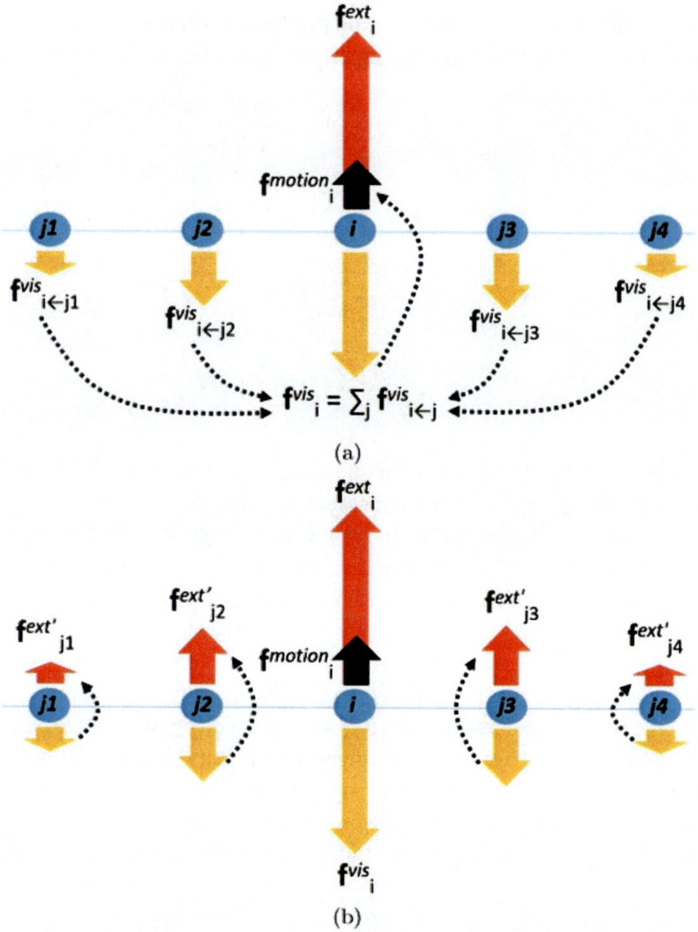

Fig. 1 When an external force $\mathbf{f}_i^{ext}$ is applied on the ith particle of a series of particles arranged in a row, not only the ith one gets the viscosity force $\mathbf{f}_i^{vis}$ but also the particles $j1 \sim j4$ are subject to new external forces $\mathbf{f}_{j1}^{ext'} \sim \mathbf{f}_{j4}^{ext'}$ according to Newton's third law. Red, yellow, and black arrows indicate external, viscosity, and net motion forces, respectively

$$\frac{\Delta \mathbf{v}_i}{\Delta t} \simeq \frac{1}{\rho_i}\mu \sum_j \frac{m_j}{\rho_j}(\mathbf{v}_j - (\mathbf{v}_i + \Delta \mathbf{v}_i))\, \nabla^2 W_{ij} + \mathbf{a}_i, \tag{16}$$

which is equivalent to

$$\frac{\Delta \mathbf{v}_i}{\Delta t} \simeq -\frac{1}{\rho_i}\mu\, \Delta \mathbf{v}_i \sum_j \frac{m_j}{\rho_j}\, \nabla^2 W_{ij} + \underbrace{\left(\frac{1}{\rho_i}\mu \sum_j \frac{m_j}{\rho_j}(\mathbf{v}_j - \mathbf{v}_i)\, \nabla^2 W_{ij} + \mathbf{a}_i\right)}_{\mathbf{f}_i^{ext'} = \mathbf{a}_i'}. \tag{17}$$

Note that, considering the combination of the last two terms in the right side of the above equation as a new external force $\mathbf{f}_i^{ext'} = \mathbf{a}_i'$, then equation (17) becomes equivalent to (12).

In view of the law of conservation of energy and the condition that the kernel Laplacian is $\nabla^2 W \geq 0$, the following expressions hold

$$\begin{aligned} \mathbf{f}_i^{motion} &= \mathbf{f}_i^{vis} + \mathbf{f}_i^{ext} = \sum_j \mathbf{f}_{i\leftarrow j}^{vis} + \mathbf{f}_i^{ext} \\ \Rightarrow \quad \mathbf{f}_i^{ext} &= \mathbf{f}_i^{motion} + \sum_j \mathbf{f}_{j\leftarrow i}^{vis} \\ \Rightarrow \quad \left\|\mathbf{f}_i^{ext}\right\| &\geq \sum_j \left\|\mathbf{f}_{j\leftarrow i}^{vis}\right\| = \sum_j \left\|\mathbf{f}_j^{ext'}\right\|. \end{aligned} \tag{18}$$

Therefore, the magnitude of the sum of the external forces, $\left\|\sum \mathbf{f}^{ext}\right\|$, always decreases and converges to zero as the procedure is iterated.

So far, a viscosity equation has been retrieved from the knowledge of an applied external force, and such a viscosity term has been used as a further external force as well. There is not so much difference with conventional explicit viscosity methods. However, imagine the sequential procedure with a very short time-step $\delta t \ll \Delta t$ in an explicit approach. The external force $\mathbf{f}_i^{ext}$ affects the particle i: after δt, the particle i changes its velocity accordingly to $\delta \mathbf{v}_i$ and, subsequently, any particle $j \in \mathcal{N}_i$ in the neighbours of i gets an external force $\mathbf{f}_j^{ext}$ from i. During the next δt, the particle j subject to the external force $\mathbf{f}_j^{ext}$ changes its velocity accordingly to $\delta \mathbf{v}_j$ and, subsequently, any particle $k \in \mathcal{N}_j$ in its neighbour, including the particle $i \in \mathcal{N}_j$, gets an additional external force $\mathbf{f}_k^{ext}$ from the particle j again. This procedure is iterated for $\Delta t/\delta t$ times, and thus the number of interactions tends to infinite as $\delta t \to 0$.

For a Newtonian fluid, since the dynamic viscosity coefficient μ is constant, it is possible to separate the computation of the new external forces, $\mathbf{f}^{ext'}$, from the sequential processing without increasing the time-step. In this way, it is possible to take a longer time-step Δt for viscosity calculation like other implicit viscosity methods.

A schematic diagram is shown in Fig. 2. On the one hand, Fig. 2a shows that the velocity changes $\delta \mathbf{v}$ are accumulated linearly during a short time δt as in the existing explicit viscosity methods. To use these approaches, the short time-step δt should be selected efficiently. Furthermore, suppose there is an SPH routine including neighbour searching and incompressible solving for each iteration as the conventional explicit viscosity methods do. In that case, the computation time of the SPH algorithm also drastically increases. On the other hand, Fig. 2b illustrates the approach proposed in this paper, which calculates the velocity changes $\Delta \mathbf{v}$ during a given time-step $\Delta t \gg \delta t$, and then the induced external force $\mathbf{f}_i^{ext'} = \sum_j \mathbf{f}_{i\leftarrow j}^{vis}$ is applied to the surrounding particles subsequently without adding a time-step. Moreover, the magnitude of the change of velocity $\|\Delta \mathbf{v}\|$ decreases as the routine is repeated: hence, the viscosity

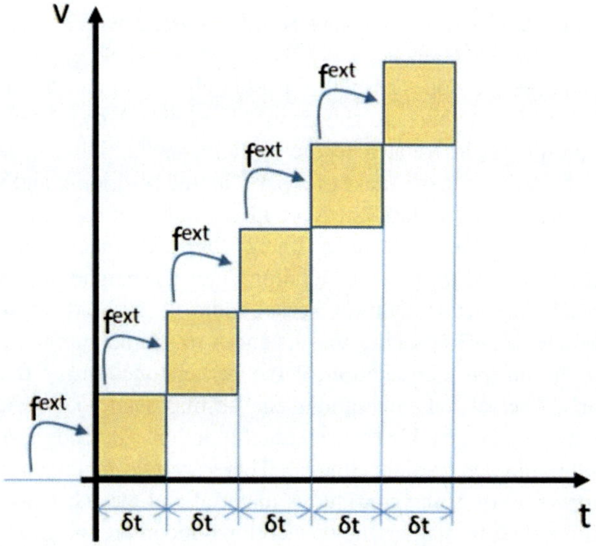

(a) conventional explicit approach

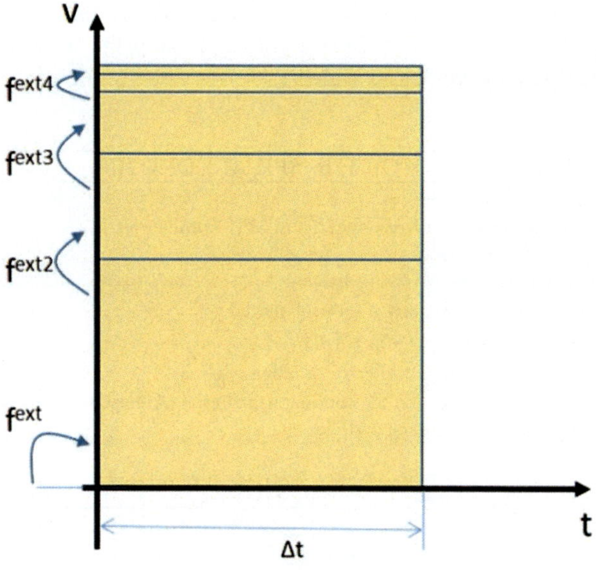

(b) proposed method

Fig. 2 While the conventional explicit approaches require a short time-step δt for sequential calculations, on the top, the proposed method uses a given time-step $\Delta t \gg \delta t$ for calculating the velocity changes and the additional external forces, on the bottom error can be controlled, as well as the number of iterations, with a trade-off between them.

The schematic pseudo-codes of both the SPH algorithm and the one related to the proposed viscosity method are shown in Algorithms 1 and 2, respectively. In Algorithm 1, the properties related to the particles are initialized, like the space dimension, the fluid density, the mass of a particle, the position **x** and velocity **v** of each particle, as well as the kernel function with the radius of its domain h, the time-step, the total simulation time, and so on (line 2). Then, the main routine (lines 4–9) is repeated until the end of the given simulation time. The neighbour search for each particle is carried out (line 4), then the kernel value W, its gradient, its Laplacian, and its density are calculated using the distances to the neighbours (line 5). The neighbour search and the computation of the kernel function are the most time-consuming parts. The related calculations can be improved with various methods like implementing a look-up table for the kernel function. The body force $\mathbf{f}^{body}$, like gravity, is applied to each particle (line 6). There are two main sub-routines: one for the incompressibility of the continuous material and the other one to calculate the viscosity force. In this chapter, the conventional incompressibility methods, like the IISPH approach in [38] and the DFSPH method in [29], are employed. They guarantee incompressibility of the continuous materials through $\rho = \rho_0$, or at least through $\frac{D\rho}{Dt} = 0$ (line 7). Under the assumption that a given continuous material is incompressible, the proposed viscosity method is run (Algorithm 2). It calculates the velocity of a particle **v** satisfying viscosity (line 8). Finally, the position of a particle **x** is updated based on the velocity **v** (line 9).

ALGORITHM 1: OUTLINE OF THE SPH ALGORITHM

1 **begin**
2   initialize particles of a continuous material in SPH formulation
3   **repeat**
4   search neighbourhood for each particle
5   compute W, ∇W, $\nabla^2 W$, and ρ for each particle
6   apply the body force $\mathbf{f}^{body}$ *e.g.* gravity
7   correct the density error s.t. $\rho = \rho_0$ or at least $\frac{D\rho}{Dt} = 0$
8   apply the viscosity effect on the continuous material (Algorithm 2)
9   update the position **x** of each particle
10   **until**
11 **end**

In detail, Algorithm 2 consists of an initial part (line 2–6) and a loop (line 8–16). Firstly, the external force $\rho\,\mathbf{a}'$ is re-calculated, including not only the original external body force, like gravity, but also the viscosity force $\mathbf{f}^{vis}$ occurred by the relative velocity between a particle i and its neighbours $\mathcal{N}_i$ using the equation (17). Within the loop part, two termination conditions are checked: the former to verify when the number of iterations is less than the maximum number *max_iter*; the latter to prove that the maximum magnitude of the accelerations of the particles $\|\mathbf{a}\|$ is greater than the minimum threshold keeping a reasonable computation time and preventing the computation of negligible small external forces (line 7). For these two termination conditions, *max_acc* and *cur_iter* are calculated (line 5–6 and line 15–16). Within the loop, two main parts can be recognized: the former is the update of the velocity of each particle (line 8–11) using (13); the latter is the calculation of the new external force $\mathbf{f}^{ext'} = \mathbf{a}'$ using (15). When each external force is negligible or the max iterations number is overcome, the viscosity method terminates.

ALGORITHM 2: PROPOSED VISCOSITY METHOD FOR INCOMPRESSIBLE NEWTONIAN FLUIDS

1 **begin**
2   **for** *each particle i of a continuous material in SPH formulation* **do**
3     $\mathbf{a}'_i = \mathbf{a}_i + \frac{\mu}{\rho_i} \sum_j \frac{m_j}{\rho_j} (\mathbf{v}_j - \mathbf{v}_i) \nabla^2 W_{ij}$ --- Eq. (17)
4   **end**
5   max_acc = $\max\{ \|\mathbf{a}'_1\|, \cdots, \|\mathbf{a}'_N\| \}$
6   cur_iter = 0
7   **while** *(max_acc > threshold)* $\wedge$ *(cur_iter < max_iter)* **do**
8   **for** *each particle i* **do**
9       $\Delta \mathbf{v}_i = (\Delta t \cdot \mathbf{a}'_i)/(1 + \Delta t \cdot \frac{\mu}{\rho_i} \sum_j \frac{m_j}{\rho_j} \nabla^2 W_{ij})$ --- Eq. (13)
10       $\mathbf{v}_i = \mathbf{v}_i + \Delta \mathbf{v}_i$
11   **end**
12   **for** *each particle i* **do**
13       $\mathbf{a}'_i = \frac{\mu}{\rho_i} \sum_j \frac{m_j}{\rho_j} (\Delta \mathbf{v}_j) \nabla^2 W_{ij}$ --- Eq. (15)
14   **end**
15   max_acc = $\max\{ \|\mathbf{a}'_1\|, \cdots, \|\mathbf{a}'_N\| \}$
16   cur_iter = cur_iter + 1
17   **end**
18 **end**

4 Other Components for the SPH-based Modelling

4.1 Kernel Functions

The smoothing kernel function W is crucial for the accuracy of the SPH-based continuum mechanic simulation. As mentioned, there is no necessary condition for the kernel except that

$$\int_{-\infty}^{\infty} W(\mathbf{x} - \mathbf{x}', h)\, d\mathbf{x}' = 1, \tag{19}$$

or the discretized approximated version

$$\sum_j \frac{m_j}{\rho_j} W(\mathbf{x}_i - \mathbf{x}_j, h) \simeq 1, \tag{20}$$

for the center position $\mathbf{x}_i$ of the kernel. Its gradient function is given by

$$\nabla_i W(\mathbf{x}_i - \mathbf{x}_j, h) = \frac{\partial W}{\partial q} \nabla_i q, \tag{21}$$

while its Laplace function is

$$\nabla_i^2 W(\mathbf{x}_i - \mathbf{x}_j, h) = \frac{\partial^2 W}{\partial q^2} \|\nabla_i q\|^2 + \frac{\partial W}{\partial q} \nabla_i^2 q, \tag{22}$$

where $q = \frac{\|\mathbf{x}_i - \mathbf{x}_j\|}{h}$, $\nabla_i q = \frac{\mathbf{x}_i - \mathbf{x}_j}{\|\mathbf{x}_i - \mathbf{x}_j\| h}$, and $\nabla_i^2 q = \frac{1}{\|\mathbf{x}_i - \mathbf{x}_j\| h} - \frac{\sqrt{\|\mathbf{x}_i - \mathbf{x}_j\|}}{h}$.

Usually, the radius of the kernel domain h is set to twice the default particle spacing, that is, four times longer than the particle radius for maintaining sufficient but not too much neighbour particles. The following Gaussian kernel is commonly used

$$W(\mathbf{x}_i - \mathbf{x}_j, h) = \frac{1}{(2\pi h^2)^{\frac{r}{2}}} \exp\left(- \frac{\|\mathbf{x}_i - \mathbf{x}_j\|^2}{2h^2} \right). \tag{23}$$

Even though a Gaussian kernel is highly recommended statistically, the computational cost for evaluating the exponential function is expensive, and it does not have compact support, that is, the range spans $-\infty$ to ∞. Therefore, an approximated cubic spline kernel function is employed in this work

$$W(\mathbf{x}_i - \mathbf{x}_j, h) = \frac{1}{\sigma} \begin{cases} \frac{4}{3} - 2q^2 + q^3 & \text{if } 0 \le q < 1 \\ \frac{1}{3}(2 - q)^3 & \text{if } 1 \le q < 2 \\ 0 & \text{otherwise,} \end{cases} \tag{24}$$

where $q = \frac{\|\mathbf{x}_i - \mathbf{x}_j\|}{h/2}$, σ is a normalization constant equal to $\{2h, \pi h^2, \frac{4}{3}\pi h^3\}$ for $r = \{1, 2, 3\}$, respectively [7, 20]. A particular kernel to handle viscosity effect, whose Laplacian is always non-negative, $\nabla^2 W \ge 0$, so that the viscosity force does not increase the relative velocity and avoid to create unstable status, was introduced in [50].

An additional method to limit the computational burden concerning the kernel function is to use a look-up table [39]. The kernel values are pre-computed for sample distances, called keys, between two particles with a specific span. The keys and the computed values are saved into the look-up table. Using the closest key to the requested one or calculating the linear interpolation using two narrowest points, finding out the desired approximated value of the kernel function is possible.

A cubic spline kernel function has been employed within the performed simulations, while its Laplacian value is set to zero if it is negative. Furthermore, when convenient, the look-up tables storing the kernel value, its gradient value, and its Laplacian value of the kernel domain, $h = 0.005$ m and $h = 0.010$ m with a span 10^{-7} m, have been exploited in some simulations using the particle spacing $1.25 \cdot 10^{-3}$ m or $02.5 \cdot 10^{-3}$ m, respectively.

4.2 *Incompressible Fluid*

Particle-based continuum mechanic methods like SPH have incompressibility problem, that is, they need some special treatments for modelling incompressible continuous materials. To address the incompressibility problem, various methods, like the PSPH [23], the ISPH [24], the WCSPH [25], the PCISPH [26], and the PBFM [12] were proposed. Recently, methods like the IISPH [28] and the DFSPH [29] have shown outstanding performance in addressing incompressible fluid. The IISPH approach tries to find out a proper pressure p so that it satisfies the derivative of the density $\frac{\mathrm{D}\rho}{\mathrm{D}t} = 0$ using the continuity equation $\frac{\mathrm{D}\rho}{\mathrm{D}t} = -\rho\nabla \cdot \mathbf{v}$ and the velocity update equation $\mathbf{v}(t + \Delta t) = \mathbf{v}(t) + \Delta t \frac{\mathbf{f}^{adv} + \mathbf{f}^{p}}{m}$, where $\mathbf{f}^{adv} \in \mathbb{R}^r$ is an advection force and $\mathbf{f}^p \in \mathbb{R}^r$ is a pressure force computed as

$$\mathbf{f}_i^p = -m_i \sum_j m_j \left(\frac{p_i}{\rho_i^2} + \frac{p_j}{\rho_j^2} \right) \nabla W_{ij}. \tag{25}$$

The DFSPH method, instead, does not find out the pressure directly but it retrieves an impulse pressure force defined as

$$\mathbf{f}_i^p = \kappa_i^v \sum_j m_j \nabla W_{ij}, \tag{26}$$

where κ^v is the stiffness parameter, so that it satisfies $\frac{\mathrm{D}\rho_i}{\mathrm{D}t} + \Delta \frac{\mathrm{D}\rho_i}{\mathrm{D}t} = 0$, where $\Delta \frac{\mathrm{D}\rho_i}{\mathrm{D}t} = \Delta t \sum_j m_j \left(\frac{\mathbf{f}_i^p}{\rho_i} - \frac{\mathbf{f}_{j \leftarrow i}^p}{\rho_i} \right) \nabla W_{ij}$.

In the proposed simulations dealing with viscosity, the IISPH method has been employed: nevertheless, other incompressibility approaches are doable. As a contribution, instead of the pressure force Eq. (25) in the original IISPH method, the one from [33] has been employed

$$\mathbf{f}_i^p = -\frac{m_i}{\rho_i} \sum_j \frac{m_j}{\rho_j} \left(\frac{p_i + p_j}{2} \right) \nabla W_{ij}, \tag{27}$$

since more stable performance are obtained within the carried out simuatlions.

5 Simulations

The set-up employed to test the proposed method is composed of an IntelCore i7-6500U CPU@2.50GHz, 8.0 Gb of memory with Windows 10 × 64 OS, equipped with MFC of Microsoft, Eigen,[1] and OpenCV[2] for 2D graphic or Open Scene Graph[3] for 3D graphics libraries based on C++ programming language. Houdini software from SideFX[4] has been employed to reconstruct the mesh from particles, and Blender[5] for the graphical rendering. A CUDA version has been implemented to compare both the proposed algorithm and the existing conventional viscosity methods. They have been tested on an NVIDIA GeForce 940MX graphic card. It has been possible to verify that CUDA implementation has sped up the execution from the 20% to the 50% compared to OpenMP-based[6] implementation.

5.1 Accuracy and Time Analysis

The accuracy of the proposed algorithm is firstly addressed. A free-fall experiment is considered, where the velocity of the particles of the continuous material is measured after applying a uniform external force like gravity. Because there is no degradation for the gravity during the free-fall, the movement of the particles is the same as the product of the acceleration with the delay time, $\Delta t\mathbf{a}$. The error introduced by the proposed viscosity method is due to the threshold η, the viscosity μ, and $\mathbf{a}$. Two values are considered for the external force, namely $\mathbf{a} = 10$ m/s^2 and $\mathbf{a} = 100$ m/s^2. The velocity of the continuous materials is measured after 0.001 s. The particle spacing and the density ρ of the SPH particles are set as 0.0025 m and $\rho = 100$ kg/m^3, respectively. The results are summarized in Table 2 and Table 3.

Based on these data, the graph in Fig. 3 is depicted with a threshold rate, $TR = \frac{\eta}{\|\mathbf{a}\|}$. From the data, it is possible to appreciate that, as long as the viscosity coefficient increases, the error rate, $ERR = \frac{\|\mathbf{a}\| - \|\mathbf{a}'\|}{\|\mathbf{a}\|}$, where $\mathbf{a}$ is the input acceleration and $\mathbf{a}'$ is the measured one, converges to the threshold rate given by the external force, that is,

$$ERR \leq \frac{\eta}{\|\mathbf{a}\|}. \tag{28}$$

[1] https://eigen.tuxfamily.org/.

[2] https://opencv.org.

[3] http://www.openscenegraph.org.

[4] https://www.sidefx.com.

[5] https://www.blender.org.

[6] https://www.openmp.org.

Table 2 Error rates (%) of free-fall experiment with $\mathbf{a} = 10$ m/s^2 and various viscosity μ and threshold η at time 0.001 s in 3D space

	$\eta = 0.01$	0.1	0.2	0.4	0.6	0.8	1.0
$\mu = 0.1$	0.01	0.05	0.05	0.50	0.50	0.50	0.50
1	0.02	0.20	0.34	0.98	1.69	1.69	2.95
10	0.05	0.56	1.16	2.40	3.69	5.02	6.05
100	0.09	0.92	1.83	3.66	5.49	7.30	9.14
200	0.10	0.96	1.91	3.82	5.73	7.65	9.54
300	0.10	0.97	1.94	3.88	5.82	7.76	9.71
400	0.10	0.98	1.96	3.91	5.86	7.82	9.78
500	0.10	0.98	1.96	3.93	5.89	7.85	9.82

Table 3 Error rates (%) of free-fall experiment with $\mathbf{a} = 100$ m/s^2 and various viscosity μ and threshold η at time 0.001 s in 3D space

	$\eta = 0.1$	1.0	2.0	4.0	6.0	8.0	10.0
$\mu = 0.1$	0.01	0.05	0.05	0.50	0.50	0.50	0.50
1	0.02	0.20	0.34	0.98	1.69	1.69	2.95
10	0.05	0.56	1.16	2.40	3.69	5.02	6.05
100	0.09	0.92	1.83	3.66	5.49	7.30	9.14
200	0.10	0.96	1.91	3.82	5.73	7.65	9.54
300	0.10	0.97	1.94	3.88	5.82	7.76	9.71
400	0.10	0.98	1.96	3.91	5.86	7.82	9.78
500	0.10	0.98	1.96	3.93	5.89	7.85	9.82

Besides, it is possible to estimate the execution time of the proposed viscosity method, which is relative to the number of iterations, $ITER$, because the time of each iteration can be considered as a constant. The number of iterations are measured from the previous simulations: the results are summarized in Tables 4 and 5.

We used the threshold rate, $TR = \frac{\eta}{\|\mathbf{a}\|}$, to display the graphical relation in Fig. 4. Furthermore, we display the gradient of each relation equation at each threshold rate, $TR = \frac{\eta}{\|\mathbf{a}\|}$, in Fig. 5.

Here, we derived the following equation between them

$$ITER \simeq -A \ln\left(\frac{\eta}{\|\mathbf{a}\|}\right) \cdot \mu, \tag{29}$$

with the coefficients $A = 1.646$ in this experiment.

Based on these results, it is possible to conclude that the proposed viscosity method has a predictable accuracy and computation time.

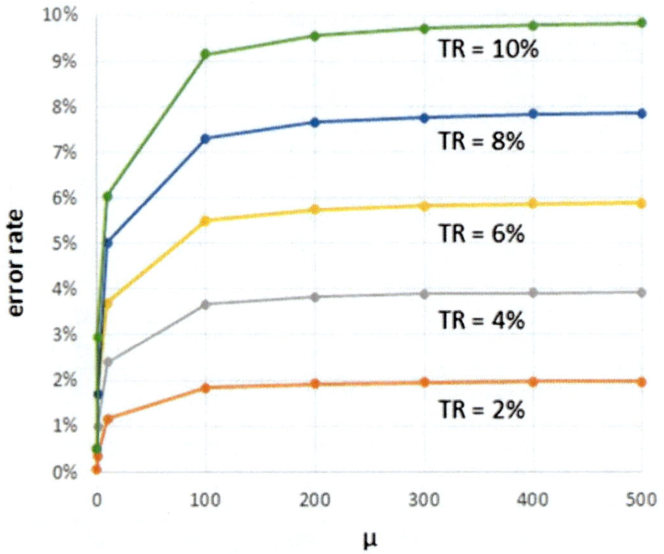

Fig. 3 The relation between μ and error rate with various threshold rates, $TR = \dfrac{\eta}{\|\mathbf{a}\|}$

Table 4 The number of iterations of the free-fall simulation with $\mathbf{a} = 10\ \text{m/s}^2$ and various viscosity μ and threshold η at time 0.001 s in 3D space

	$\eta = 0.01$	0.1	0.2	0.4	0.6	0.8	1.0
$\mu = 0.1$	4	3	3	2	2	2	2
1	15	10	9	7	6	6	5
10	118	79	67	55	48	43	40
100	1142	761	647	532	465	418	381
200	2278	1519	1290	1062	928	833	760
300	3415	2277	1934	1591	1391	1249	1138
400	4551	3034	2577	2121	1854	1664	1517
500	5688	3792	3221	2650	2317	2080	1896

5.2 *Couette Flow Experiment*

The Couette flow appears within a viscous fluid between two parallel plates. The upper plate is moving with a constant velocity while the lower is stationary: because of the difference of the velocities between the two plates, the speed of each layer of the flow is different. As time goes to infinity, the gradient of the velocity to the vertical coordinate, $\dfrac{\partial \mathbf{v}_x}{\partial y}$, is constant, where y is the axis orthogonal to the fluid movement that is along the x direction. The viscosity of the fluid affects how fast the gradient

Table 5 The number of iterations of the free-fall simulation with $\mathbf{a} = 100\,\text{m/s}^2$ and various viscosity μ and threshold η at time 0.001 s in 3D space

	$\eta = 0.1$	1.0	2.0	4.0	6.0	8.0	10.0
$\mu = 0.1$	4	3	3	2	2	2	2
1	15	10	9	7	6	6	5
10	118	79	67	55	48	43	40
100	1142	761	647	532	465	418	381
200	2278	1519	1290	1062	928	833	760
300	3415	2277	1934	1591	1391	1249	1138
400	4551	3034	2577	2121	1854	1664	1517
500	5688	3792	3221	2650	2317	2080	1896

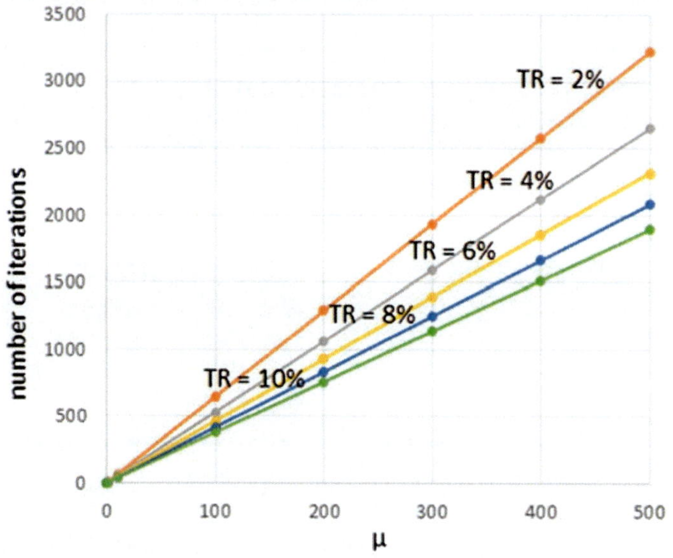

Fig. 4 The relation between μ and the number of iterations with various threshold rates, $TR = \frac{\eta}{\|\mathbf{a}\|}$

becomes constant. The following equation is an analytic solution for the Couette flow [30]:

$$\mathbf{v}_x(y, t) = \mathbf{v}_0 \frac{y}{y_0} + \sum_{n=1}^{\infty} \frac{2\,\mathbf{v}_0}{n\pi}(-1)^n \sin\left(\frac{n\pi}{y_0} y\right) \exp\left(-\nu \left(\frac{n\pi}{y_0}\right)^2 t\right), \tag{30}$$

where $\nu = \frac{\mu}{\rho}$ is the kinematic viscosity, $\mathbf{v}_0 \in \mathbb{R}^r$ is the velocity of the top flow, and $y_0 \in \mathbb{R}$ is the height of the top flow. By comparing the analytic solution with the experimental results, it is possible to verify the accuracy of the viscosity algorithm.

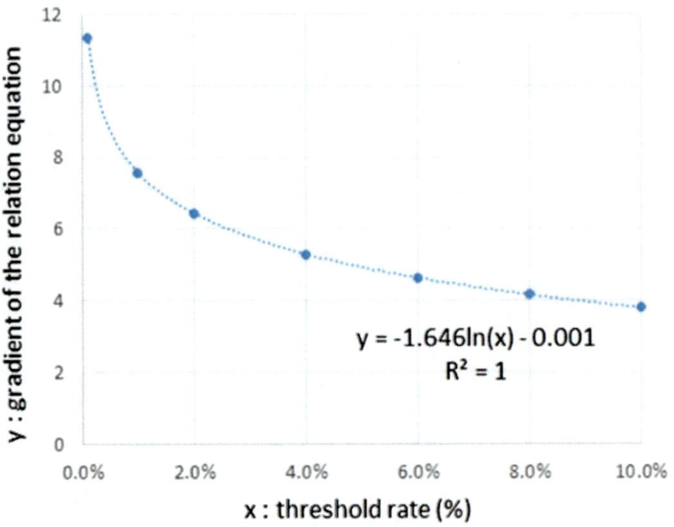

Fig. 5 The relation between threshold rates, $TR = \frac{\eta}{\|\mathbf{a}\|}$, and the gradients of the relation equations in Fig. 4

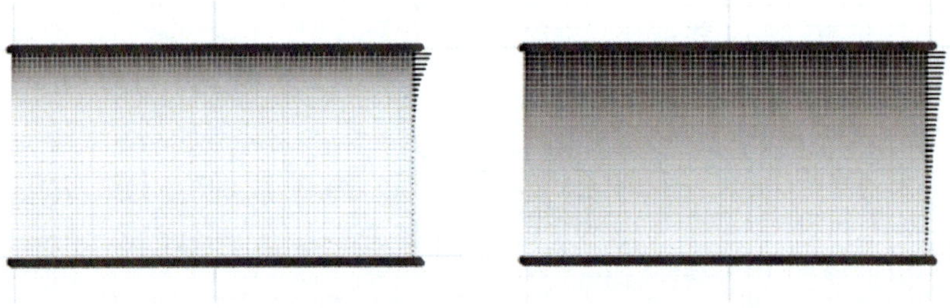

Fig. 6 Couette flow simulation with the gap between two plates $y_0 = 0.1$ m, the velocity of the upper plate $\mathbf{v}_0 = 1.0$ m/s, the total simulation time of 0.01 s, the density of the fluid $\rho = 100$, the particle spacing $1.25 \cdot 10^{-3}$ m, time-step $\Delta t = 0.1 \cdot 10^{-3}$ s, and dynamic viscosity coefficients: **a** $\mu = 0.1$, Re $= 25$, **b** $\mu = 1.0$, Re $= 2.5$. The right side bar indicates the magnitude of the velocity of each layer

The carried out simulation about the Couette flow has height $y_0 = 0.1$ m, velocity of the upper plate $\mathbf{v}_0 = 1$ m/s, and time $t0.01$ s, The parameters relative to the SPH formulation are: $\rho = 100$ for the fluid density, the particle spacing is equal to $1.25 \cdot 10^{-3}$ m, and the time-step is $\Delta t = 0.1 \cdot 10^{-3}$ s.

The results of two flows with $\mu = 0.1$ and $\mu = 1.0$ after $t = 0.01$ s are depicted in Fig. 6, in which the gray colour shows the velocity magnitudes of each layer in the fluid and lines in the right side. Reynolds number for Couette flow is defined [53] as

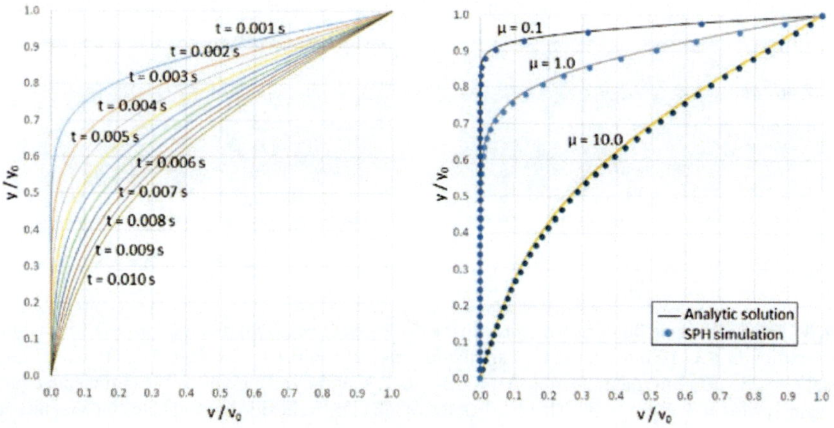

(a) (b)

Fig. 7 Couette flow experiment. **a** The results from time 0.001 s to 0.01 s with viscosity coefficient $\mu = 10$. **b** The results from time 0.01 s with various viscosity coefficients, namely: $\mu = 0.1$, $\mu = 1.0$, and $\mu = 10.0$. Lines and dots indicate the analytic solution and SPH simulation results, respectively

$$\text{Re} := \frac{\rho\, \mathbf{v}_0\, \frac{1}{2} y_0}{\mu}. \tag{31}$$

Hence, Reynolds numbers for $\mu = 0.1$ and $\mu = 1.0$ are $\text{Re} = 25$ and $\text{Re} = 2.5$, respectively.

The results are more easily collected in Fig. 7. It is possible to verify that the top and bottom fluid layers' speeds are almost identical to the velocity of the upper and lower plates, respectively. The curves in Fig. 7a indicate the velocities of each fluid layer with time t: as time goes, the gradient of the velocity equation becomes constant. The velocity results of the proposed algorithm has been compared with the analytic solution of Eq. (30) in Fig. 7b for three different viscosity fluids, namely $\mu = 0.1$, $\mu = 1.0$, and $\mu = 10.0$. The velocity decreases smoothly as the height decreases in the higher viscosity fluid than the lower one. Hence, the proposed algorithm has almost the same results as the analytic solutions.

5.3 *Poiseuille Plane Flow Experiment*

Next simulation test deals with the Poiseuille plate flow [30, 54]. Like the Couette flow, there are two parallel plates with distance y_0, but both of them are stationary. A constant force is applied into the fluid between the two plates. Depending on how strong the viscosity of the fluid is, the velocities of each layer of the fluid are

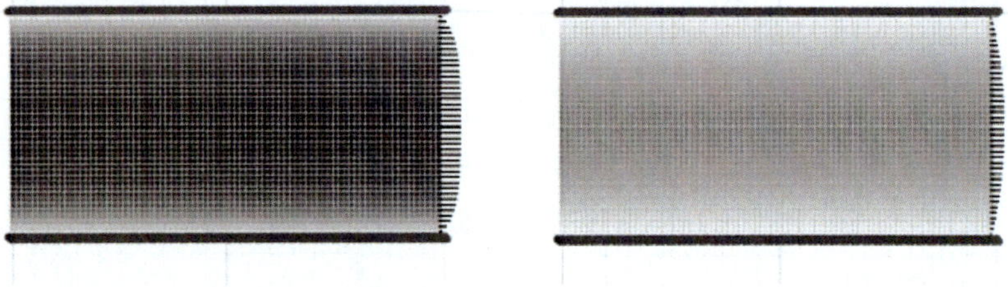

(a) (b)

Fig. 8 Poiseuille plane flow experiment with the distance between two plates of $y_0 = 0.1$ m, a fluid acceleration of $\mathbf{a} = 10.0$ m/s^2, the total simulation time of $t = 0.1$ s, $\rho = 100$, the particle spacing equal to $1.25 \cdot 10^{-3}$ m, the time-step $\Delta t = 0.1 \cdot 10^{-3}$ s, and dynamic viscosity coefficients equal to, namely, **a** $\mu = 0.1$, Re = 33.3 **b** $\mu = 1.0$, Re = 3.3. The right side bar indicates the magnitude of the velocity of each layer

determined at a certain time. The following equation is the analytic solution for the Poiseuille plate flow [30]

$$\mathbf{v}_x(y,t) = -\frac{\mathbf{a}}{2\nu} y(y - y_0) \\ - \sum_{n=0}^{\infty} \frac{4\,\mathbf{a}\,y_0^2}{\nu\pi^3(2n+1)^3} \sin\left(\frac{(2n+1)\pi}{y_0} y\right) \exp\left(-\nu\left(\frac{(2n+1)\pi}{y_0}\right)^2 t\right).$$

The simulation test is carried out with height $y_0 = 0.1$ m, acceleration $\mathbf{a} = 10.0$ m/s^2, time $t = 0.1$ s, and the SPH relative parameters equal to $\rho = 100$, the particle spacing equal to $1.25 \cdot 10^{-3}$ m, and a time-step $\Delta t = 0.0001$ s.

The experimental results of the two flows with $\mu = 0.1$ and $\mu = 1.0$ are shown in Fig. 8. Reynolds number for Poiseuille plane flow is defined as [53]

$$\text{Re} := \frac{\rho\,\bar{\mathbf{v}}\,y_0}{\mu}, \tag{32}$$

where $\bar{\mathbf{v}}$ is the average velocity. Another definition of Reynolds number is also used, $\text{Re} := \dfrac{\rho\,\mathbf{v}_0 \frac{1}{2} y_0}{\mu}$, where $\mathbf{v}_0$ is the velocity at the mid-plane of the channel [55]. The average velocity for Poiseuille plane flow is $\bar{\mathbf{v}} = \frac{2}{3}\mathbf{v}_0$. Hence, the Reynolds numbers for $\mu = 0.1$ and $\mu = 1.0$ are Re = 33.3 and Re = 3.3, respectively.

Because the upper and lower plates are stationary, the speeds of the top and bottom fluid layer are zero, while the speed of the center fluid is maximum. The average velocity of the fluid is faster if $\mu = 0.1$, rather than in case of $\mu = 1.0$.

The results with three different viscosity fluids $\mu = 0.1$, $\mu = 1.0$, and $\mu = 10.0$ are represented in Fig. 9. Since the acceleration is $\mathbf{a} = 10.0$ m/s^2, and the time $t =$

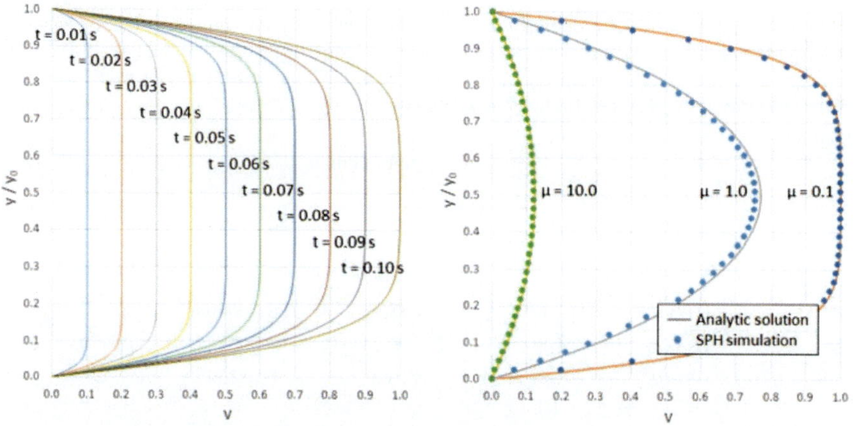

(a) (b)

Fig. 9 Poiseuille flow experiment. **a** The results from time 0.01 to 0.10 s with viscosity coefficient $\mu = 0.1$. **b** The results from time 0.1 s with various viscosity coefficients, namely, $\mu = 0.1$, $\mu = 1.0$, and $\mu = 10.0$. Lines and dots indicates the analytic solution and SPH simulation results, respectively

0.1 s, the maximum speed is 1 m/s. It is possible to appreciate that the higher the viscosity of the fluid, the lower the maximum reached speed. Simulation results are verified to be similar to the analytic solutions.

5.4 Comparison with Conventional Viscosity Methods

A comparison with the existing viscosity approaches mentioned in Sect. 1 is carried out to prove the efficiency of the proposed method. The Couette flow experimentation presented in Sect. 5.2 is chosen as a test-bed for such a comparison. It is worth remarking that the comparison results of the viscosity methods might differ depending on what kind of test-bed is used.

The simulations are configured with the height and the velocity of the top plate as $y_0 = 0.1$ m and $v_0 = 1$ m/s, respectively. The SPH particles for the incompressible Newtonian fluid have particle spacing equal to $1.25 \cdot 10^{-3}$ m and rest density $\rho_0 = 100$. Then, the velocity of each particle along the vertical direction is measured at time 0.01 s and compared with the analytic solution for a fluid of viscosity 1 using an RMSE method. Because each viscosity method has its parameters, it is impossible to use the same numbers for them. Hence, the best parameters matching the analytic solution have been selected: such a best matching induces longer time-step and less RMSE.

It is worth saying that, within the conventional methods, adjusting the parameters has frequently caused explosion phenomena of the fluid. This can be explained by the fact that too close particles have a strong repulsive force. Therefore, they move very

Table 6 Comparison of the proposed viscosity method with explicit/implicit methods. Couette flow experiments [30] for simulating a physical fluid of dynamic viscosity 1 at simulation time 0.01 s, particle spacing 0.00125 m, and flow rest density $\rho_0 = 100$. No optimization like parallelization or precompiled kernel. PM stands for proposed method

Method	max Δt (s)	iterations	real time/simul time	RMSE	parameters
[20]	0.00001	1000	6209.3	0.026075	$\alpha = 0.01$, $\beta = 0$, $c = 1484$
[30]	0.0001	100	635.9	0.006468	$\mu = 0.4452$
[24]	0.0001	100	706.2	0.006747	$\mu = 0.4452$
[36]	0.0001	100	710.8	0.082348	$\nu_0 = 0.000005$, $k = 0$
[37]	0.001	44	223.4	0.009944	$\mu = 1.2$, PCG tol. = 1%
[38]	0.001	434	81.2	0.072388	$\xi = 0.9$, PCG tol. = 1%
[39]	0.0001	265686	37863.9	0.155345	$\gamma = 0.89$, $\eta^{visco} = 0.1$
PM	0.001	259	90.6	0.007486	$\mu = 0.88$, $\eta = 0.1$
PM	0.001	278	92.4	0.012060	$\mu = 1.00$, $\eta = 0.1$

fast in opposite directions. Usually, a fast movement triggers subsequent explosions of nearby particles. Within the comparison experiments, the parameters for which an explosion occurs are neglected. The best settings corresponding to stable behaviour are thus selected.

The obtained results are summarized in Table 6 and their velocity curves are shown in Fig. 10 and Fig. 11. In these figures, the gray solid lines with dynamic viscosity coefficients μ indicate the reference viscosity curve calculated by the analytic solution in (30).

In the following, conventional explicit and implicit viscosity methods are compared separately.

The first four methods shown in Table 6 are explicit viscosity methods. The viscosity force is calculated according to the related viscosity equations at each time step Δt. Therefore the total iteration number is equal to $t/\Delta t$. The classic viscosity equation proposed in [20] requires the shortest time step, $\Delta t = 0.1 \cdot 10^{-4}$ s. Both of the viscosity equations proposed in [30] and in [24] produce very accurate results. The viscosity equation proposed in [36] generates a zigzag velocity curve. Such a zigzag velocity curve occurs because the calculation of the shear stress and the calculation of its derivative are separated so that the particle updated by the shear stress

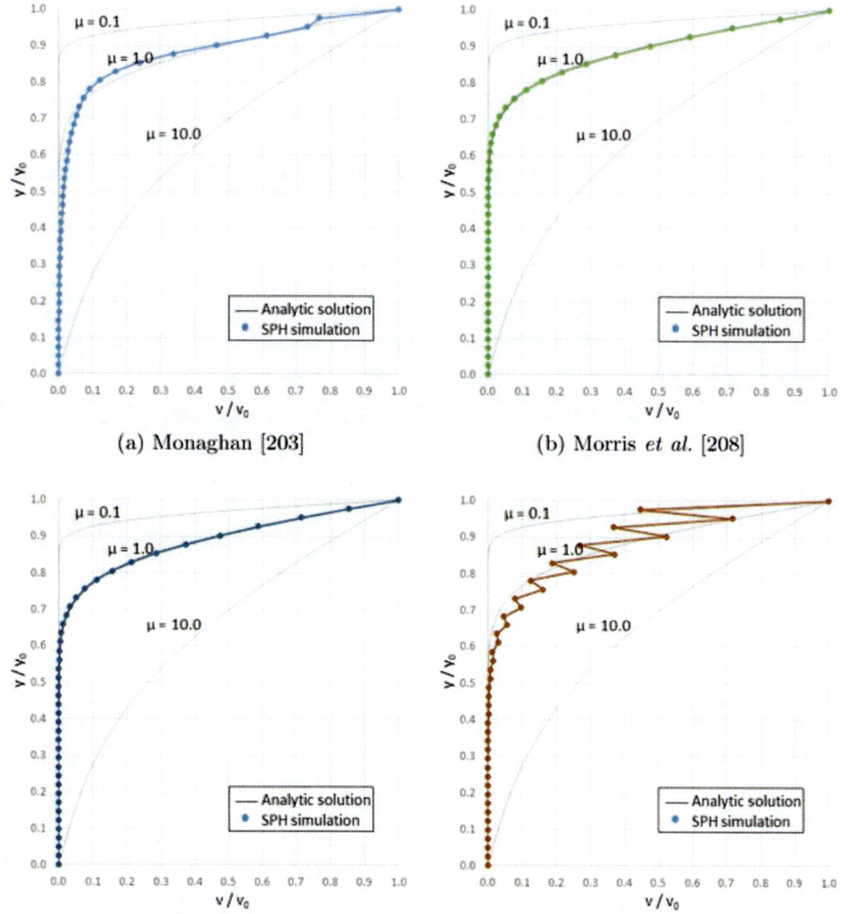

(a) Monaghan [203]

(b) Morris *et al.* [208]

(c) Shao & Lo [279]

(d) Andrade *et al.* [6]

Fig. 10 Couette flow experiments with various conventional viscosity methods and the proposed one for simulating a physical fluid of viscosity 1 at simulation time 0.01 s, particle spacing of $1.25 \cdot 10^{-3}$ m, and flow rest density $\rho_0 = 100$. The relative data are shown in Table 6

is different from the particle updated by the viscosity force. However, the overall pattern of the velocity curve is similar to the one of the analytic solution.

The following three viscosity methods shown in Table 6 are categorized into implicit viscosity methods. They update their velocities iteratively so that the relation equation about viscosity for all the fluid particles is solved. Therefore, viscosity diffusion can occur within a time step, and these implicit viscosity methods usually work well with relatively long time steps. In the comparison experiment, the viscosity algorithm proposed in [37] has recorded the fewest iterations among the conventional methods. The viscosity algorithm proposed in [38] has finished the simulation with the shortest running time, while the velocity curve is somewhat different from the

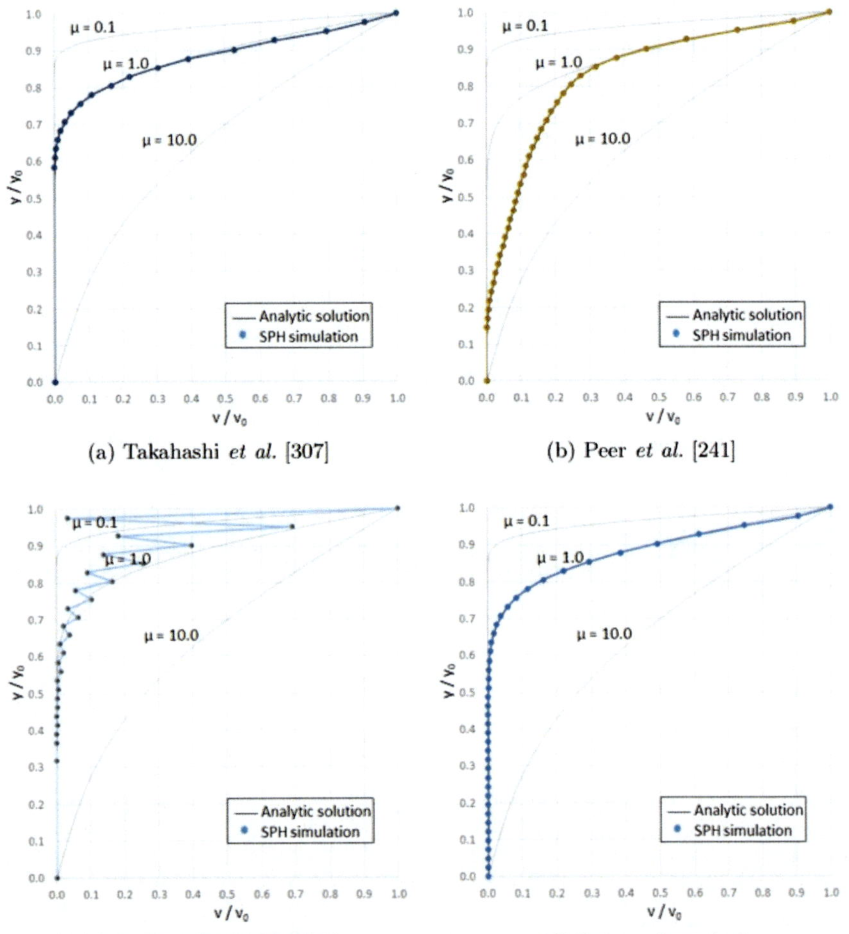

(a) Takahashi *et al.* [307]

(b) Peer *et al.* [241]

(c) Bender & Koschier [23]

(d) Proposed method

Fig. 11 Couette flow experiments with various conventional viscosity methods and the proposed one for simulating a physical fluid of viscosity 1.0 at simulation time $t = 0.01\,s$, particle spacing $r = 0.00125\,m$, and flow rest density $\rho_0 = 100$. Relative data are shown in Table 6

one of the analytic solution. The viscosity algorithm in [39] has also shown a zigzag curve, with reasons similar to the ones mentioned above.

The method proposed in this paper can work with a long time-step $\Delta t = 10^{-3}$ s. The time ratio between the real-time to simulation time of the proposed approach ranked second among the compared viscosity methods. Even though the best time performance is recorded by the viscosity method proposed in [38], the proposed method has higher accuracy. Moreover, it has been appreciated that the proposed method is more stable when there is a change in the viscosity parameter. Based on these results, it is possible to affirm that the proposed viscosity is comparable to the existing viscosity methods regarding accuracy, speed, and stability.

5.5 *Additional Simulations*

Two further simulations are presented. The former is about free-falls of several viscous fluids (water, honey, ketchup, and shortening) over a bunny sculpture. The dynamic viscosity coefficients of these fluids are $\mu = 8.94 \times 10^{-4}, 2{-}10, 50{-}100$, and ≈ 250, respectively. The following parameters have been considered to simulate these fluids, having in mind the above range $\mu = 0.0,\ 5.0,\ 70.0$, and 250.0. The simulation scenes for each case study are shown in Fig. 12. Because the time gap between the viscous fluids is big, i.e., the movement of shortening is languid compared to other fluids, the best scene for each liquid is illustrated. With the help of additional graphical treatment, it is possible to verify how the simulated case is close to the real one.

The latter simulation is about the stretching of a bread dough. Doughs are a non-Newtonian fluid, and they have complex properties including viscosity, elasticity, and plasticity [56–58]. Hence, the proposed viscosity method for Newtonian fluids does not perfectly match the simulation of a dough, but it can approximately simulate its behaviour. Taking data from physical experiments, the shear stress of general a dough has a relation equation $\tau = 298.76 + 5177\,\dot{\gamma}^{0.417}$ [57], where $\dot{\gamma}$ is the strain rate. The dynamic viscosity coefficient $\mu = 2500.0$ has been selected for a simulation with 6000 particles with spacing $1.25 \cdot 10^{-3}$ m. Some results are shown in Fig. 13 at time 0 s, 1.5 s, and 3 s, respectively. The dough is pushed, stretched, and torn: the simulated behaviours are similar to a real dough.

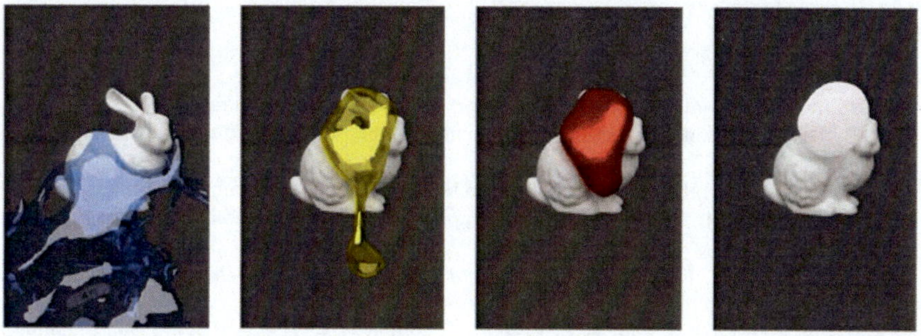

Fig. 12 Simulations of viscous fluids: water $\mu = 8.94 \times 10^{-4}\,(0.0)$, honey $\mu = 2 - 10\,(5.0)$, ketchup $\mu = 50 - 100\,(70.0)$, and shortening $\mu \approx 250\,(250.0)$. The numbers indicate physical dynamic viscosity coefficients and the numbers in parenthesis are the coefficients for these simulations

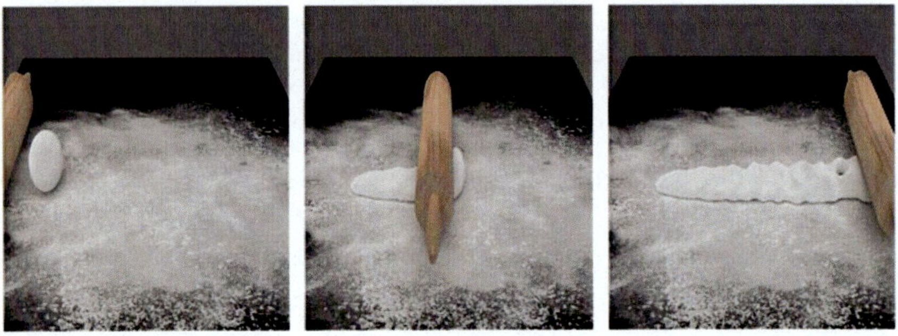

Fig. 13 Simulation of stretching of a bread dough with the dynamic viscosity coefficient $\mu = 2500.0$ at simulation time (left) 0 s, (middle) 1.5 s, and (right) 3 s

6 Discussion and Conclusion

A novel implicit viscosity method based on the SPH formulation for incompressible Newtonian fluids has been proposed. Such formulation can simulate wide-ranged viscosity behaviours in continuous materials.

Concerning the contributions, the proposed method exhibits a shorter computation time; it employs a physical viscosity coefficient, which helps predict the viscous behaviour of the continuous material more accurately. The carried out simulations show that the computational time is predictable. Finally, the accuracy is similar or higher than conventional viscosity methods in modelling continuous and viscous materials. Additionally, some graphical simulations show that the approach might be suitable for graphics animations.

References

1. T.J. Chung. *Computational Fluid Dynamics* (Cambridge University Press, 2010)
2. R. Eymard, T. Gallouët, R. Herbin, Finite volume methods. Handb. Numer. Anal. **7**, 713–1018 (2000)
3. G. Irving, J. Teran, R. Fedkiw, Invertible finite elements for robust simulation of large deformation, in *2004 ACM SIGGRAPH/Eurographics Symposium on Computer Animation*, pp. 131–140 (2004)
4. O.C. Zienkiewicz, R.L. Taylor, J.Z. Zhu. *The Finite Element Method Set* (Butterworth-Heinemann, 2005)
5. G.R. Liu, J. Zhang, K.Y. Lam, H. Li, G. Xu, Z.H. Zhong, G.Y. Li, X. Han, A gradient smoothing method (GSM) with directional correction for solid mechanics problems. Comput. Mech. **41**(3), 457–472 (2008)
6. M.B. Liu, G.R. Liu, Smoothed Particle Hydrodynamics (SPH): an overview and recent developments. Arch. Comput. Methods Eng. **17**(1), 25–76 (2010)
7. J.J. Monaghan, Smoothed particle hydrodynamics. Rep. Prog. Phys. **68**(1), 1703–1759 (2005)
8. J.U. Brackbill, H.M. Ruppel, FLIP: A method for adaptively zoned, particle-in-cell calculations of fluid flows in two dimensions. J. Comput. Phys. **65**(2), 314–343 (1986)

9. J. Cornelis, M. Ihmsen, A. Peer, M. Teschner, IISPH-FLIP for incompressible fluids. Comput. Gr. Forum **33**(2), 255–262 (2014)
10. A. Stomakhin, C. Schroeder, L. Chai, J. Teran, A. Selle, A material point method for snow simulation. ACM Trans. Gr. **32**(4), 102:1–102:10 (2013)
11. D. Sulsky, S.-J. Zhou, H.L. Schreyer, Application of a particle-in-cell method to solid mechanics. Comput. Phys. Commun. **87**(1), 236–252 (1995)
12. M. Macklin, M. Müller, Position based fluids. ACM Trans. Graph. **32**(4), 104:1–104:12 (2013)
13. M. Müller, B. Heidelberger, M. Hennix, J. Ratcliff, Position based dynamics. J. Vis. Commun. Image Represent. **18**(2), 109–118 (2007)
14. F.H. Harlow, *The particle-in-cell method for numerical solution of problems in fluid dynamics* Los Alamos Scientific Lab., N. Mex, Technical report (1962)
15. Y. Zhu, R. Bridson, Animating sand as a fluid. ACM Trans. Gr. **24**(3), 965–972 (2005)
16. T. Takahashi, Y. Dobashi, I. Fujishiro, T. Nishita, Volume preserving viscoelastic fluids with large deformations using position-based velocity corrections. Vis. Comput. **32**(1), 57–66 (2016)
17. R.A. Gingold, J.J. Monaghan, Smoothed particle hydrodynamics-theory and application to non-spherical stars. Mon. Not. R. Astron. Soc. **181**, 375–389 (1977)
18. L.B. Lucy, A numerical approach to the testing of the fission hypothesis. Astron. J. **82**, 1013–1024 (1977)
19. J.J. Monaghan, R.A. Gingold, Shock simulation by the particle method SPH. J. Comput. Phys. **52**(2), 374–389 (1983)
20. J.J. Monaghan, Smoothed particle hydrodynamics. Ann. Rev. Astron. Astrophys. **30**(1), 543–574 (1992)
21. J.J. Monaghan, Simulating free surface flows with SPH. J. Comput. Phys. **110**(2), 399–406 (1994)
22. M. Desbrun, M.-P. Gascuel, Smoothed particles: A new paradigm for animating highly deformable bodies, in *Eurographics Workshop on Computer Animation and Simulation '96*, pp. 61–76 (1996)
23. S.J. Cummins, M. Rudman, An SPH projection method. J. Comput. Phys. **152**(2), 584–607 (1999)
24. S. Shao, E.Y.M. Lo, Incompressible SPH method for simulating Newtonian and non-Newtonian flows with a free surface. Adv. Water Res. **26**(7), 787–800 (2003)
25. M. Becker, M. Teschner, Weakly compressible SPH for free surface flows, in *2007 ACM SIGGRAPH/Eurographics Symposium on Computer Animation*, pp. 209–217 (2007)
26. B. Solenthaler, R. Pajarola, Predictive-corrective incompressible SPH. ACM Trans. Gr..s, **28**(3), 40:1–40:6 (2009)
27. B. Solenthaler, R. Pajarola, Predictive-corrective incompressible SPH, in *ACM SIGGRAPH 2009 Papers*, pp. 40:1–40:6 (2009)
28. M. Ihmsen, J. Cornelis, B. Solenthaler, C. Horvath, M. Teschner, Implicit incompressible SPH. IEEE Trans. Vis. Comput. Gr. **20**(3), 426–435 (2014)
29. J. Bender, D. Koschier, Divergence-free smoothed particle hydrodynamics, in *14th ACM SIGGRAPH/Eurographics Symposium on Computer Animation*, pp. 147–155 (2015)
30. J.P. Morris, P.J. Fox, Y. Zhu, Modeling low Reynolds number incompressible flows using SPH. J. Comput. Phys. **136**(1), 214–226 (1997)
31. S. Clavet, P. Beaudoin, P. Poulin, Particle-based viscoelastic fluid simulation, in *2005 ACM SIGGRAPH/Eurographics Symposium on Computer Animation*, pp. 219–228 (2005)
32. Y. Chang, K. Bao, J. Zhu, E. Wu, High viscosity fluid simulation using particle-based method, in *2011 IEEE International Symposium on VR Innovation*, pp. 199–205 (2011)
33. M. Müller, D. Charypar, M. Gross, Particle-based fluid simulation for interactive applications, in *ACM Eurographics/SIGGRAPH Symposium on Computer Animation*, pp. 154–159 (2003)
34. M.M. Cross, Rheology of non-Newtonian fluids: A new flow equation for pseudoplastic systems. J. Colloid Sci. **20**, 417–437 (1965)
35. L.F.d.S. Andrade, M. Sandim, F. Petronetto, P. Pagliosa, A. Paiva, SPH fluids for viscous jet buckling, in *27th SIBGRAPI Conference on Graphics, Patterns and Images*, pp. 65–72 (2014)

36. L.F.d.S. Andrade, M. Sandim, F. Petronetto, P. Pagliosa, A. Paiva. Particle-based fluids for viscous jet buckling. Comput. Gr. **52**, 106–115 (2015)
37. T. Takahashi, Y. Dobashi, I. Fujishiro, T. Nishita, M.C. Lin, Implicit formulation for SPH-based viscous fluids. Comput. Gr. Forum **34**(2), 493–502 (2015)
38. A. Peer, M. Ihmsen, J. Cornelis, M. Teschner, An implicit viscosity formulation for SPH fluids. ACM Trans. Gr. **34**(4), 114:1–114:10 (2015)
39. J. Bender, D. Koschier, Divergence-free SPH for incompressible and viscous fluids. IEEE Trans. Visual Comput. Gr. **23**, 1193–1206 (2017)
40. H. Barreiro, I. García-Fernández, I. Alduán, M.A. Otaduy, Conformation constraints for efficient viscoelastic fluid simulation. ACM Trans. Gr. **36**(6), 221:1–221:11 (2017)
41. M. Weiler, D. Koschier, M. Brand, J. Bender, A physically consistent implicit viscosity solver for SPH fluids. Comput. Gr. Forum **37**(2), 145–155 (2018)
42. M. Carlson, P.J. Mucha, R.B. Van Horn III, G. Turk, Melting and flowing, in *2002 ACM SIGGRAPH/Eurographics Symposium on Computer Animation*, pp. 167–174 (2002)
43. C. Batty, R. Bridson, Accurate viscous free surfaces for buckling, coiling, and rotating liquids, in *2008 ACM SIGGRAPH/Eurographics Symposium on Computer Animation*, pp. 219–228 (2008)
44. A. Peer, M. Ihmsen, Prescribed velocity gradients for highly viscous SPH fluids with vorticity diffusion. IEEE Trans. Visual Comput. Gr. **23**(12), 2656–2662 (2016)
45. E. Rossi, A. Colagrossi, D. Durante, G. Graziani, Simulating 2D viscous flow around geometries with vertices through the diffused vortex hydrodynamics method. Comput. Methods Appl. Mech. Eng. **302**, 147–169 (2016)
46. A.E. Chorin. *A Mathematical Introduction to Fluid Mechanics* (Springer, 1992)
47. A. Powell, MIT materials science and engineering. 3.21 Lectures on Fluid Flow and Kinectics (2003)
48. R. Temam, *Navier-Stokes Equations: Theory and Numerical Analysis* (AMS Chelsea Publishing, 1977)
49. M. Ihmsen, J. Orthmann, B. Solenthaler, A. Kolb, M. Teschner, SPH fluids in computer graphics, in *EUROGRAPHICS 2014 — State of The Art Report*, pp. 21–42 (2014)
50. M. Kelager, Lagrangian fluid dynamics using smoothed particle hydrodynamics. Master's thesis, University of Copenhagen (2006)
51. M. Gomez-Gesteira, B.D. Rogers, R.A. Dalrymple, A. Crespo, State-of-the-art of classical SPH for free-surface flows. J. Hydraul. Res. **48**, 6–27 (2010)
52. F. Colin, R. Egli, F.Y. Lin, Computing a null divergence velocity field using smoothed particle hydrodynamics. J. Comput. Phys. **217**(2), 680–692 (2006)
53. H.-S. Dou, B. Khoo, Investigation of turbulent transition in plane Couette flows using energy gradient method. Adv. Appl. Math. Mech. **3**, 165–180 (2005)
54. X. X, J. Ouyang, W. Li, Q. Liu, SPH simulations of 2D transient viscoelastic flows using brownian configuration fields. J. Non-Newtonian Fluid Mech. **208**, 59–71 (2014)
55. L. Trefethen, A.E. Trefethen, S.C. Reddy, T. Driscoll, Hydrodynamic stability without eigenvalues. Science **261**(5121), 578–584 (1993)
56. E. Mitsoulis, Numerical simulation of calendering viscoplastic fluids. J. Nonnewton. Fluid Mech. **154**(2–3), 77–88 (2008)
57. E. Mitsoulis, S.G. Hatzikiriakos, Rolling of bread dough: Experiments and simulations. Food Bioprod. Process. **87**(2), 124–138 (2009)
58. S. Sofou, E.B. Muliawan, S.G. Hatzikiriakos, E. Mitsoulis, Rheological characterization and constitutive modeling of bread dough. Rheol. Acta **47**(4), 369–381 (2008)

Perception and Motion Planning for Unknotting/untangling of Ropes of Finite Thickness

Midhun S. Menon and Vincenzo Lippiello

Abstract Handling physical ropes/strings is an important capability to enable robots for human interaction, surgery/suturing, and so on. This chapter will discuss the specific problem of untangling strings/ropes with finite thickness. The chapter discusses the motivation for this problem, the problem statement, the state of the art in this area, and the specific direction of this research. Subsequently, an outline of the rope simulator being developed, which is a prerequisite, is provided. A brief discussion on the employed rope model and handling of self-collisions is provided.

1 Brief Introduction

Algorithms for modelling/manipulating ropes and string-like objects by robots is an area of research that is witnessing a resurgence in interest thanks to cheaper and easier access to computing power and a growing number of applications in different fields ranging from medical robotics, search-and-rescue robots, protein folding and drug-discovery to applications in computer graphics.

The handling of ropes, strings, and clothes presents an interesting case of prehensile and non-prehensile manipulation. Though such objects look simple and can be modelled, they pose challenging questions regarding various physical aspects like knotting-unknotting operations.

However, a primary requirement for approaching the manipulation and motion planning of objects like rope/string is to develop a virtual simulator. This chapter primarily deals with developing a rope simulator, followed by the future direction of research.

M. S. Menon (✉)
OLA Electric, Bangalore, India

V. Lippiello
Department of Electrical Engineering and Information Technology, PRISMA Lab, CREATE Consortium & University of Naples Federico II, Via Claudio 21, 80125 Naples, Italy
e-mail: vincenzo.lippiello@unina.it

B. Siciliano and F. Ruggiero (eds.), *Robot Dynamic Manipulation*,
Springer Tracts in Advanced Robotics 144,
https://doi.org/10.1007/978-3-030-93290-9_4

2 Problem Statement

The rope is represented mathematically using an arc length parametrized curve $\mathcal{C}(s,t) \in \mathbb{R}^3$, with $s \in \mathbb{R}$ the arc length parameter. A manipulation step consists of a vector of arc-length parameter values $S = \{s_1, s_2, \ldots, s_k\}$ and a set of corresponding 3D displacement vectors $P_i = \{\mathbf{p}_{s_1}, \mathbf{p}_{s_2}, \ldots, \mathbf{p}_{s_k}\}$ with $\mathbf{p}_{s_i} \in \mathbb{R}^3$, and $i = 1, \ldots, k$. It may be noted that k is not a constant but rather can vary in each time step based on the type of manipulation move, i.e., single($k = 1$), double($k = 2$), or multi-point manipulation.

For a given rope of finite diameter $d > 0$ and length $L > 0$ resting on horizontal plane, with initial configuration $\mathcal{C}_S(s, 0)$ and goal configuration $\mathcal{C}_G(s, T)$, an objective is to develop a motion planner to compute finite number of discrete quasi-static single/multi-point manipulation steps for times $0 \leq t_1 \leq t_2 \leq \cdots \leq T$ to achieve the demanded configuration change.

By calling the moves quasi-static, the idea is that the 3D shape of the rope at the end of each time step t_i is derived by post-processing of the image data from an RGB-D perception sensor mounted on the robot. In our case, the visual sensor used is Intel RealSense SR300 structured light camera. The result of stereo-processing is an arc length parametrized curve $\mathcal{C}_I(s, t_i)$ which gives the intermediate shape of the rope. The motion planner takes in $\mathcal{C}_I(s, t_i)$ as input and derives the next incremental manipulation step to take the rope to incremental configuration $\mathcal{C}_I(s, t_{i+1})$ which is closer to the goal configuration $\mathcal{C}_G(s, T)$ than $\mathcal{C}_I(s, t_i)$, the goal configuration.

The idea to implement this motion planner is to have a virtual model of a rope or a rope simulator. The motion planner will test out various trial moves before arriving at the candidate move. This will be based on the minimisation of some metric, which is a functional measuring how close or how far is the current configuration from the goal configuration.

3 Preliminaries and Associated State-of-the-art

A rope simulator simulates the motion of the rope in response to an input motion given at generic point(s) on the rope. The simulator has a rope model and an interference avoidance algorithm that avoids self and environmental obstacle interference/penetration. These combine to generate a virtual model of a rope that behaves like a standard physical rope or string. Using this model, the 3D configuration of the rope model in response to any generic single/multi-point perturbation is computed and analysed to evaluate if the new configuration is closer to the target configuration using a synthetic health metric functional evaluated on the configuration.

Ropes and strings are hyper-redundant systems due to their flexible nature. This poses the conflicting challenges of handling many DoFs and resolving the redundancy in the fastest possible refresh rates.

A large body of research already exists in simulating ropes and strings, many of which are from the computer graphics community and some from the medical robotics community. Significant contributions in these papers are in two main areas: computational models of such objects and interference detection/prevention algorithms. These are detailed in the following subsections.

It may be mentioned here that there are many off-the-shelf rope simulators available as part of commercial physics engines like MuJoCo [1]. However, they all have minor shortcomings: for example, some of them are closed source. This makes it difficult to understand the form of redundancy resolution used in the algorithm and practically challenging to modify them easily. Henceforth, all discussions will be not considering such commercial simulator platforms.

3.1 Rope Models

Solid mechanics dealt with small deformations of solid bodies for a long time [2]. However, it does not capture the significant deformation behaviour of objects like rope and string. The shape assumed by a weighted thread in a gravitational field was studied in the calculus of variations [3] and the shape was found to be the Catenary curve [4]. Subsequently, the Catenary curves were used in cloth modelling [5]. However, these ignore bend rigidity in the formulation. In contrast, Feynman used the plate flexure model in [6] to model cloth with in-plane and out-of-plane stiffness.

Synthetic physical deformation models for elastic objects arose in computer graphics to rapidly simulate the realistic motion of deformable bodies. Some of the earliest works in this area include [7–13]. The survey papers [14] and [15] give a good overview of the research in this field over subsequent years.

The research on models for one-dimensional flexible objects was preceded by problem of using minimal curvature-energy splines for non-linear interpolation [16–20] and generation of smooth curves for ease of manufacturing [21]. Subsequent works have addresses on adding boundary conditions and length constraints to this minimal-energy curve [22–25].

The recent years have witnessed a surge in requirement for such special purpose models for elastic and topologically one-dimensional bodies due to better and cheaper availability of computational power and the consequent rise of demand in fields like computer graphics and animation, snake robots, and robotic surgical simulators. This resulted in a new classification for such objects called Deformable Linear Objects (DLO). Some researchers try to simulate snake/worm motion as DLO using spring-mass system [26] while others try to use energy-minimisation techniques in contour extraction from images [27]. Subsequently, computation on such spring-mass systems by adaptive refinement has been accelerated [28]. The work in [29] generalises the minimal energy curve to include twist and bend energies which have been extended using differential geometry to dynamics [30]. There have also been works on accelerating techniques. Another popular modelling method was using Cosserat rod elements [31–33]. A more detailed overview can be found in [34].

An equally important area of development has been in interference detection and collision avoidance algorithms for DLOs. In this case, it has to detect the DLO's self-collision and a collision of DLO with surroundings. Self collisions are essential because any physically realistic motion simulation should always avoid the rope from passing through itself and respect the rope's finite thickness nature. The work in [35] uses a spline of springs with collision avoidance and adaptive subdivision to create a virtual knot tying simulation. The collision detection scheme is a brute force check for interference between spheres. Finite-Element Methods (FEM) and an optimisation-based scheme were employed to incorporate collision avoidance and solve for the associated configuration of the DLO [36].

While most of the aforementioned works focused on approaching the problem from a dynamics perspective, some research has also focused on kinematic approaches. The Follow The Leader (FTL) approach is used in [37] and [38] for computing rope configurations.

Clearly, all the above-mentioned works are taken from the field of computer graphics to simulate the motion of ropes, strings, hair, or fur. Another research branch tackling the same problem is from the robotics community working on hyper-redundant serial manipulators and inverse kinematics. The early works in [39] and [40] utilizes the concept of backbone curves and modal decomposition to model the manipulator's inverse kinematics. Another group of researchers attempt to resolve this redundancy by minimising specific metrics [41, 42]. Working on these lines, These works were extended using the Tractrix curve for redundancy resolution [43], an optimisation-based approach [44], and adding obstacle avoidance to the framework [45].

4 Proposed Algorithm

The majority of the solvers simulate the system's dynamics by solving the governing recursive Newton-Euler equations. Others solve the system's dynamics by minimising the Lagrangian of the system. However, the problem being solved in the given scenario is only for discrete states and as movements are quasi-static. In between two such quasi-static manipulation steps, the current state/configuration of the DLO is taken as feedback to re-plan the subsequent manipulation move. The criticality of dynamic terms is minimised. Moreover, there are additional advantages of confining to just kinematic or quasi-static simulations. The rope system is a badly scaled system due to its high longitudinal stiffness and low-lateral stiffness, making the associated dynamical system ill-conditioned and prone to numerical errors and instabilities. Restricting to kinematic simulation can help in avoiding these issues.

Given the above-mentioned considerations, it is decided to utilise the tractrix algorithm [45] for the simulator. The following subsections detail the approach taken to model the rope.

4.1 Rope Model

The rope is modelled as a serial kinematic chain of rigid links connected by spherical joints. From any given state of the rope $C_I(s, t_i)$, the first step is to discretize it into a rigid chain by sampling at arc length parameter values $s = \{s_0 = 0, s_1, s_2, \ldots, s_n = L\}$ (generic spacing) and convert it into an n-link rigid kinematic chain with joint coordinates denoted by $P_I = \{\mathbf{r}_0, \mathbf{r}_1, \mathbf{r}_2, \ldots, \mathbf{r}_n\}$ with $\mathbf{r}_i \in \mathbb{R}^3$, with $i = 1, \ldots, n$. The links are denoted by $l_1, l_2, \ldots, l_n$.

Without loss of generality, consider the case of a single point manipulation where a prescribed displacement vector $\mathbf{m}_i \in \mathcal{R}^3$ at the i-th joint on the kinematic chain introduces a configuration change. To find the new configuration of the rope, the tractrix algorithm is run on the chain to solve for a new P_I. It may be noted that, if the manipulation point is not on the ends of the rope ($i \neq 0$ or $i \neq n$), the kinematic chain can be treated as two sub-chains with head at i and solving for the new configuration P_I.

4.2 Self-collision Avoidance Model

For the interference model, most of the papers mentioned above use an explicit interference detection step wherein different methods are employed to detect solid-solid interference between different rope parts. Subsequently, the interference is removed by brute force corrections. However, most interference detectors use the shortest perpendicular distance(signed/unsigned), which is an inherently discontinuous measure and is numerically not robust when the objects are too close to each other, which can potentially cause blowup errors.

The tractrix algorithm solves the manipulation step, one joint at a time. Hence, all links except those connected to the i-th considered joint may be considered as obstacles to be avoided. Along the lines of [45], these obstacles are modelled as superquadrics [46, 47]. Subsequently, the obstacle avoidance problem is set up as a motion minimisation problem with obstacles as geometric constraints, similar to the artificial potential force approach in [48]. The difference is that, here, the obstacles generate constraint avoidance velocities and not constraint avoidance forces.

The approach guarantees that the obstacles are avoided optimally by ensuring that the rope velocities are tangential, which causes the obstacles to graze on top of each other. This simulates the actual physical behaviour of the rope, which further substantiates the choice of this algorithm.

5 Motion Planner

In this section, the motion planner will be discussed. The motion planner is based on optimising an objective function.

Physical knot theory is an upcoming area in mathematics, which tries to view the knots in their canonical forms as a configuration minimising specially designed energy functionals. The field came up from the need to check ambient isotropy between two configurations of a knot [49, 50]). Over the years, it has found growing application specifically in the area of protein confirmation study. Various energy functionals have been suggested, of which the first functional for polygonal knots were done in [51]. Subsequently, other energy functionals were also proposed [52].

These energy functionals have the usual property of energy, namely positive definiteness and infinite energy barrier to the self-crossing of the polygonal chain, which guarantees topologically consistent motion plans.

6 Discussion and Conclusion

To date, the energy mentioned above functionals were used for only mathematical knots, which are closed curves in space and have been able to successfully un-knot configurations in non-canonical states to canonical energy stationary/minimal configurations. This research will attempt to use these functions to un-knot physical knots by minimising these energy functionals in their original form or slightly modified form as required.

As these energy functionals possess multiple local minima and inflection points, sub-optimal convergence will cause the gradient-based optimisation approaches to fail. Hence, the idea is to use approaches that are not based on the gradient and heuristic ones, like metaheuristic approaches, simulated annealing, or perturbation methods, to solve this problem.

References

1. E. Todorov, T. Erez, Y. Tassa, MuJoCo: A physics engine for model-based control, in *2012 IEEE/RSJ International Conference on Intelligent Robots and Systems*, pp. 5026–5033 (2012)
2. Y.-C. Fung, *Foundations of Solid Mechanics (Book on deformation and motion of elastic and plastic solids including variational calculus and tensor analysis)* (Prentice-Hall, 1965)
3. L.D. Elsgolc, *Calculus of Variations* (Courier Corporation, 2012)
4. H.M. Irvine, *Cable Structures*, vol. 17 (MIT Press, 1981)
5. J. Weil, The synthesis of cloth objects. ACM Siggraph Comput. Gr. **20**(4), 49–54 (1986)
6. C.R. Feynman, *Modeling the Appearance of cloth*. PhD thesis, Massachusetts Institute of Technology (1986)
7. W.W. Armstrong, M.W. Green, The dynamics of articulated rigid bodies for purposes of animation. Vis. comput. **1**(4), 231–240 (1985)

8. A.H. Barr, Global and local deformations of solid primitives. *Readings in Computer Vision*, pp. 661–670 (1987)
9. R. Barzel, A.H. Barr, A modeling system based on dynamic constraints, in *5th Annual Conference on Computer Graphics and Interactive Techniques*, pp. 179–188 (1988)
10. T.W. Sederberg, S.R. Parry, Free-form deformation of solid geometric models, in *13th Annual Conference on Computer Graphics and Interactive Techniques*, pp. 151–160 (1986)
11. D. Terzopoulos, J. Platt, A. Barr, K. Fleischer, Elastically deformable models. ACM Siggraph Comput. Gr. **21**(4), 205–214 (1987)
12. J.P. Wilhelms, B.A. Barsky, Using dynamic analysis to animate articulated bodies such as humans and robots, in *Computer-Generated Images* (Springer, 1985), pp. 209–229
13. A. Witkin, K. Fleischer, A. Barr, Energy constraints on parameterized models. ACM SIGGRAPH Comput. Gr. **21**(4), 225–232 (1987)
14. S.F.F. Gibson, B. Mirtich, *A survey of deformable modeling in computer graphics* (Technical report, Citeseer, 1997)
15. D.N. Metaxas, *Physics-based Deformable Models: Applications to Computer Vision, Graphics and Medical Imaging*, vol. 389 (Springer Science & Business Media, 2012)
16. G. Birkhoff, H. Burchard, D. Thomas, *Nonlinear Interpolation by Splines* (General Motors Corporation, Pseudosplines and Elastica. Research Laboratories, 1965)
17. G. Brunnett, A new characterization of plane elastica, in *Mathematical methods in computer aided geometric design II* (Elsevier, 1992), pp. 43–56
18. S.D. Fisher, J.W. Jerome, Stable and unstable elastica equilibrium and the problem of minimum curvature, in *Minimum Norm Extremals in Function Spaces* (Springer, 1975), pp. 90–106
19. E. Jou, W. Han, Elastica and minimal–energy splines, in *Curves and Surfaces* (Elsevier, 1991), pp. 247–250
20. M.A. Malcolm, On the computation of nonlinear spline functions. SIAM J. Numer. Anal. **14**(2), 254–282 (1977)
21. C. de Boor, *A Practical Guide to Splines* (Springer, 1978)
22. B.K.P. Horn, The curve of least energy. ACM Trans. Math. Softw. (TOMS) **9**(4), 441–460 (1983)
23. E. Jou, Han, Minimal-energy splines with various end constraints, in *Curve and Surface Design* (SIAM, 1992), pp. 23–40
24. M. Kallay, Plane curves of minimal energy. ACM Trans. Math. Softw. (TOMS) **12**(3), 219–222 (1986)
25. M. Kallay, Method to approximate the space curve of least energy and prescribed length. Comput. Aided Des. **19**(2), 73–76 (1987)
26. G.S.P. Miller, The motion dynamics of snakes and worms, in *15th Annual Conference on Computer Graphics and Interactive Techniques*, pp. 169–173 (1988)
27. M. Kass, A. Witkin, D. Terzopoulos, Snakes: Active contour models. Int. J. Comput. Vis. **1**(4), 321–331 (1988)
28. D. Hutchinson, M. Preston, T. Hewitt, Adaptive refinement for mass/spring simulations, in *Computer Animation and Simulation'96* (Springer, 1996), pp. 31–45
29. H. Wakamatsu, S. Hirai, K. Iwata, Modeling of linear objects considering bend, twist, and extensional deformations, in *1995 IEEE International Conference on Robotics and Automation*, pp. 433–438 (1995)
30. H. Wakamatsu, K. Takahashi, S. Hirai, Dynamic modeling of linear object deformation based on differential geometry coordinates, in *2005 IEEE International Conference on Robotics and Automation*, pp. 1028–1033 (2005)
31. M. Grégoire, E. Schömer, Interactive simulation of one-dimensional flexible parts. Comput. Aided Des. **39**(8), 694–707 (2007)
32. D.K. Pai, Strands: Interactive simulation of thin solids using cosserat models. Comput. Gr. Forum **21**(3), 347–352 (2002)
33. J. Spillmann, M. Teschner, CoRdE: Cosserat rod elements for the dynamic simulation of one-dimensional elastic objects, in *2007 ACM SIGGRAPH/Eurographics Symposium on Computer Animation*, pp. 63–72 (2007)

34. K. Ward, F. Bertails, T.-Y. Kim, S.R. Marschner, M.-P. Cani, M.C. Lin, A survey on hair modeling: Styling, simulation, and rendering. IEEE Trans. Vis. Comput. Gr. **13**(2), 213–234 (2007)
35. J. Phillips, A. Ladd, L.E. Kavraki., Simulated knot tying, in *2002 IEEE International Conference on Robotics and Automation*, pp. 841–846 (2002)
36. J. Spillmann, M. Teschner, An adaptive contact model for the robust simulation of knots. Comput. Gr. Forum **27**(2), 497–506 (2008)
37. J. Brown, J.-C. Latombe, K. Montgomery, Real-time knot-tying simulation. Vis. Comput. **20**(2–3), 165–179 (2004)
38. M. Müller, T.-Y. Kim, N. Chentanez, Fast simulation of inextensible hair and fur, in *Eurographics/ACM SIGGRAPH Symposium on Computer Animation*, pp. 39–44 (2012)
39. G.S. Chirikjian, J.W. Burdick, Kinematically optimal hyper-redundant manipulator configurations. IEEE Trans. Robot. Autom. **11**(6), 794–806 (1995)
40. K.E. Zanganeh, J. Angeles, The inverse kinematics of hyper-redundant manipulators using splines, in *1995 IEEE International Conference on Robotics and Automation*, pp. 2797–2802 (1995)
41. D.N. Nenchev, Redundancy resolution through local optimization: A review. J. Robot. Syst. **6**(6), 769–798 (1989)
42. T. Yoshikawa, Analysis and control of robot manipulators with redundancy, in *Robotics Research: The First International Symposium*, pp. 735–747 (1984)
43. V.C. Ravi, S. Rakshit, A. Ghosal. Redundancy resolution using tractrix—simulations and experiments. J. Mech. Robot. **2**(3), 031013–1–031013–7 (2010)
44. M.S. Menon, G.K. Ananthasuresh, A. Ghosal, Natural motion of one-dimensional flexible objects using minimization approaches. Mech. Mach. Theory **67**, 64–76 (2013)
45. M.S. Menon, V.C. Ravi, A. Ghosal, Trajectory planning and obstacle avoidance for hyper-redundant serial robots. J. Mech. Robot. **9**(4), 041010–1–041010–9 (2017)
46. A.H. Barr, Superquadrics and angle-preserving transformations. IEEE Comput. Gr. Appl. **1**(1), 11–23 (1981)
47. M. Gardner, *Mathematical Carnival: A New Round-up Tantalizers and Puzzles from Scientific American* (Vintage Books, 1977)
48. O. Khatib, Real-time obstacle avoidance for manipulators and mobile robots, in *Autonomous Robot Vehicles* (Springer, 1986), pp. 396–404
49. S. Fukuhara, Energy of a knot, in *A fête of Topology* (Elsevier, 1988), pp. 443–451
50. J. O'hara, Energy of a knot. Topology **30**(2), 241–247 (1991)
51. G. Buck, J. Simon, Knots as dynamical systems. Topol. Appl. **51**(3), 229–246 (1993)
52. J.K. Simon, Energy functions for polygonal knots. J. Knot Theory Ramificat. **3**(3), 299–320 (1994)

Nonprehensile Manipulation Planning and Control

Pizza-Peel Handling Through a Sliding Nonprehensile Manipulation Primitive

Alejandro Gutierrez-Giles, Fabio Ruggiero, Vincenzo Lippiello, and Bruno Siciliano

Abstract The sliding primitive is a ubiquitous nonprehensile manipulation task, generally performed by mechanical systems represented by underactuated nonlinear models. A literature review of the recent works dealing with this task is first introduced. Then, a particular nonprehensile manipulation task that has arisen in the framework of the RoDyMan project, i.e., a pizza and peel mechanical system, is addressed. A more in-depth study is presented for this system, including modelling, control, and stability analysis. Finally, a discussion on the current achievements and some directions for future work is provided. This chapter is based on the work presented in [1].

1 Brief Introduction

Nonprehensile manipulation through the sliding primitive is a critical control application in the industry, particularly in the so-called *part feeders*. Accordingly, much research has been carried out over the last years in various directions. For example, optimisation in terms of the time required to positioning and orienting a part on a plane is one of the most pursued objectives. Another direction of research deals with the minimum number of actuators/degrees of freedom required to complete a given sliding motion task.

A. Gutierrez-Giles (✉)
CECAv-UNAM, Av. Universidad 3000, Ciudad Universitaria, CDMX 04510, Mexico
e-mail: alejandro_giles@cecav.unam.mx

F. Ruggiero · V. Lippiello · B. Siciliano
Department of Electrical Engineering and Information Technology, PRISMA Lab, CREATE Consortium & University of Naples Federico II, Via Claudio 21, 80125 Naples, Italy
e-mail: fabio.ruggiero@unina.it

V. Lippiello
e-mail: vincenzo.lippiello@unina.it

B. Siciliano
e-mail: bruno.siciliano@unina.it

B. Siciliano and F. Ruggiero (eds.), *Robot Dynamic Manipulation*,
Springer Tracts in Advanced Robotics 144,
https://doi.org/10.1007/978-3-030-93290-9_5

Table 1 Main symbols used in this chapter

Definition	Symbol
World frame	$\mathcal{W}$
Peel (hand) frame	$\mathcal{H}$
Pizza (object) frame	$\mathcal{O}$
Position of the origin of the frame $\mathcal{H}$ in $\mathcal{W}$	$\mathbf{o}_h = \begin{bmatrix} x_h & y_h & z_h \end{bmatrix}^T \in \mathbb{R}^3$
Position of the origin of the frame $\mathcal{O}$ in $\mathcal{H}$	$\mathbf{o}_o^h = \begin{bmatrix} x_o & y_o & z_o \end{bmatrix}^T \in \mathbb{R}^3$
Rotation matrix of $\mathcal{H}$ with respect to $\mathcal{W}$	$\mathbf{R}_h \in SO(3)$
Rotation matrix of $\mathcal{O}$ with respect to $\mathcal{H}$	$\mathbf{R}_o^h \in SO(3)$
Rotation angle of $\mathcal{H}$ with respect to $\mathcal{W}$ around x_h	$\theta \in \mathbb{R}$
Rotation angle of $\mathcal{O}$ with respect to $\mathcal{H}$ around z_o	$\phi \in \mathbb{R}$
Mass of the peel	$m_h > 0$
Mass of the pizza	$m_o > 0$
First row-first column element of the peel's inertia matrix in $\mathcal{H}$	$I_{hx} \in \mathbb{R}$
Inertia matrix of the pizza in $\mathcal{O}$	$\mathbf{I}_o \in \mathbb{R}^{3\times 3}$
First row-first column element of $\mathbf{I}_o$	$I_{ox} \in \mathbb{R}$
Third row-third column element of $\mathbf{I}_o$	$I_{oz} \in \mathbb{R}$
Gravity acceleration	$g = 9.81$ m/s^2
Force applied over the peel along x_h	$u_h \in \mathbb{R}$
Torque applied around the axis x_h	$\tau_h \in \mathbb{R}$
Linear Coulomb friction coefficient between the pizza and the peel	$\mu_o > 0$
Angular Coulomb friction coefficient between the pizza and the peel	$\mu_\phi > 0$
Desired pizza rotation speed	$\dot{\phi}_d \in \mathbb{R}$
Controller gains	$k_i \in \mathbb{R}$, with $i = 1, \ldots, 9$

From a theoretical point of view, the specialists have made a great effort to understand the dynamic effects of all the forces that interact during a sliding manipulation task. In particular, it has been noticed that friction, both static and dynamic, plays a fundamental role in this kind of task. Although there are very precise mechanisms in the actuality that can efficiently move a given part to an arbitrary position and orientation in the plane, most of them carry out this task employing open-loop controllers.

The advantages of the feedback control are well-known inside and outside the control community, and thus there are some efforts to incorporate feedback for non-prehensile sliding manipulation. This chapter aims to design a model-based feedback controller for a specific sliding manipulation task, which consists of an unactuated disk and a 2-DoF manipulator.

First, a review of the current state of research is introduced, related to moving parts through force fields generated by the friction forces arising when two surfaces in contact are in relative motion. Then, a particular application of the sliding primitive in the context of the RoDyMan project is described, i.e., a pizza and peel manipulation system. A mathematical model is developed for this system, which in turn serves as a base for designing the feedback control strategy. The notation employed throughout this chapter is summarized in Table 1 for the convenience of the reader.

2 State of the Art

Manipulating objects through a vibrating surface has been a handy application in the industrial context, in particular for *part feeders*. This application has motivated much research in the last decades. From a theoretical point of view, in the late 1990s, the static equilibrium positions and orientations of two-dimensional parts under planar vector fields are investigated in [2], without specifying which mechanism is employed to create these vector fields. On the other hand, the mechanisms to create such vector fields are analysed in [3], where a pixel-wise array is proposed as a universal planar mechanism. In this same work, the authors establish a theoretical tradeoff between mechanical design and motion planning complexities. The universal parts feeders are studied in [4] for non-symmetric parts in the philosophy of minimalist robotics, i.e., the minimal configuration of resources required to solve a given task. Later, a fascinating result is stated in [5]: a horizontal vibrating plate is a universal planar manipulator, including the construction of the proposed mechanism. This mechanism is improved in [6], where the authors are capable of moving a specific part among several others over the plate without moving the remaining ones. A variation of the universal planar manipulator, similar to the one developed in [3], is constructed in [7] by employing three orthogonal actuators. All the approaches mentioned above use open-loop controllers.

To the best of the authors' knowledge, one of the first attempts to introduce feedback is presented in [8], where the authors utilise non-smooth analysis and discontinuous control to stabilise an arbitrary position and orientation of the part. Interestingly, in the same article, it is shown that the open-loop controllers are unstable in the orientation coordinate near the static equilibrium points.

In [9], it is noted that the existing universal planar manipulators always generate a force vector field with zero divergence, which is a practical disadvantage. The authors design a universal planar manipulator up to 6 DoFs that can generate force fields with non-zero divergence to solve this problem. This mechanism is employed in [10] in conjunction with a bang-bang control strategy, contrarily to the commonly employed sinusoidal, to generate movement in the parts towards and away predefined nodal lines. Later, the same authors proposed an optimal controller to generate a twist field that can position and orient a part using their previously designed universal planar manipulator. The effect of both dynamic and static friction on the velocity of the part is studied in [11], where a strategy is proposed to maximise the transport velocity of the sliding parts.

A task-specific mechanism is designed in [12] to translate and rotate a disk to an arbitrary position and orientation. The Neapolitan pizza chefs inspired this mechanism, and the design philosophy is that of minimalist robotics. The authors analyse several configurations for completing the task, arriving at the solution of a two-DoF manipulator. The mechanism and the control strategy are further improved in [13]. The authors also show in [14] that the same strategy can be applied to rotate a deformable disk. Later, the same authors proposed a model of the deformable object in [15]. At the same time, they also find a correlation between the plate frequency and different kinds of motion, i.e. sliding, walking, and running. Finally, in [16], the same authors find that a deformable disk rotates faster than a rigid one. The effects of the friction between the plate and the disk are investigated in the same article.

3 Pizza-Peel Manipulation Task

In this section, a dynamic model is first developed based on the Euler-Lagrange equations of motion and some friction properties found in the literature. Then, a feedback control strategy is designed to rotate the pizza. A stability analysis is later presented to show the boundedness of both pizza and peel coordinates and a limit cycle that rotates the pizza at the desired speed. Finally, a numerical simulation is presented to illustrate the validity of the approach.

3.1 *Dynamic Model*

Recall the relevant frames as the world frame, $\mathcal{W}$, the frame attached to the peel (hand), $\mathcal{H}$, and the frame attached to the pizza (object), $\mathcal{O}$, as shown in Fig. 1. In this chapter, a superscript is used to denote to which frame is the vector or matrix referenced. When referred to $\mathcal{W}$, this superscript is omitted. It is assumed that that

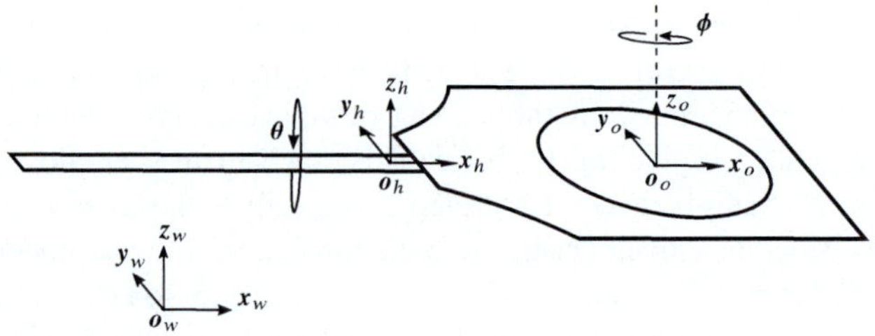

Fig. 1 Peel and pizza system

the peel is driven by a robotic manipulator. The generalised coordinates for the peel are chosen to be $x_h \in \mathbb{R}$ and $\theta \in \mathbb{R}$. The generalised coordinates for the pizza are chosen as $x_o, y_o, \phi \in \mathbb{R}$. Therefore, the configuration of the system is completely described by the vector

$$\mathbf{q} = \left[x_h \; \theta \; x_o \; y_o \; \phi\right]^T \in \mathbb{R}^5. \tag{1}$$

The pizza centre of mass with respect to $\mathcal{W}$ is given by

$$\mathbf{p}_o = \mathbf{p}_h + \mathbf{R}_h \mathbf{p}_o^h = \left[x_h + x_o \; y_o c_\theta \; y_o s_\theta\right]^T , \tag{2}$$

where s_x and c_x are shorthand notations for $\sin(x)$ and $\cos(x)$, respectively. The orientation of $\mathcal{O}$ with respect to $\mathcal{W}$ is described by the rotation matrix

$$\mathbf{R}_o = \mathbf{R}_h \mathbf{R}_o^h . \tag{3}$$

From this last matrix, one can obtain the pizza angular velocity vector $\boldsymbol{\omega}_o \in \mathbb{R}^3$ through the relation

$$\dot{\mathbf{R}}_o = \mathbf{S}(\boldsymbol{\omega}_o)\mathbf{R}_o \tag{4}$$

The kinetic energy, in terms of the generalised coordinates and velocities, is given by

$$\mathcal{T}(\mathbf{q}, \dot{\mathbf{q}}) = \frac{1}{2} m_h \dot{x}_h^2 + \frac{1}{2} I_{hx} \dot{\theta}^2 + \frac{1}{2} m_o \dot{\mathbf{p}}_o^{\mathrm{T}} \dot{\mathbf{p}}_o + \frac{1}{2} \boldsymbol{\omega}_o^T \mathbf{R}_o \mathbf{I}_o \mathbf{R}_o^T \boldsymbol{\omega}_o. \tag{5}$$

The potential energy can be computed by

$$\mathcal{U}(\mathbf{q}) = m_o g s_\theta y_o . \tag{6}$$

For obtaining the dynamic model, consider the Euler–Lagrange equations of motion

$$\frac{\mathrm{d}}{\mathrm{d}t}\left(\frac{\partial \mathcal{L}}{\partial \dot{\mathbf{q}}}\right)^T - \left(\frac{\partial \mathcal{L}}{\partial \mathbf{q}}\right)^T = \boldsymbol{\xi} \tag{7}$$

with Lagrangian $\mathcal{L}(\mathbf{q}, \dot{\mathbf{q}}) = \mathcal{T}(\mathbf{q}, \dot{\mathbf{q}}) - \mathcal{U}(\mathbf{q})$ and the non-conservative and external forces are represented by the vector

$$\boldsymbol{\xi} = \left[u_h \; \tau_\theta \; f_x \; f_y \; \tau_\phi\right]^T , \tag{8}$$

where f_x, f_y, and τ_ϕ are explained below.

The Coulomb friction terms play an important role for this particular task. These terms are defined as functions of the relative velocities between the peel and the pizza $\dot{x}_o$ and $\dot{y}_o$, and are described by [13]

$$f_x = m_o g \mu_o \text{sign}(\dot{x}_o) \tag{9}$$

$$f_y = m_o g \mu_o \text{sign}(\dot{y}_o) , \tag{10}$$

where sign(x) is the function defined by

$$\text{sign}(x) = \begin{cases} 1 & \text{if } x > 0 \\ 0 & \text{if } x = 0 \\ -1 & \text{if } x < 0 \end{cases} .$$

As explained in [13], there is a torque over the $\mathbf{z}_o^h$ axis produced by the movement of the x_h coordinate and the change of the pressure distribution which in turn is induced by the acceleration on the θ coordinate and is given by

$$\tau_\phi = \mu_\phi I_{ox} \text{sign}(\dot{x}_o)\ddot{\theta}. \tag{11}$$

By Newton's third law of motion, there must be a reaction torque acting on the θ coordinate. However, this torque can be neglected by assuming that the inertia of the peel is much bigger than that of the pizza.

A further simplification can be made if the linear and angular accelerations of the peel are assumed to be the inputs, i.e.,

$$\mathbf{u} \triangleq \left[u_h \; u_\theta\right]^T \triangleq \left[\ddot{x}_h \; \ddot{\theta}\right]^T . \tag{12}$$

The following approximation of the sign function is made in order to employ continuous tools to analyse the system dynamics

$$\text{sign}(x_i) \approx \tanh(k_i x_i) , \tag{13}$$

where each $k_i > 0$ is a constant. The objective is to control the pizza rotation speed, for which the regulation error is defined as

$$\dot{\tilde{\phi}} = \dot{\phi} - \dot{\phi}_d \in \mathbb{R} , \tag{14}$$

Next, let the state space vector $\mathbf{x} \in \mathbb{R}^9$ be defined by

$$\begin{aligned} \mathbf{x} &= \left[x_1 \; x_2 \; \cdots \; x_9\right]^T \\ &\triangleq \left[x_h \; \theta_h \; x_o \; y_o \; \dot{x}_h \; \dot{\theta}_h \; \dot{x}_o \; \dot{y}_o \; \dot{\tilde{\phi}}\right]^T . \end{aligned} \tag{15}$$

Thus, the system dynamics can be put in the form

$$\dot{\mathbf{x}} = \mathbf{f}(\mathbf{x}) + \mathbf{g}_1 u_1 + \mathbf{g}_2(\mathbf{x}) u_2 , \tag{16}$$

where

$$\mathbf{f}(\mathbf{x}) = \begin{bmatrix} x_5 \\ x_6 \\ x_7 \\ x_8 \\ \mathbf{f}_2(\mathbf{x}) \end{bmatrix} \quad \mathbf{g}_1 = \begin{bmatrix} \mathbf{0}_4 \\ \mathbf{g}_{12} \end{bmatrix} \quad \mathbf{g}_2(\mathbf{x}) = \begin{bmatrix} \mathbf{0}_4 \\ \mathbf{g}_{22}(\mathbf{x}) \end{bmatrix}, \tag{17}$$

with

$$\mathbf{f}_2(\mathbf{x}) = \begin{bmatrix} 0 \\ 0 \\ -\frac{b_o}{m_o} x_7 - g\mu_o \tanh(k_7 x_7) \\ -\frac{b_o}{m_o} x_8 - g\mu_o \tanh(k_8 x_8) - g\sin(x_2) + x_4 x_6^2 \\ 0 \end{bmatrix} \tag{18}$$

$$\mathbf{g}_{12} = \begin{bmatrix} 1 \; 0 \; -1 \; 0 \; 0 \end{bmatrix}^T \tag{19}$$

$$\mathbf{g}_{22}(\mathbf{x}) = \begin{bmatrix} 0 \; 1 \; 0 \; 0 \; -\mu_\phi \left(I_{ox}/I_{oz} \right) \tanh(k_7 x_7) \end{bmatrix}^T. \tag{20}$$

By computing the Philip Hall basis [17, p. 344] with the vector fields $\mathbf{f}$, $\mathbf{g}_1$, and $\mathbf{g}_2$, it can be verified that the *accessibility distribution* is of dimension 9 in the set $\mathcal{D} = \left\{ \mathbf{x} \in \mathbb{R}^9 : \dot{\theta} \neq 0, \dot{x}_o \neq 0, y_o \neq 0 \right\}$, and therefore the system is *accessible*. Furthermore, by computing the base

$$\{\mathbf{g}_1, \mathbf{g}_2, [\mathbf{g}_1, \mathbf{f}], [\mathbf{g}_2, \mathbf{f}], [\mathbf{g}_1, \mathbf{g}_2], [\mathbf{f}, [\mathbf{g}_1, \mathbf{f},]], [\mathbf{f}, [\mathbf{g}_2, \mathbf{f}]], \\ [\mathbf{f}, [\mathbf{g}_1, \mathbf{g}_2]], [\mathbf{g}_1, [\mathbf{g}_1, \mathbf{f}]], [\mathbf{g}_2, [\mathbf{g}_2, \mathbf{f}]]\}$$

it can be proven that the system is *strongly accessible* in $\mathcal{D}$ [18, p. 180]. Nevertheless, if the centripetal force term $x_4 x_6^2$ is neglected, as it is commonly assumed in the related literature (see for example [13, 19]), the strong accessibility condition is no longer fulfilled, but only the accessibility one, restricted to $\mathcal{D}_a = \left\{ \mathbf{x} \in \mathbb{R}^9 : \dot{x}_o \neq 0, \dot{y}_o \neq 0 \right\}$. The following assumption is made in this work.

Assumption 1 The centrifugal force term $x_4 x_6^2$ in (16) can be neglected.

This assumption is made to simplify the controller design and the stability analysis, but the term $x_4 x_6^2$ is kept for the numerical simulation of the system dynamics.

3.2 *Controller Design and Stability Analysis*

The control objective of this work is to induce a rotating movement on the pizza dough at a desired angular speed $\dot{\phi}_d$ while keeping the remaining coordinates as close to zero as possible. The contribution of this work is the design of a closed-loop control law for achieving the task mentioned above. The first control law, intended for the linear motion of the peel, is composed of a feedforward term, very similar to the reported open-loop controllers in the literature, plus a PD control to stabilise the linear peel direction, that is

$$u_1 = -k_1 x_1 - k_5 x_5 + a_h \sin(\omega_h t) \tag{21}$$

$$u_2 = \frac{\tanh(k_7 x_7) I_{oz}}{\mu_\phi I_{ox}} k_9 x_9 - k_2 x_2 - k_6 x_6 \,, \tag{22}$$

with $a_h \in \mathbb{R}$ and $\omega_h > 0$. The feedforward term $a_h \sin(\omega_h t)$ ensures the required condition $\dot{x}_o \neq 0$. On the other hand, the control law (22) is a PD control that stabilises the peel orientation, plus a nonlinear term employed to induce a rotation in the pizza by exploiting the torque generated from (11).

In order to carry out a mathematical analysis, consider the closed-loop dynamics

$$\begin{aligned}
\dot{x}_1 &= x_5, \\
\dot{x}_2 &= x_6, \\
\dot{x}_3 &= x_7, \\
\dot{x}_4 &= x_8, \\
\dot{x}_5 &= -k_1 x_1 - k_5 x_5 + a_h \sin(\omega_h t)\,, \\
\dot{x}_6 &= -k_2 x_2 - k_6 x_6 + \frac{I_{oz}}{\mu_\phi I_{ox}} \tanh(k_7 x_7) k_9 x_9, \\
\dot{x}_7 &= -g\mu_o \tanh(k_7 x_7) + k_1 x_1 + k_5 x_5 - a_h \sin(\omega_h t)\,, \\
\dot{x}_8 &= -g\mu_o \tanh(k_8 x_8) - g \sin(x_2), \\
\dot{x}_9 &= -k_9 x_9 \tanh^2(k_7 x_7) + \frac{\mu_\phi I_{ox}}{I_{oz}} \tanh(k_7 x_7)\,(k_2 x_2 + k_6 x_6)\,.
\end{aligned} \tag{23}$$

The following analysis is developed considering that stationary state has reached. For a more straightforward exposition of the following development, the closed-loop dynamics is divided into the four subsystems shown in Fig. 2.

(1) The subsystem Σ_1, comprised of the states x_1 and x_5, is a linear stable system with arbitrarily chosen poles by means of the gains k_1 and k_5, under the effect of the sinusoidal input $a_h \sin(\omega_h t)$. The amplitude of the states x_1 and x_5 in stationary

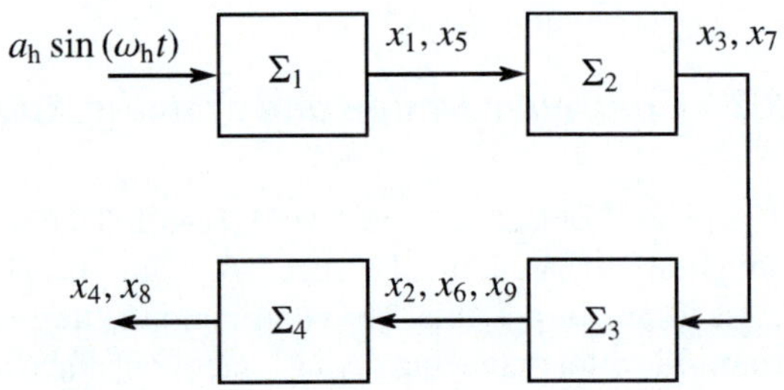

Fig. 2 Closed-loop dynamics

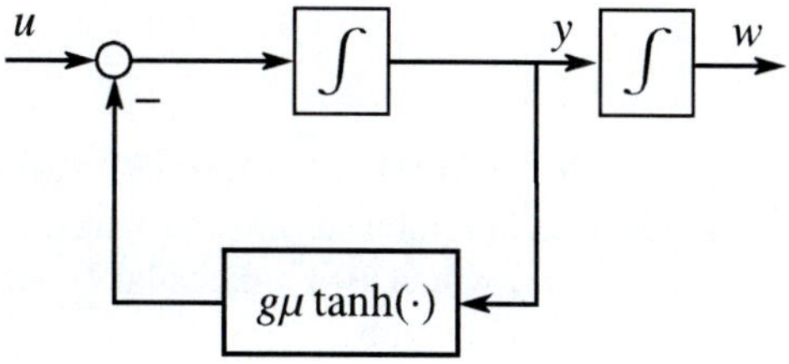

Fig. 3 Nonlinear feedback loop

state are sinusoidals with frequency ω_h and amplitude easily computed to be

$$|x_1| = \frac{a_h}{D(k_1, k_5, \omega_\mathrm{h})} \tag{24}$$

$$|x_5| = \frac{a_\mathrm{h}\omega_h}{D(k_1, k_5, \omega_h)}, \tag{25}$$

where

$$D(k_1, k_5, \omega_\mathrm{h}) = \sqrt{\left(k_1 - \omega_h^2\right)^2 + k_5^2\omega_\mathrm{h}^2}. \tag{26}$$

(2) The second subsystem, which corresponds to the states x_3 and x_7, is analysed by employing the *describing function* method [20, p. 157], for which the configuration shown in Fig. 3 is considered with $u = -u_1$ and $y = x_7$. To approximate the nonlinearity in the Σ_2 subsystem, the following describing function is used

$$g\mu_\mathrm{o} \tanh(\cdot) \approx \frac{4\, g\mu_\mathrm{o}}{\pi A(\cdot)}, \tag{27}$$

where $A(\cdot)$ is the input amplitude of the nonlinear block signal, which in turn is assumed to be sinusoidal. The approximate closed-loop transfer function of the subsystem Σ_2 is

$$y(s) = \left(\frac{\pi A(\cdot)}{s\pi A(\cdot) + 4\, g\mu_o}\right) u(s) = h(s)u(s). \tag{28}$$

The output of this system is the sum of a self-oscillatory response and a forced response [21, Chap. 3]. The frequency of the self oscillatory component ω_s is obtained by solving the equation

$$j\omega_\mathrm{s} = \frac{4\, g\mu_o}{\pi A(\cdot)}. \tag{29}$$

Because $A(\cdot)$ is a positive and real function, no unforced periodic response is present. On the other hand, the gain for the closed-loop pseudo-transfer function can be computed as

$$|h(s)| = \frac{\pi^2 A^2(\cdot)}{\sqrt{\omega_h^2 \pi^2 A^2(\cdot) + 16\, g^2 \mu_o^2}}, \tag{30}$$

which has a minimum at zero and a maximum at ω_h^{-1}. The input to this subsystem is $u = -u_1$, which is a sinusoidal signal with zero mean, frequency ω_h and amplitude bounded by

$$|u_1| \leq \frac{a_h \omega_h^2}{D(k_1, k_5, \omega_h)}. \tag{31}$$

Therefore, the approximate steady state output of this subsystem is a sinusoidal with zero mean given by

$$x_7 = a_7 \sin(\omega_h t + \phi_7), \tag{32}$$

where

$$|a_7| \leq \frac{a_h \omega_h}{D(k_1, k_5, \omega_h)} \tag{33}$$

and $\phi_7 \in \mathbb{R}$ is the phase shift given by

$$\phi_7 = \text{atan2}(-a_h \omega_h, 4\, g\mu_o/(\pi A(\cdot))), \tag{34}$$

which can be bounded by $-\pi/2 \leq \phi_7 \leq 0$. The response in steady state for x_3 can be approximated by

$$x_3 = a_3 \sin(\omega_h t + \phi_3) + c_3, \tag{35}$$

where $c_3 \in \mathbb{R}$ is a bias constant, $\phi_3 \in \mathbb{R}$ is a phase shift, and

$$|a_3| \leq \frac{a_\text{h}}{D(k_1, k_5, \omega_h)}. \tag{36}$$

(3) For showing the stability properties of the subsystem Σ_3, it is first recalled the following [22, Theorem 10.3].

Theorem 1 *Consider the system*

$$\dot{\mathbf{x}} = \mathbf{f}(\mathbf{x}) + \epsilon \mathbf{g}(t, \mathbf{x}, \epsilon). \tag{37}$$

Suppose

- $\mathbf{f}$, $\mathbf{g}$, *and their first partial derivatives with respect to* $\mathbf{x}$ *are continuous and bounded for all* $(t, \mathbf{x}, \epsilon) \in [0, \infty) \times D_0 \times [-\epsilon_0, \epsilon_0]$, *for every compact set* $D_0 \subset D$, *where* $D \subset \mathbb{R}^n$ *is a domain that contains the origin.*
- *The origin is an exponentially stable equilibrium point of the autonomous system*

$$\dot{\mathbf{x}} = \mathbf{f}(\mathbf{x})\,; \tag{38}$$

- $\mathbf{g}(t, \mathbf{x}, \epsilon)$ *is* T*-periodic in* t.

Then, there exist positive constants ϵ^* *and* k *such that for all* $|\epsilon| < \epsilon^*$*, Eq. (37) has a unique T-periodic solution* $\bar{\mathbf{x}}(t, \epsilon)$ *with the property that* $\|\bar{\mathbf{x}}(t, \epsilon)\| \leq k|\epsilon|$*. Moreover, this solution is exponentially stable.*

After employing the identity $\tanh^2(x) = 1 - \operatorname{sech}^2(x)$, the dynamics of the subsystem Σ_3 can be written as Eq. (37) with $\mathbf{x} = \begin{bmatrix} x_2 \; x_6 \; x_9 \end{bmatrix}^T$ and

$$\mathbf{f} = \begin{bmatrix} x_6 \\ -k_2 x_2 - k_6 x_6 \\ -k_9 x_9 \end{bmatrix} \tag{39}$$

$$\mathbf{g} = \begin{bmatrix} 0 \\ \dfrac{1}{c_\mu} \tanh(k_7 x_7) k_9 x_9 \\ \operatorname{sech}^2(k_7 x_7) k_9 x_9 + c_\mu \tanh(k_7 x_7)\,(k_2 x_2 + k_6 x_6) \end{bmatrix} \tag{40}$$

where $c_\mu = \mu_\phi I_{\mathrm{ox}} I_{\mathrm{oz}}^{-1}$, $\mathbf{f}(\mathbf{x}) = \mathbf{f}$ and $\mathbf{g}(t, \mathbf{x}, \epsilon) = \mathbf{g}$, with $\epsilon = 1$. As proven in the item 2), the steady state solution for x_7 is a T-periodic function of time t with period $T = 2\pi/\omega_{\mathrm{h}}$, and thus it is $\mathbf{g}$ in (40). It is not difficult to prove that the autonomous subsystem $\dot{\mathbf{x}} = \mathbf{f}(\mathbf{x})$, with $\mathbf{f}(\mathbf{x})$ defined in (39), is exponentially stable. By applying Theorem 1 restricted to the region $\mathbb{D} \triangleq \{\mathbf{x} : \|\mathbf{x}\| \leq \rho\}$, with $\rho > 0$, one can conclude that the solution trajectories for the states $\mathbf{x} = \begin{bmatrix} x_2 \; x_6 \; x_9 \end{bmatrix}^T$ are T-periodic functions of time, and the states converge exponentially to these solutions. As stated in Theorem 1, the periodic solution is bounded by $\|\bar{\mathbf{x}}(t, \epsilon)\| \leq k|\epsilon|$. Furthermore, from the theory of linear bounded input-bounded output systems, the ultimate bound $k|\epsilon|$ can be made arbitrarily small by making the gains k_2, k_6, and k_9 arbitrarily large.

(4) For the fourth subsystem, consider again the block diagram depicted in Fig. 3, with $u = g \sin(x_2)$, $y = x_8$, and $w = x_4$. The transfer function employed to approximate this system is given by (28). By the same arguments as the given in the item 2), the solution trajectories for this system can be approximated by

$$x_4 = a_4 \sin(\omega_h t + \phi_4) + c_4 \tag{41}$$

$$x_8 = a_8 \sin(\omega_h t + \phi_8)\,, \tag{42}$$

where $c_4 \in \mathbb{R}$ is a bias constant and

$$a_4 \leq \frac{c_8 g}{\omega_h^2} \tag{43}$$

$$a_8 \leq \frac{c_8 g}{\omega_h}\,, \tag{44}$$

with $c_8 = \sup_t(|\sin(x_2(t))|)$.

In summary, the above analysis shows that the approximate solutions of the closed-loop system are bounded and periodic. Furthermore, the ultimate bound for the states x_2, x_6, and x_9 can be driven arbitrarily close to zero, which means that the peel will be as close as desired to the horizontal position and that the pizza rotation speed will be arbitrarily close to the desired value, i.e. $\dot{\phi} \approx \dot{\phi}_d$.

3.3 Numerical Simulation

A numerical simulation is proposed to validate the results of Sect. 3.1. The centrifugal term $x_4x_6^2$ in (20) is kept for the simulation to test the robustness of the controller design. The parameters considered for the system are displayed in Table 2.

The parameters of the controller (21)–(22) were chosen empirically as $\omega_\mathrm{h} = 18\,\mathrm{rad/s}$, $a_\mathrm{h} = 2$, $k_1 = 10$, $k_2 = 10$, $k_5 = 10$, $k_6 = 50$, $k_7 = 20$, and $k_9 = 40$. It is considered that the sample time for the control loop is $T = 5\,\mathrm{ms}$. The desired speed for the pizza rotation is $1\,\mathrm{rad/s}$ in counter-clockwise direction.

The actual and desired rotation speed are shown in Fig. 4. It can be appreciated that the actual speed is very close to the desired one in the steady-state.

Table 2 Parameters for the numerical simulation

Meaning	Parameter	Value
Pizza mass	m_o	0.25 kg
Pizza x-inertia moment	I_ox	$0.01\,\mathrm{kg\,m^2}$
Pizza z-inertia moment	I_oz	$0.028\,\mathrm{kg\,m^2}$
Linear Coulomb friction coefficient	μ_o	0.5
Rotational Coulomb friction coefficient	μ_ϕ	0.5
Gravity acceleration constant	g	$9.81\,\mathrm{m/s^2}$

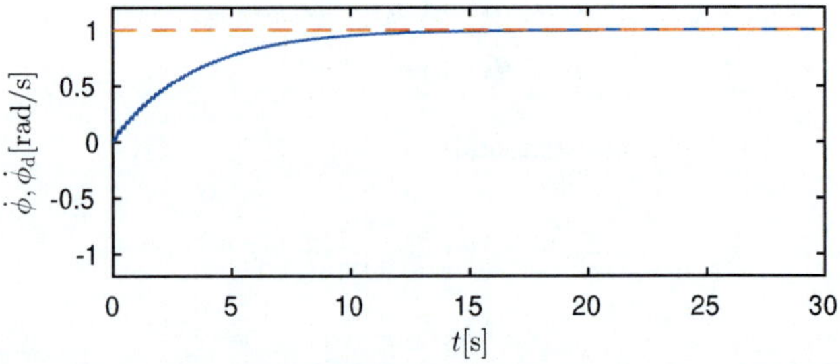

Fig. 4 Pizza rotation speed: real (—), desired (—)

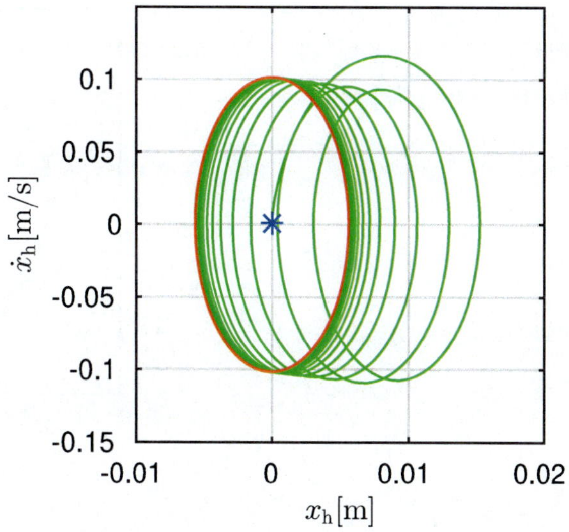

Fig. 5 Phase portrait of $x_{\rm h}$ and $\dot{x}_{\rm h}$: $t = 0$ s ($*$), $0 < t \leq 15$ s (—), $t > 15$ s (—)

The phase portrait of $x_{\rm h}$ is shown in Fig. 5. In this plot and the following ones, a blue marker indicates the initial point, a green line denotes the first 15 seconds (transient response), and a red line indicates the last 15 seconds (steady-state). It can be appreciated that the $x_{\rm h}$ and $\dot{x}_{\rm h}$ coordinates keep oscillating around zero. Indeed, the amplitude of $x_{\rm h}$ coincides with the one predicted by (24), i.e. , $|x_h| = 5.5 \cdot 10^{-3}$ m. On the other hand, the phase portrait for the $\theta_{\rm h}$ coordinate is shown in Fig. 6. In

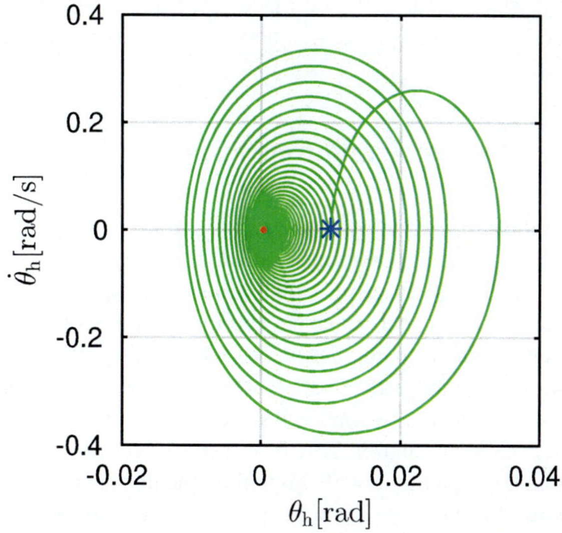

Fig. 6 Phase portrait of $\theta_{\rm h}$ and $\dot{\theta}_{\rm h}$: $t = 0$ s ($*$), $0 < t \leq 15$ s (—), $t > 15$ s (—)

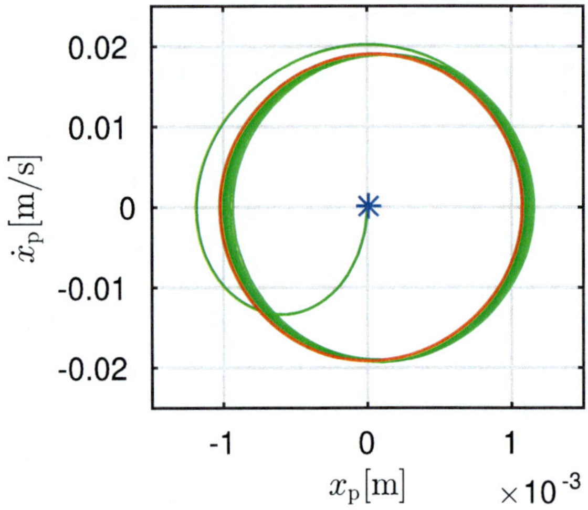

Fig. 7 Phase portrait of x_p and $\dot{x}_p$: $t = 0$ s (∗), $0 < t \leq 15$ s (—), $t > 15$ s (—)

this figure, it can be appreciated that in the stationary state, this coordinate remains oscillating very close to zero, following the stability analysis of Sect. 3.1.

The phase portrait for the x_p coordinate is shown in Fig. 7. The predicted value for the amplitude of the oscillations for x_p computed in (36) has as an upper bound $|a_3| \leq 5.5 \cdot 10^{-3}$, which is a very conservative one, as can be seen in the figure. The reason behind this large margin is because we are taking the worst case for the pseudo-transfer function gain in (30). Finally, the phase portrait for the y_p coordinate is displayed in Fig. 8. As it can be appreciated in the figure, the oscillation amplitude of this coordinate can be made arbitrarily small, as it depends on the amplitude of θ_h, which in turn can be made arbitrarily small. However, in this case, there is a non-negligible bias term of about $6 \cdot 10^{-3}$ m as stated in (41).

4 Discussion and Conclusion

In this chapter, an overview of the sliding manipulation primitive is first given, and then a particular case was studied. The main intention was to underline the potential of moving parts using sliding and the complication arising in the control design and mathematical analysis. The literature review showed that although within the nonprehensile manipulation tasks, the sliding primitive is one of the most commonly used in the industry, most controllers are designed in an open-loop fashion, and they lack a formal mathematical analysis. This chapter attempted to solve a particular sliding manipulation task from the model-based feedback control perspective. By employing a numerical simulation, the designed controller satisfies the desired task with the stability properties given by the mathematical analysis carried out in this chapter.

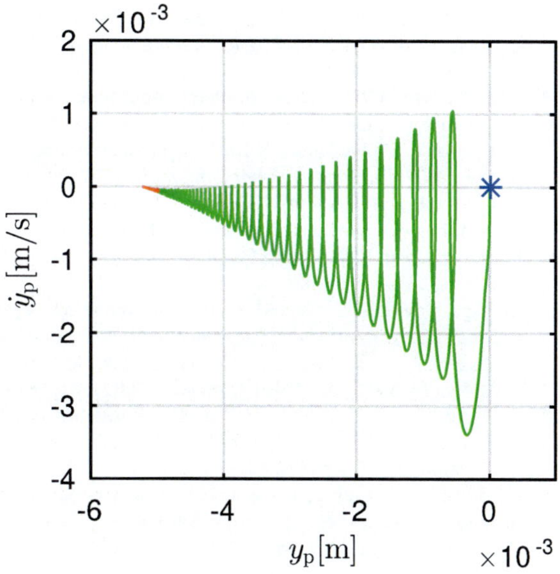

Fig. 8 Phase portrait of y_p and $\dot{y}_p$: $t = 0$ s (∗), $0 < t \leq 15$ s (—), $t > 15$ s (—)

Although there are some nice features of the designed controller over the existing ones in the literature, it still has some room for improvement. In this direction, the design could be improved to obtain asymptotic stability in all the coordinates or practical asymptotic stability. The experimental evaluation of the controller is also left as remaining work. Furthermore, the generalisation of the results presented in this chapter to a more extensive set of sliding manipulation tasks is the ultimate goal pursued in this work.

References

1. A. Gutierrez-Giles, F. Ruggiero, V. Lippiello, B. Siciliano, Closed-loop control of a nonprehensile manipulation system inspired by a pizza-peel mechanism, in *European Control Conference*, pp. 1580–1585, Naples, I (2019)
2. M.G. Coutinho, P.M. Will, A general theory for positioning and orienting 2D polygonal or curved parts using intelligent motion surfaces, in *1998 IEEE International Conference on Robotics and Automation*, pp. 856–862 (1998)
3. K.-F. Bohringer, B.R. Donald, N.C. MacDonald, Programmable force fields for distributed manipulation, with applications to MEMS actuator arrays and vibratory parts feeders. Int. J. Robot. Res. **18**(2), 168–200 (1999)
4. K.F. Bohringer, B.R. Donald, L.E. Kavraki, F. Lamiraux, Part orientation with one or two stable equilibria using programmable force fields. IEEE Trans. Robot. Autom. **16**(2), 157–170 (2000)
5. D. Reznik, E. Moshkovich, J. Canny, Building a universal planar manipulator, in *Distributed Manipulation*, pp. 147–171 (Springer, 2000)
6. D.S. Reznik, J.F. Canny, C'mon part, do the local motion! in *2001 IEEE International Conference on Robotics and Automation*, pp. 2235–2242 (2001)

7. P.U. Frei, An intelligent vibratory conveyor for the individual object transportation in two dimensions, in *2002 IEEE/RSJ International Conference on Intelligent Robots and Systems*, pp. 1832–1837 (2002)
8. T.D. Murphey, J.W. Burdick, Feedback control methods for distributed manipulation systems that involve mechanical contacts. Int. J. Robot. Res. **23**(7–8), 763–781 (2004)
9. T.H. Vose, P. Umbanhowar, K.M. Lynch, Vibration-induced frictional force fields on a rigid plate, in *2007 IEEE International Conference on Robotics and Automation*, pp. 660–667 (2007)
10. T.H. Vose, P. Umbanhowar, K.M. Lynch, Friction-induced lines of attraction and repulsion for parts sliding on an oscillated plate. IEEE Trans. Autom. Sci. Eng. **6**(4), 685–699 (2009)
11. P. Umbanhowar, K.M. Lynch, Optimal vibratory stick-slip transport. IEEE Trans. Autom. Sci. Eng. **5**(3), 537–544 (2008)
12. M. Higashimori, K. Utsumi, M. Kaneko, Dexterous hyper plate inspired by pizza manipulation, in *2008 IEEE International Conference on Robotics and Automation*, pp. 399–406 (2008)
13. M. Higashimori, K. Utsumi, Y. Omoto, M. Kaneko, Dynamic manipulation inspired by the handling of a pizza peel. IEEE Trans. Rob. **25**(4), 829–838 (2009)
14. M. Higashimori, Y. Omoto, M. Kaneko, Non-grasp manipulation of deformable object by using pizza handling mechanism, in *2009 IEEE International Conference on Robotics and Automation*, pp. 120–125 (2009)
15. I.G. Ramirez-Alpizar, M. Higashimori, M. Kaneko, C.-H. Tsai, I. Kao, Nonprehensile dynamic manipulation of a sheet-like viscoelastic object, in *2011 IEEE International Conference on Robotics and Automation*, pp. 5103–5108 (2011)
16. I.G. Ramirez-Alpizar, M. Higashimori, M. Kaneko, C.-H.D. Tsai, I. Kao, Dynamic nonprehensile manipulation for rotating a thin deformable object: An analogy to bipedal gaits. IEEE Trans. Rob. **28**(3), 607–618 (2012)
17. R.M. Murray, Z. Li, S.S. Sastry, *A Mathematical Introduction to Robotic Manipulation* (CRC press, 1994)
18. A. Bloch, J. Baillieul, P. Crouch, J.E. Marsden, D. Zenkov, P.S. Krishnaprasad, R.M. Murray, *Nonholonomic Mechanics and Control*, vol. 24 (Springer, 2003)
19. T.H. Vose, P. Umbanhowar, K.M. Lynch, Sliding manipulation of rigid bodies on a controlled 6-DoF plate. Int. J. Robot. Res. **31**(7), 819–838 (2012)
20. J.-J. Slotine, W. Li, *Applied Nonlinear Control*, vol. 199 (Prentice-Hall, 1991)
21. I. Boiko, *Discontinuous Control Systems: Frequency-domain Analysis and Design* (Springer Science & Business Media, 2008)
22. H.K. Khalil, *Nonlinear Systems* (Prentice Hall, 2002)

Holonomic Rolling Nonprehensile Manipulation Primitive

Alejandro Donaire, Fabio Ruggiero, Vincenzo Lippiello, and Bruno Siciliano

Abstract In this chapter, the design of nonlinear controllers for non-prehensile holonomic rolling system is reviewed. A general model for the class of non-prehensile rolling system considered in this work is first formulated. Then, both the input-state linearisation approach and the interconnection and damping assignment passivity-based control technique for rolling systems are addressed. The class of control designs presented in this chapter make use of energy concepts and physical properties. Three benchmark examples are used to illustrate the control design presented, namely the disk-on-disk, the ball-and-beam, and the eccentric disk-on-disk This chapter is based on the works presented in [1–3].

1 Brief Introduction

The manipulation problem of nonprehensile planar rolling concerns tasks that involve an actuated manipulator, which in this chapter is called the hand, and an object that is manipulated without form or force closure grasps [4]. There exist many examples of such robotic systems, the *disk-on-disk* [5, 6], the *ball-and-beam* [7, 8], and the *butterfly robot* [9–11] among the benchmarks presented in the literature. In detail, DoD is composed of an upper disk (object) that can roll without slipping on the rim of a lower actuated disk (hand). The BnB consists of a beam (hand) actuated

A. Donaire (✉)
The University of Newcastle, University Drive, Callaghan 2308, NSW, Australia
e-mail: alejandro.donaire@newcastle.edu.au

F. Ruggiero · V. Lippiello · B. Siciliano
Department of Electrical Engineering and Information Technology, PRISMA Lab, CREATE Consortium & University of Naples Federico II, Via Claudio 21, 80125 Naples, Italy
e-mail: fabio.ruggiero@unina.it

V. Lippiello
e-mail: vincenzo.lippiello@unina.it

B. Siciliano
e-mail: bruno.siciliano@unina.it

B. Siciliano and F. Ruggiero (eds.), *Robot Dynamic Manipulation*,
Springer Tracts in Advanced Robotics 144,
https://doi.org/10.1007/978-3-030-93290-9_6

by a torque around its CoM together with a ball (object) rolling on it. The butterfly robot is composed of an actuated butterfly-shaped link (hand) and a ball (object) that can freely roll on the rim of the link. The control objective in these examples is to balance the object and drive the hand towards the desired configuration. The shared characteristic of these planar systems is the holonomic nature of the pure rolling constraint. A constraint is said to be holonomic if it comes from an integrable kinematic constraint that can be expressed in the so-called Pfaffian form [12]. The integrability condition allows instantaneous motion in every admissible direction, that is just one in the case of planar systems like the ones addressed in this chapter (Table 1).

This chapter presents control designs for nonprehensile planar rolling systems based on two techniques: input-state FLin and IDA-PBC. The designs are illustrated on two benchmarks, namely the disk-on-disk and the ball-and-beam systems. In addition, a different system, referred to as eccentric DoD, is used to demonstrate a novel procedure to design IDA-PBC controllers. This last example is a variant of the DoD system, where the center of rotation of the hand and its geometric center are not coincident.

This chapter is organised as follows. First, a general model of nonprehensile planar rolling systems is presented. Then, a control design based on FLin is presented to design controllers for nonprehensile planar rolling systems. In a subsequence section, the IDA-PBC technique is used to design controllers for nonprehensile planar rolling systems. In each section, the control designs are applied to all or some of the benchmark examples. The chapter is wrapped up with discussion and conclusions.

2 Dynamic Model of Nonprehensile Holonomic Rolling Manipulation Systems

In this section, the model of a general class of nonprehensile planar rolling systems shown in Fig. 1 is formulated. Previous models are derived under the assumption that the hand can only rotate around its center of mass [2, 5, 6]. However, the development in this section relax this assumption and allows the formulation of the dynamics of a larger class of rolling tasks [3]. Consider the inertial world fixed frame $\mathcal{W}$, which is without loss of generality attached to the holder where the hand is actuated (i.e., the center of rotation of the hand).

At least locally, the shapes should be of class $\mathcal{C}^2$. Any point of the hand shape is given by the chart $\mathbf{c}_h^h(s_h) = \begin{bmatrix} u_h(s_h) & v_h(s_h) \end{bmatrix}^T \in \mathbb{R}^2$, expressed with respect to $\mathcal{H}$, while any point of the object shape is given by $\mathbf{c}_o^o(s_o) = \begin{bmatrix} u_o(s_o) & v_o(s_o) \end{bmatrix}^T \in \mathbb{R}^2$, expressed with respect to $\mathcal{O}$. Notice that s_h increases counterclockwise along the hand, while s_o increases clockwise along the object. With this choice, the pure rolling assumption is $\dot{s}_h = \dot{s}_o$. Without loss of generality, the frames $\mathcal{W}$ and $\mathcal{H}$ coincide at $\theta_h = 0$, the point $s_h = 0$ is at the intersection between the vertical (gravitational) axis of $\mathcal{W}$ and the hand shape (i.e., $\mathbf{c}_h(0) = \begin{bmatrix} 0 & v_h(0) \end{bmatrix}^T$ in $\mathcal{W}$), and thus $s_h = s_o$ at

Table 1 Main symbols used in this chapter

Definition	Symbol
Frame attached to the CoM of the hand	$\mathcal{H}$
Frame attached to the CoM of the object	$\mathcal{O}$
Angle of the hand in $\mathcal{W}$	$\theta_h \in \mathbb{R}$
Position of $\mathcal{H}$ in $\mathcal{W}$	$\mathbf{p}_h(\theta_h) \in \mathbb{R}^2$
Position of $\mathcal{O}$ in $\mathcal{W}$	$\mathbf{p}_o \in \mathbb{R}^2$
Angle of the object in $\mathcal{W}$	$\theta_o \in \mathbb{R}$
Arclength parameter for the hand	$s_h \in \mathbb{R}$
Arclength parameter for the object	$s_o \in \mathbb{R}$
Rotation of $\mathcal{H}$ with respect to $\mathcal{W}$	$\mathbf{R}(\theta_h) \in SO(2)$
Rotation of $\mathcal{O}$ with respect to $\mathcal{W}$	$\mathbf{R}(\theta_o) \in SO(2)$
Configuration vector	$\mathbf{q} = \begin{bmatrix} q_1 & q_2 \end{bmatrix}^T \in \mathbb{R}^2$
Mass matrix	$\mathbf{M}(q) = \begin{bmatrix} m_{11}(\mathbf{q}) & m_{12}(\mathbf{q}) \\ m_{12}(\mathbf{q}) & m_{22}(\mathbf{q}) \end{bmatrix} \in \mathbb{R}^{2\times 2}$
Coriolis matrix	$\mathbf{C}(q, \dot{q}) = \begin{bmatrix} c_{11} & c_{12} \\ c_{21} & c_{22} \end{bmatrix} \in \mathbb{R}^{2\times 2}$
Gravity vector	$\mathbf{g}(\mathbf{q}) = \begin{bmatrix} g_1(\mathbf{q}) & g_2(\mathbf{q}) \end{bmatrix}^T \in \mathbb{R}^2$
Inertia of the hand	$I_h \in \mathbb{R}$
Mass of the hand	$m_h > 0$
Inertia of the object	$I_o \in \mathbb{R}$
Mass of the object	$m_o > 0$
Gravity acceleration	$g = 9.81$ m/s^2
Selection vector	$\mathbf{e}_{=} \begin{bmatrix} 0 & 1 \end{bmatrix}^T \in \mathbb{R}^2$
Momentum vector	$\mathbf{p} = \mathbf{M}(\mathbf{q})\dot{\mathbf{q}} \in \mathbb{R}^n$
Hand torque applied at the hand's center of rotation	$\tau_h \in \mathbb{R}$
Hand acceleration	$a_h \in \mathbb{R}$
Radius of the actuated disk (DoD and eccentric DoD setups)	$\rho_h > 0$
Radius of the upper disk (DoD setup and eccentric DoD setups) and radius of the rolling ball (BnB setup)	$\rho > 0$
Distance between the beam's CoM and its surface where the ball rolls (BnB setup)	$d_h > 0$
Hamiltonian function	$H : \mathbb{R}^4 \to \mathbb{R}$
Desired Hamiltonian function	$H_d : \mathbb{R}^4 \to \mathbb{R}$
Desired mass matrix	$\mathbf{M}_d(q) \in \mathbb{R}^{2\times 2}$

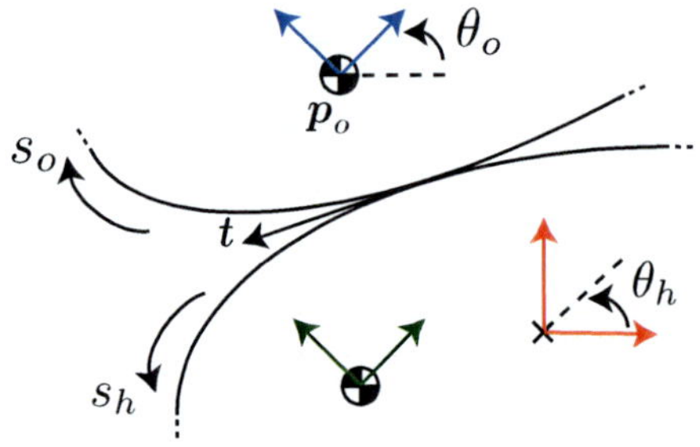

Fig. 1 A general nonprehensile planar rolling manipulation system with the center of rotation of the hand (indicated by the × symbol) not corresponding to its center of mass. The world fixed frame $\mathcal{W}$ is in red. The hand frame $\mathcal{H}$ and the object frame $\mathcal{O}$, in green and blue, respectively, are placed at their respective CoM

all times during rolling. Therefore, the contact location will be specified only by s_h throughout the remainder of the paper. As the first assumption, the hand and the object maintain pure rolling contact for all time. The arclength parametrization implies the property $\|\mathbf{c}_h^{h\prime}\| = 1$, with the symbol $'$ indicating the derivative with respect to the parameter s_h. The same holds for $\mathbf{c}_o^o(s_h)$. At the contact point $\mathbf{c}_h^h(s_h)$, the tangent vector to the shapes is expressed as $\mathbf{t}^h(s_h) = \mathbf{c}_h^{h\prime} \in \mathbb{R}^2$ forming an angle $\phi_h(s_h) = \text{atan2}(v_h'(s_h), u_h'(s_h))$ in $\mathcal{H}$. The same tangent can be expressed with respect to $\mathcal{O}$ with an angle $\phi_o(s_h) = \text{atan2}(v_o'(s_h), u_o'(s_h))$. The signed curvatures of the shapes are defined as: $\kappa_h(s_h) = \phi_h'(s_h) = u_h'(s_h)v_h''(s_h) - u_h''(s_h)v_h'(s_h)$, $\kappa_o(s_h) = \phi_o'(s_h) = u_o'(s_h)v_o''(s_h) - u_o''(s_h)v_o'(s_h)$. The relative curvature at the contact point is given by

$$\kappa_r(s_h) = \kappa_h(s_h) - \kappa_o(s_h). \tag{1}$$

Notice that $\kappa_h(s_h) > 0$ and $\kappa_o(s_h) < 0$ denote convexity at the contact point for the hand and the object, respectively. Hence, $\kappa_r(s_h) > 0$ guarantees a single contact point at least locally [6]. The following constraint expresses the angle of the tangent $\mathbf{t}^h(s_h)$ with respect to $\mathcal{W}$: $\theta_h + \phi_h(s_h) = \theta_o + \phi_o(s_h)$. Therefore, taking into account (1), the following relations hold

$$\theta_o = \theta_h + \phi_h(s_h) - \phi_o(s_h), \tag{2a}$$

$$\dot{\theta}_o = \dot{\theta}_h + \kappa_r(s_h)\dot{s}_h. \tag{2b}$$

Assuming that $\mathbf{R}(\theta) \in SO(2)$ is the rotation matrix in the 2D space, notice that the relation $\dot{\mathbf{R}}(\theta) = \mathbf{R}(\bar{\theta})\dot{\theta}$ holds with $\bar{\theta} = \theta + \frac{\pi}{2}$. The position of the CoM of the hand in $\mathcal{W}$ is denoted by $\mathbf{p}_h(\theta_h) = \begin{bmatrix} u_w(\theta_h) & v_w(\theta_h) \end{bmatrix}^T$. The coincidence between the contact points on both the hand and the object is expressed by $\mathbf{p}_h(\theta_h) + \mathbf{R}(\theta_h)\mathbf{c}_h^h(s_h) = \mathbf{p}_o + \mathbf{R}(\theta_o)\mathbf{c}_o^o(s_h)$, yielding to the equation $\mathbf{p}_o = \mathbf{p}_h(\theta_h) + \mathbf{R}(\theta_h)\mathbf{c}_h^h(s_h) - \mathbf{R}(\theta_o)\mathbf{c}_o^o(s_h)$, and, consequently, $\dot{\mathbf{p}}_o = \boldsymbol{\gamma}(\mathbf{q})\dot{\theta}_h + \boldsymbol{\eta}(\mathbf{q})\dot{s}_h = \begin{bmatrix} \boldsymbol{\gamma}(\mathbf{q}) & \boldsymbol{\eta}(\mathbf{q}) \end{bmatrix} \dot{\mathbf{q}}$, with $\mathbf{q} = \begin{bmatrix} \theta_h & s_h \end{bmatrix}^T$ and

$$\gamma = \mathbf{p}_h^{\backslash} + \mathbf{R}(\bar{\theta}_h)\mathbf{c}_h^h - \mathbf{R}(\bar{\theta}_o)\mathbf{c}_o^o, \tag{3a}$$

$$\eta = \mathbf{R}(\theta_h)\mathbf{c}_h^{h\prime} - \mathbf{R}(\theta_o)\mathbf{c}_o^{o\prime} - \kappa_r \mathbf{R}(\bar{\theta}_o)\mathbf{c}_o^o, \tag{3b}$$

in which dependencies have been dropped, while (2b) is included and (2a) has to be plugged in. The symbol $^{\backslash}$ indicates the derivative with respect to θ_h. For this class of systems the kinetic energy is given by

$$T = \frac{1}{2}\left(I_h\dot{\theta}_h^2 + m_h\dot{\mathbf{p}}_h^T(\theta_h)\dot{\mathbf{p}}_h(\theta_h) + m_o\dot{\mathbf{p}}_o^T\dot{\mathbf{p}}_o + I_o\dot{\theta}_o^2\right) = \frac{1}{2}\mathbf{p}^T\mathbf{M}^{-1}(\mathbf{q})\mathbf{p},$$

where the elements of the mass matrix are given by $m_{11}(\mathbf{q}) = I_h + I_o + m_h\mathbf{p}_h^{\backslash T}\mathbf{p}_h^{\backslash} + m_o\gamma^T(\mathbf{q})\gamma(\mathbf{q})$, $m_{12}(\mathbf{q}) = I_o\kappa_r(s_h) + m_o\gamma(\mathbf{q})^T\eta(\mathbf{q})$, and $m_{22}(\mathbf{q}) = I_o\kappa_r^2(s_h) + m_o\eta(\mathbf{q})^T\eta(\mathbf{q})$. The potential energy is, instead, given by

$$V(\mathbf{q}) = g\mathbf{e}_2^T(m_o\mathbf{p}_o(\mathbf{q}) + m_h\mathbf{p}_h(\mathbf{q})). \tag{4}$$

Given the kinetic and potential energy functions, then the dynamics of a mechanical system can be readily derived using the Euler-Lagrange equations or equivalently the Hamilton canonical equation of motion [13].

3 Input-State Feedback Linearisation

The scope of this section is to find and apply a general diffeomorphism to achieve an input-state FLin of the whole dynamics. Such state transformation renders the system in the so-called *normal form* (i.e., a chain of integrators) without internal dynamics. Given some assumptions on the shapes of both the object and the hand, EFL is employed to stabilize the system.

For a brief mathematical background regarding input-state linearisation and differential flatness, please see [2].

3.1 Hypotheses on the Shapes and Input-State Linearisation

Having in mind the derivation in Sect. 2, the dynamic model is derived through the *Euler-Lagrange* formalism. The so-called Lagrange function is given by $\mathcal{L} = T - V$. The dynamic model equations are then given by $\frac{\mathrm{d}}{\mathrm{d}t}\frac{\partial\mathcal{L}}{\partial\dot{q}_i} - \frac{\partial\mathcal{L}}{\partial q_i} = \tau_i$, with $i = 1, 2$ and τ_i the associated generalized force acting on the ith generalized coordinate. Therefore, by computing the Lagrange equations and considering the Christoffel symbols of the first type [12], the dynamic model can be written as $\mathbf{M}(\mathbf{q})\ddot{\mathbf{q}} + \mathbf{C}(\mathbf{q}, \dot{\mathbf{q}})\dot{\mathbf{q}} + \mathbf{g}(\mathbf{q}) = \boldsymbol{\tau}$, where $\boldsymbol{\tau} = \begin{bmatrix}\tau_h & 0\end{bmatrix}^T$ represents the generalized input force, $\mathbf{g}(\mathbf{q}) = \left(\frac{\partial \mathcal{V}(\mathbf{q})}{\partial \mathbf{q}}\right)^T$, and $\mathbf{C}(\mathbf{q}, \dot{\mathbf{q}})$ is a suitable matrix whose generic element is given by

$$c_{ij}(\mathbf{q}, \dot{\mathbf{q}}) = \frac{1}{2}\sum_{k=1}^{2}\left(\frac{\partial m_{ij}(\mathbf{q})}{\partial q_k} + \frac{\partial m_{ik}(\mathbf{q})}{\partial q_j} - \frac{\partial m_{jk}(\mathbf{q})}{\partial q_i}\right)\dot{q}_k, \tag{5}$$

with $i, j = 1, 2$. By neglecting dependencies, the dynamic model can be written in the following extended form

$$m_{11}\ddot{\theta}_h + m_{12}\ddot{s}_h + c_{11}\dot{\theta}_h + c_{12}\dot{s}_h + g_1 = \tau_h, \tag{6a}$$
$$m_{12}\ddot{\theta}_h + m_{22}\ddot{s}_h + c_{21}\dot{\theta}_h + c_{22}\dot{s}_h + g_2 = 0. \tag{6b}$$

During experimentation, when highly-geared harmonic drive plus DC motors are present, the hand's angular acceleration is more convenient than the hand's torque [6]. It is thus suitable to rewrite (6) with $\ddot{\theta}_h = a_h$ as input

$$\ddot{\theta}_h = a_h, \tag{7a}$$
$$\ddot{s}_h = -m_{22}^{-1}(m_{12}a_h + c_{21}\dot{\theta}_h + c_{22}\dot{s}_h + g_2), \tag{7b}$$

where dependencies have been neglected. The equation relating τ_h and a_h is given by

$$\tau_h = \xi(\mathbf{q}, \dot{\mathbf{q}}) + \sigma(\mathbf{q})a_h, \tag{8}$$

with $\xi(\mathbf{q}, \dot{\mathbf{q}}) = g_1 + c_{11}\dot{\theta}_h + c_{12}\dot{s}_h - \frac{m_{12}}{m_{22}}(g_2 + c_{21}\dot{\theta}_h + c_{22}\dot{s}_h)$ and $\sigma(\mathbf{q}) = m_{11} - \frac{m_{12}^2}{m_{22}}$.

Assumption 3.1 The Coriolis terms $c_{21}(\mathbf{q}, \dot{\mathbf{q}})$ and $c_{22}(\mathbf{q}, \dot{\mathbf{q}})$ are equal to zero.

Remark 3.1 Looking at (5), Assumption 3.1 is verified when terms $m_{12} = m_{21}$ and m_{22} do not depend on $\mathbf{q}$, and when m_{11} depends only on θ_h. Looking at the particular expressions of m_{ij}, this means that κ_r has to be constant, *i.e.* $\kappa_r' = 0$, the combination of the products $\boldsymbol{\gamma}(\mathbf{q})^T\boldsymbol{\eta}(\mathbf{q})$ and $\boldsymbol{\eta}(\mathbf{q})^T\boldsymbol{\eta}(\mathbf{q})$ do not depend on $\mathbf{q}$, and the product $\boldsymbol{\gamma}(\mathbf{q})^T\boldsymbol{\gamma}(\mathbf{q})$ depends only on θ_h. Considering (3) and the expressions of $\kappa_h(s_h)$ and $\kappa_o(s_h)$, the aforementioned properties are thus governed by the shapes of both the hand and the object.

Assumption 3.1 simplifies (7) as follows

$$\ddot{\theta}_h = a_h, \tag{9a}$$
$$\ddot{s}_h = -\frac{1}{m_{22}}(m12a_h + g_2(\mathbf{q})). \tag{9b}$$

By indicating the state of the system as $\mathbf{x} = \left[x_1\ x_2\ x_3\ x_4\right]^T = \left[\theta_h\ \dot{\theta}_h\ s_h\ \dot{s}_h\right]^T \in \mathbb{R}^4$, (9) can be written in the affine state-space form $\dot{\mathbf{x}} = \mathbf{f}(\mathbf{x}) + \mathbf{b}(\mathbf{x})u$, with $u = a_h$ and

$$\mathbf{f}(\mathbf{x}) = \left[x_2\ 0\ x_4\ -\frac{g_2(\mathbf{x})}{m_{22}}\right]^T \in \mathbb{R}^4, \tag{10a}$$

$$\mathbf{b} = \left[0\ 1\ 0\ -\frac{m_{12}}{m_{22}}\right]^T \in \mathbb{R}^4. \tag{10b}$$

In order to check whether (9) is input-state FLin, the controllability matrix $\mathbf{T}(\mathbf{x}) = \left[\mathbf{b}\ \mathrm{ad}_{\mathbf{f}}\mathbf{b}\ \mathrm{ad}_{\mathbf{f}}^2\mathbf{b}\ \mathrm{ad}_{\mathbf{f}}^3\mathbf{b}\right] \in \mathbb{R}^{4\times 4}$ has to be invertible in a certain region Ω, and the set given by the first three columns of $\mathbf{T}(\mathbf{x})$ has to be involutive. Taking into account (10), the detailed expression of the controllability matrix is

$$\mathbf{T}(\mathbf{x}) = \begin{bmatrix} 0 & -1 & 0 & 0 \\ 1 & 0 & 0 & 0 \\ 0 & \frac{m_{21}}{m_{22}} & 0 & t_{34} \\ -\frac{m_{21}}{b_{22}} & 0 & t_{43} & t_{44} \end{bmatrix}, \tag{11}$$

with $t_{34} = \frac{1}{m_{22}}\frac{\partial g_2(\mathbf{x})}{\partial x_1} - \frac{m_{12}}{m_{22}^2}\frac{\partial g_2(\mathbf{x})}{\partial x_3}$, $t_{43} = \frac{m_{12}}{m_{22}^2}\frac{\partial g_2(\mathbf{x})}{\partial x_3} - \frac{1}{m_{22}}\frac{\partial g_2(\mathbf{x})}{\partial x_1}$, and $t_{44} = x_4\frac{m_{12}}{m_{22}}\frac{\partial^2 g_2(\mathbf{x})}{\partial x_3^2} - \frac{x_2}{m_{22}}\frac{\partial^2 g_2(\mathbf{x})}{\partial x_1^2}$. Defining the region $\Omega = \left\{\mathbf{x} \in \mathbb{R}^4 : \frac{\partial g_2(\mathbf{x})}{\partial x_1} \neq \frac{m_{12}}{m_{22}}\frac{\partial g_2(\mathbf{x})}{\partial x_3}\right\}$, it is possible to prove that $\mathbf{T}(\mathbf{x})$ in (11) is made by linearly independent columns: the first three of them build an involutive set (proofs are omitted for brevity). The system (9) is then input-state FLin in Ω.

To bring (9) in the following normal form

$$\begin{cases} \dot{z}_i = z_{i+1} \\ \dot{z}_n = v \end{cases}, \tag{12}$$

with $i = 1, \ldots, n-1$, and $v \in \mathbb{R}$ and $z_i \in \mathbb{R}$ a new input and state variable, respectively, a diffeomorphism

$$\mathbf{z} = \left[z_1\ z_2 \cdots z_n\right]^T = \boldsymbol{\phi}(\mathbf{x}) = \left[z_1\ L_{\mathbf{f}}z_1\ \ldots\ L_{\mathbf{f}}^{n-1}z_1\right]^T \tag{13}$$

has to be found. Hence, in order to compute the first component z_1, the following equations

$$\frac{\partial z_1}{\partial \mathbf{x}} \mathrm{ad}_{\mathbf{f}}^{i} \mathbf{b} = 0, \tag{14a}$$

$$\frac{\partial z_1}{\partial \mathbf{x}} \mathrm{ad}_{\mathbf{f}}^{n-1} \mathbf{b} \neq 0, \tag{14b}$$

with $i = 0, \ldots, n-2$, must hold for the vector fields (10). In particular, looking at the expression of the first three columns of $\mathbf{T}(\mathbf{x})$, condition (14a) yields $\frac{\partial z_1}{\partial x_2} - \frac{m_{12}}{m_{22}} \frac{\partial z_1}{\partial x_4} = 0$, $\frac{m_{12}}{m_{22}} \frac{\partial z_1}{\partial x_3} - \frac{\partial z_1}{\partial x_1} = 0$, and $\frac{\partial z_1}{\partial x_4} t_{43} = 0$. The solution to this system is then given by $z_1 = \frac{m_{12}}{m_{22}} x_1 + x_3$. It is easy to verify that such a choice for z_1 also satisfies (14b). Therefore, the complete diffeomorphism is given by

$$\boldsymbol{\phi}(\mathbf{x}) = \begin{bmatrix} z_1 \\ z_2 \\ z_3 \\ z_4 \end{bmatrix} = \begin{bmatrix} y \\ \dot{y} \\ \ddot{y} \\ y^{(3)} \end{bmatrix} = \begin{bmatrix} \frac{m_{12}}{m_{22}} x_1 + x_3 \\ \frac{m_{12}}{m_{22}} x_2 + x_4 \\ -\frac{g_2(\mathbf{x})}{m_{22}} \\ -\frac{1}{m_{22}} \left(\frac{\partial g_2(\mathbf{x})}{\partial x_1} x_2 + \frac{\partial g_2(\mathbf{x})}{\partial x_3} x_4 \right) \end{bmatrix}, \tag{15}$$

where $y^{(j)}$ is the jth-order derivative, with $j \geq 3$. The input transformation $a_h = \alpha(\mathbf{x}) + \beta(\mathbf{x})v$ finally renders (9) in the normal form (12), with

$$\alpha(\mathbf{x}) = -\frac{\left(\frac{\partial^2 g_2(\mathbf{x})}{\partial x_1^2} x_2 - \frac{g_2(\mathbf{x})}{m_{22}} \frac{\partial g_2(\mathbf{x})}{\partial x_3} + \frac{\partial^2 g_2(\mathbf{x})}{\partial x_3^2} x_4 \right)}{\left(\frac{\partial g_2(\mathbf{x})}{\partial x_1} - \frac{m_{12}}{m_{22}} \frac{\partial g_2(\mathbf{x})}{\partial x_3} \right)}, \tag{16a}$$

$$\beta(\mathbf{x}) = -m_{22} \left(\frac{\partial g_2(\mathbf{x})}{\partial x_1} - \frac{m_{12}}{m_{22}} \frac{\partial g_2(\mathbf{x})}{\partial x_3} \right)^{-1}. \tag{16b}$$

This is the core result since, under Assumption 3.1, a general diffeomorphism is found to change a nonprehensile 2D rolling manipulation system into a normal form where simple linear controllers can be applied.

Therefore, in general, any suitable approach can be employed to control the normal form (12). The EFL technique [14] is here considered. In detail, a change of coordinates is applied to (9)–(15). To get the normal form, the EFL technique does not use the feedback transformation $a_h = \alpha(\mathbf{x}) + \beta(\mathbf{x})v$, but $a_h = \alpha(\mathbf{x}^\star) + \beta(\mathbf{x}^\star)v$, where $\mathbf{x}^\star$ is the desired state[1] (in feedforward). The new virtual input v is instead designed as an extended PID^{n-1} plus a feedforward action

[1] Eventually retrieved from $\mathbf{z}^\star$ through $\boldsymbol{\phi}^{-1}$.

$$v = z_4^\star + \sum_{i=0}^{4} k_i e_i, \tag{17a}$$

$$e_0 = \int_o^t e_1(\tau)\mathrm{d}\tau, \tag{17b}$$

$$e_i = z_i^\star - z_i, \tag{17c}$$

with k_i positive gains such that the resulting characteristic polynomial is Hurwitz.

3.2 *Case Studies*

3.2.1 Disk-on-disk

This case study considers the balancing of a disk free to roll on an actuated disk. Referring to Fig. 2, the shape of the hand, i.e. the actuated disk, is parametrized by $\mathbf{c}_h^h(s_h) = \rho_h \left[-\sin\frac{s_h}{\rho_h} \ \cos\frac{s_h}{\rho_h} \right]^T$. The upper disk's shape is parametrized by $\mathbf{c}_o^o(s_h) = -\rho_o \left[\sin\frac{s_h}{\rho_o} \ \cos\frac{s_h}{\rho_o} \right]^T$. Considering (1), the relative curvature is given by $\kappa_r = \frac{\rho_h + \rho_o}{\rho_h \rho_o}$. The upper disk angular velocity is given by $\dot{\theta}_o = \dot{\theta}_h + \kappa_r \dot{s}_h$. The vectors $\boldsymbol{\gamma}(\mathbf{q})$ and $\boldsymbol{\eta}(\mathbf{q})$ are computed like in (3): $\boldsymbol{\gamma}(\mathbf{q}) = -(\rho_h + \rho_o) \left[\cos\left(\theta_h + \frac{s_h}{\rho_h}\right) \ \sin\left(\theta_h + \frac{s_h}{\rho_h}\right) \right]^T$, and $\boldsymbol{\eta}(\mathbf{q}) = -\rho_o \kappa_r \left[\cos\left(\theta_h + \frac{s_h}{\rho_h}\right) \ \sin\left(\theta_h + \frac{s_h}{\rho_h}\right) \right]^T$. Therefore, the parameters of the DoD dynamic model are $m_{11} = I_h + I_o + m_o(\rho_h + \rho_o)^2$, $m_{12} = m_{21} = I_o\kappa_r + m_o\frac{(\rho_h + \rho_o)^2}{\rho_h}$, $m_{22} = I_o\kappa_r^2 + m_o\rho_o^2\kappa_r^2$, $c_{11} = c_{12} = c_{21} =$

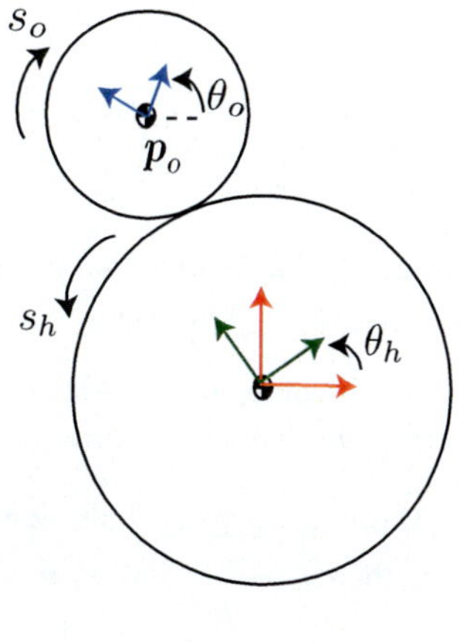

Fig. 2 A representation of the DoD system. In red the world fixed frame $\mathcal{W}$. In green the hand frame $\mathcal{H}$, while in blue the object frame $\mathcal{O}$, placed at the respective CoMs

$c_{22} = 0$, $g_1 = -m_o g(\rho_h + \rho_o)\sin\left(\theta_h + \frac{s_h}{\rho_h}\right)$ and $g_2 = -m_o g \rho_o \kappa_r \sin\left(\theta_h + \frac{s_h}{\rho_h}\right)$. Notice that the quantity $\theta_h + \frac{s_h}{\rho_h}$ is the angle of the object's CoM with respect to the vertical axis of $\mathcal{W}$. It is possible to verify that the DoD dynamic model fully verifies Assumption 3.1. Hence, considering the acceleration a_h of the actuated disk as input, the DoD dynamics can be written as in (9) with τ_h as in (8). The affine state space form assumes the following vector fields (10) $\mathbf{f}(\mathbf{x}) = \left[x_2 \; 0 \; x_4 \; \frac{m_o g \rho_o \sin\left(x_1 + \frac{x_3}{\rho_h}\right)}{I_o\kappa_r + m_o\rho_o^2\kappa_r}\right]^T$ and $\mathbf{b} = \left[0 \; 1 \; 0 \; -\frac{m_o\rho_o^2 + m_o\rho_h\rho_o + I_o}{\kappa_r(m_o\rho_o^2 + I_o)}\right]^T$. Computing the matrix $\mathbf{T}(\mathbf{x})$ as in (11), it is possible to verify that the approximate dynamic model is input-state linearizable in the region $\Omega = \left\{\mathbf{x} \in \mathbb{R}^4 : \cos\left(x_1 + \frac{x_3}{\rho_h}\right) \neq 0 \Rightarrow |x_1 + \frac{x_3}{\rho_h}| < \frac{\pi}{2}\right\}$. Notice that such a region is not restrictive because, with no bound on other states, Ω covers all practical situations since outside it the disk falls down from the hand. Finally, the diffeomorphism (15) is

$$\phi(\mathbf{x}) = \left[\frac{m_{12}}{m_{22}}x_1 + x_3 \;\; \frac{m_{12}}{m_{22}}x_2 + x_4 \;\; \frac{m_o g \rho_o \sin\left(x_1 + \frac{x_3}{\rho_h}\right)}{I_o\kappa_r + m_o\rho_o^2\kappa_r} \;\; \frac{m_o g \rho_o \kappa_r \left(x_2 + \frac{x_4}{\rho_h}\right)\cos\left(x_1 + \frac{x_3}{\rho_h}\right)}{b_{22}}\right]^T,$$

with

$$\alpha(\mathbf{x}) = \frac{\sin\left(x_1 + \frac{x_3}{\rho_h}\right)x_2 - \left(\frac{m_o g \rho_o \kappa_r}{b_{22}\rho_h}\sin\left(x_1 + \frac{x_3}{\rho_h}\right) + \frac{x_4}{\rho_h^2}\right)\cos\left(x_1 + \frac{x_3}{\rho_h}\right)}{\left(1 - \frac{b_{12}}{b_{22}\rho_h}\right)\cos\left(x_1 + \frac{x_3}{\rho_h}\right)}$$

and $\beta(\mathbf{x}) = m_{22}\left(m_o g \rho_o \kappa_r \left(1 - \frac{m_{12}}{m_{22}\rho_h}\right)\cos\left(x_1 + \frac{x_3}{\rho_3}\right)\right)^{-1}$. The control is again performed with the EFL technique described in Sect. 3.1.

Remark 3.2 Notice that in this case study the only possibility of balancing is with the object directly above the hand's CoM, i.e. $\theta_h + \frac{s_h}{\rho_h} = 0$. As noticed in [6], any other balancing position leads to constant angular acceleration resulting in unbounded velocities. The differential flatness loses thus some sense for the disk on disk.

Looking at $\boldsymbol{\phi}(\mathbf{x})$, stabilizing the origin $\mathbf{z} = \mathbf{0_4}$ is equivalent to stabilizing $\mathbf{x} = \mathbf{0_4}$ and then $x_1 + \frac{x_3}{\rho_h} = 0$. However, notice that through the following further change of coordinates $\bar{\mathbf{z}} = \left[z_1 - \left(\frac{m_{12} - m_{22}\rho_h}{m_{22}} \theta_h^\star \right) \; z_2 \; z_3 \; z_4 \right]^T \in \mathbb{R}^4$ it is possible to balance the object with θ_h at a desired constant angle $\theta_h^\star = x_1^\star$. It is easy to verify that such additional diffeomorphism does not change the normal form (12) expressed now in terms of $\bar{\mathbf{z}}$. With some algebra, it is possible to show that stabilizing the origin $\bar{\mathbf{z}} = \mathbf{0}_4$ yields $x_1 = x_1^\star$, $x_1 + \frac{x_3}{\rho_h} = 0$ and $x_2 = x_4 = 0$.

The input-state FLin plus the EFL controller are employed in [2] on a real-hardware DoD device.

3.2.2 Ball-and-beam

Referring to Fig. 3, the beam can rotate around its center of mass while the ball can only roll along the beam. The shape of the hand, *i.e.*, the beam, is parametrized through $\mathbf{c}_h^h(s_h) = \left[-s_h \; d_h\right]^T$. The ball's shape is parametrized by $\mathbf{c}_o^o(s_h) = -\rho_o \left[\sin \frac{s_h}{\rho_o} \; \cos \frac{s_h}{\rho_o} \right]^T$. Considering (1), the signed curvatures of the beam and the ball are $\kappa_h = 0$ and $\kappa_o = \rho_o^{-1}$, respectively. The relative curvature is thus given by $\kappa_r = \rho_o^{-1}$. The ball's angular velocity is instead given by (2b) $\dot{\theta}_o = \dot{\theta}_h + \frac{\dot{s}_h}{\rho_o}$. In order to compute the mass matrix of the system, the vectors $\boldsymbol{\gamma}(\mathbf{q})$ and $\boldsymbol{\eta}(\mathbf{q})$ in (3) are $\boldsymbol{\gamma}(\mathbf{q}) = \left[-(\rho_o + d_h)\cos\theta_h + s_h \sin\theta_h \; -(\rho_o + d_h)\sin\theta_h - s_h\cos\theta_h\right]^T$, and $\boldsymbol{\eta}(\mathbf{q}) = \left[-\cos(\theta_h) \; -\sin(\theta_h)\right]^T$. Therefore, the parameters of the BnB dynamic model are $m_{11} = I_h + I_o + m_o d_h^2 + 2m_o d_h \rho_o + m_o \rho_o^2 + m_o s_h^2$, $m_{12} = m_{21} = \frac{I_o}{\rho_o} + m_o d_h + m_o \rho_o$, $m_{22} = \frac{I_o}{\rho_o^2} + m_o$, $c_{11} = m_o s_h \dot{s}_h$, $c_{12} = m_o s_h \dot{\theta}_h$, $c_{21} = m_o s_h \dot{\theta}_h$, $c_{22} = 0$, $g_1 = -m_o g((d_h + \rho_o)\sin\theta_h + s_h\cos\theta_h)$ and $g_2 = -m_o g \sin\theta_h$. Considering the acceleration a_h of the beam as input, the system can be written as in (7), with τ_h as in (8). However, it is possible to verify that Assumption 3.1 is not verified for the ball and beam case since $c_{21} \neq 0$. Even if κ_r is constant and the products $\boldsymbol{\gamma}(\mathbf{q})^T\boldsymbol{\eta}(\mathbf{q})$ and $\boldsymbol{\eta}(\mathbf{q})^T\boldsymbol{\eta}(\mathbf{q})$ do not depend on $\mathbf{q}$, the product $\boldsymbol{\gamma}(\mathbf{q})^T\boldsymbol{\gamma}(\mathbf{q})$ does not depend only on θ_h, but it depends on the arclength parameter. Therefore, the ball and beam system is not

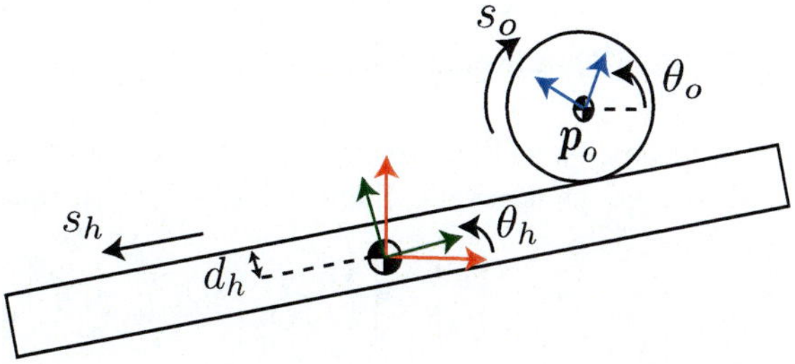

Fig. 3 A representation of the BnB system. In red the world fixed frame $\mathcal{W}$. In green the hand frame $\mathcal{H}$, while in blue the object frame $\mathcal{O}$, placed at the respective CoMs

input-state linearizable. This result is well known in the literature, nevertheless, in many cases it is possible to approximate c_{21} to zero [7]. This is true for small velocities of the beam, small masses of the ball and not so long beam. Hence, by putting $c_{21} = 0$, only for control design purposes, it is possible to write the approximated ball and beam system like in (9). The affine state space form of the approximated ball and beam system has the following vector fields (10) $\mathbf{f}(\mathbf{x}) = \left[x_2 \; 0 \; x_4 \; \dfrac{m_o \rho_o^2 g \sin x_1}{I_o + m_o \rho_o^2} \right]^T$ and $\mathbf{b} = \left[0 \; 1 \; 0 \; -\rho_o \dfrac{m_o \rho_o^2 + d_h m_o \rho_o + I_o}{I_o + m_o \rho_o^2} \right]^T$. Computing the matrix $\mathbf{T}(\mathbf{x})$ as in (11), it is possible to verify that the approximate dynamic model is input-state FLin in the region $\Omega = \left\{ \mathbf{x} \in \mathbb{R}^4 : \cos \theta_h \neq 0 \Rightarrow |\theta_h| < \dfrac{\pi}{2} \right\}$. Notice that such a region is not restrictive because, with no bound on other states, Ω covers all practical situations since outside it the ball falls down from the hand. Finally, it is possible to compute the diffeomorphism (15) $\boldsymbol{\phi}(\mathbf{x}) = \left[\dfrac{m_{12}}{m_{22}} x_1 + x_3 \; \dfrac{m_{12}}{m_{22}} x_2 + x_4 \; \dfrac{m_{og}}{b_{22}} \sin x_1 \; \dfrac{m_{og} x_2}{m_{22}} \cos x_1 \right]^T$, yielding the normal form

$$\dot{z}_1 = z_2, \tag{18a}$$
$$\dot{z}_2 = z_3, \tag{18b}$$
$$\dot{z}_3 = z_4, \tag{18c}$$
$$\dot{z}_4 = \beta(\mathbf{x})|^{-1}_{\mathbf{x}=\boldsymbol{\phi}^{-1}(\mathbf{z})} \left(a_h - \alpha(\mathbf{x})|_{\mathbf{x}=\boldsymbol{\phi}^{-1}(\mathbf{z})} \right), \tag{18d}$$

with $\alpha(\mathbf{x}) = m_o g x_2 \tan x_1$ and $\beta(\mathbf{x}) = \dfrac{m_{22}}{m_o g \cos x_1}$ from (16). The control is then performed with the EFL technique described in Sect. 3.1.

The designed controller for the BnB system is tested in simulation within [2], showing acceptable performance despite the employed assumption to make the BnB system as input-state feedback linearisable.

4 Passivity-Based Approach

4.1 Background on Passivity-Based Control

4.1.1 Port-Hamiltonian Systems

The pH framework allows modeling of mechanical systems including the information about the energy transfer explicitly. The canonical Hamiltonian equations of motion are

$$\begin{bmatrix} \dot{\mathbf{q}} \\ \dot{\mathbf{p}} \end{bmatrix} = \begin{bmatrix} \mathbf{O}_2 & \mathbf{I}_2 \\ -\mathbf{I}_2 & \mathbf{O}_2 \end{bmatrix} \nabla H(\mathbf{q}, \mathbf{p}) + \begin{bmatrix} \mathbf{0}_2 \\ \mathbf{g}_i(\mathbf{q}) \end{bmatrix} u, \tag{19}$$

where $\mathbf{g}_i(\mathbf{q}) \in \mathbb{R}^2$ is the input mapping vector. The function H is the Hamiltonian, which represents the total energy (kinetic plus potential) stored in the system, having the form

$$H(\mathbf{q}, \mathbf{p}) = \frac{1}{2}\mathbf{p}^T \mathbf{M}^{-1}(\mathbf{q})\mathbf{p} + V(\mathbf{q}).$$

A full development of the Hamilton canonical equations of motion can be found in [13].

4.1.2 Interconnection and Damping Assignment

The stabilisation problem of the dynamics (19) to the desired equilibrium $(\mathbf{q}, \mathbf{p}) = (\mathbf{q}^\star, \mathbf{0}_2)$ is achieved using the IDA-PBC by assigning the target dynamics to the closed loop [8]

$$\begin{bmatrix} \dot{\mathbf{q}} \\ \dot{\mathbf{p}} \end{bmatrix} = \begin{bmatrix} \mathbf{O}_2 & \mathbf{M}^{-1}(\mathbf{q})\mathbf{M}_d(\mathbf{q}) \\ -\mathbf{M}_d(\mathbf{q})\mathbf{M}^{-1}(\mathbf{q}) & \mathbf{J}_2(\mathbf{q}, \mathbf{p}) \end{bmatrix} \nabla H_d(\mathbf{q}, \mathbf{p}), \tag{20}$$

where $\mathbf{J}_2(\mathbf{q}, \mathbf{p}) \in \mathbb{R}^{2\times 2}$ is the desired interconnection matrix. The desired total energy function is given by

$$H_d(\mathbf{q}, \mathbf{p}) = \frac{1}{2}\mathbf{p}^T \mathbf{M}_d^{-1}(\mathbf{q})\mathbf{p} + V_d(\mathbf{q}), \tag{21}$$

with $V_d(\mathbf{q}) \in \mathbb{R}$ the desired potential energy function. Then, $(\mathbf{q}^\star, \mathbf{0}_2)$ will be a stable equilibrium configuration of the closed-loop (20) if

C.1 $\mathbf{M}_d(\mathbf{q})$ is symmetric and positive definite;
C.2 $\mathbf{q}^\star = \arg\min V_d(\mathbf{q})$;
C.3 $\mathbf{J}_2(\mathbf{q}, \mathbf{p})$ is skew-symmetric.

The stabilization of the desired equilibrium is achieved by identifying the class of Hamiltonian systems that can be obtained via feedback. The conditions under which this feedback law exists are the matching conditions, i.e., matching the original dynamic system (19) and the target dynamic system (20):

$$\begin{bmatrix} \mathbf{O}_2 & \mathbf{I}_2 \\ -\mathbf{I}_2 & \mathbf{O}_2 \end{bmatrix} \nabla H + \begin{bmatrix} \mathbf{0}_2 \\ \mathbf{g}_i(\mathbf{q}) \end{bmatrix} \mathbf{u} = \begin{bmatrix} \mathbf{0}_2 & \mathbf{M}^{-1}\mathbf{M}_d \\ -\mathbf{M}_d\mathbf{M}^{-1} & \mathbf{J}_2 \end{bmatrix} \nabla H_d, \tag{22}$$

where the dependency of the functions on their argument has been drop to simplify the notation. The first line in (22) is straightforwardly satisfied, while the second line in (22) corresponds to the following set of PDEs

$$\mathbf{g}_i^\perp(\mathbf{q}) \left(\nabla_q H(\mathbf{q}, \mathbf{p}) - \mathbf{M}_d(\mathbf{q})\mathbf{M}^{-1}(\mathbf{q})\nabla_q H_d(\mathbf{q}, \mathbf{p}) + \mathbf{J}_2(\mathbf{q}, \mathbf{p})\mathbf{M}_d^{-1}(\mathbf{q})\mathbf{p} \right) = 0, \tag{23}$$

where $\mathbf{g}_i^{\perp}(\mathbf{q}) \in \mathbb{R}^{1\times 2}$ is the full rank left annihilator of $\mathbf{g}_i(\mathbf{q})$, which satisfies $\mathbf{g}_i^{\perp}(\mathbf{q})\mathbf{g}_i(\mathbf{q}) = 0$. The PDEs (23) can be separated into the two subsets of PDEs, namely

$$\mathbf{g}_i^{\perp}\left(\nabla_q(\mathbf{p}^T\mathbf{M}^{-1}(\mathbf{q})\mathbf{p}) - \mathbf{M}_d(\mathbf{q})\mathbf{M}^{-1}(\mathbf{q})\nabla_q(\mathbf{p}^T\mathbf{M}_d^{-1}(\mathbf{q})\mathbf{p}) + 2\mathbf{J}_2(\mathbf{q},\mathbf{p})\mathbf{M}_d^{-1}(\mathbf{q})\mathbf{p}\right) = 0, \tag{24}$$

$$\mathbf{g}_i^{\perp}\left(\nabla_q V(\mathbf{q}) - \mathbf{M}_d(\mathbf{q})\mathbf{M}^{-1}(\mathbf{q})\nabla_q V_d(\mathbf{q})\right) = 0, \tag{25}$$

where (24) and (25) are the kinetic and the potential energy matching equations, respectively. By solving (24)–(25) for $\mathbf{M}_d(\mathbf{q})$, $V_d(\mathbf{q})$ and $\mathbf{J}_2(\mathbf{q},\mathbf{p})$, subject to **C.1**, **C.2**, and **C.3**, the energy shaping control is given by

$$u_{es} = (\mathbf{g}_i^T(\mathbf{q})\mathbf{g}_i(\mathbf{q}))^{-1}\mathbf{g}_i^T(\mathbf{q})(\nabla_q H(\mathbf{q},\mathbf{p}) - \mathbf{M}_d(\mathbf{q})\mathbf{M}^{-1}(\mathbf{q})\nabla_q H_d(\mathbf{q},\mathbf{p}) + \mathbf{J}_2(\mathbf{q},\mathbf{p})\mathbf{M}_d^{-1}(\mathbf{q})\mathbf{p}). \tag{26}$$

It is worth remarking that not every desired $\mathbf{M}_d(\mathbf{q})$, $V_d(\mathbf{q})$ and $\mathbf{J}_2(\mathbf{q},\mathbf{p})$ can be chosen, but only those solving (24)–(25) subject to the conditions **C.1**, **C.2**, and **C.3**.

By applying (26) to the Hamiltonian dynamics (19), the closed-loop target dynamics (20) is obtained. Damping aimed at achieving asymptotic stability is then injected through

$$u_{di} = -K_v\mathbf{g}_i^T(\mathbf{q})\nabla_p H_d(\mathbf{q},\mathbf{p}), \tag{27}$$

where $K_v > 0$ is a gain. The damping injection (27) and the energy shaping control (26) are then assembled to generate the IDA-PBC

$$u = u_{es} + u_{di}. \tag{28}$$

Therefore, through this adjustment, the closed-loop dynamics (20) takes the form

$$\begin{bmatrix}\dot{\mathbf{q}}\\ \dot{\mathbf{p}}\end{bmatrix} = \begin{bmatrix}\mathbf{O}_2 & \mathbf{M}^{-1}(\mathbf{q})\mathbf{M}_d(\mathbf{q})\\ -\mathbf{M}_d(\mathbf{q})\mathbf{M}^{-1}(\mathbf{q}) & \mathbf{J}_2(\mathbf{q},\mathbf{p}) - \mathbf{R}_d(\mathbf{q})\end{bmatrix}\nabla H_d(\mathbf{q},\mathbf{p}), \tag{29}$$

where $\mathbf{R}_d(\mathbf{q}) = \mathbf{g}_i(\mathbf{q})K_v\mathbf{g}_i^T(\mathbf{q}) \in \mathbb{R}^{2\times 2}$ is the positive-(semi)definite dissipation matrix [8, 15].

The stability of desired equilibrium is analysed by using the desired Hamiltonian (21) as a Lyapunov function and computing its time derivative along the trajectories of the closed-loop dynamics (29) as follows

$$\begin{aligned}\dot{H}_d(\mathbf{q},\mathbf{p}) &= \nabla_p H_d^T\dot{\mathbf{p}} + \nabla_q H_d^T\dot{\mathbf{q}}\\ &= -\mathbf{p}^T\mathbf{M}_d^{-1}\mathbf{g}_i K_v\mathbf{g}_i^T\mathbf{M}_d^{-1}\mathbf{p} \le 0,\end{aligned}$$

which ensures stability of the desired equilibrium. Asymptotic stability follows from LaSalle's invariance principle [16], or equivalently from detectability of the signal $\mathbf{y}_d = K_v\mathbf{g}_i^T\mathbf{M}_d^{-1}\mathbf{p}$ [17].

4.1.3 Integral Control

In many cases, the presence of disturbances deteriorates the control system's performance, or in the worst case, will produce instabilities. A typical scenario in practice is to consider constant disturbances, which account for slow-varying perturbations. The dynamics of the closed-loop (29) under the action of a matched disturbance is

$$\begin{bmatrix} \dot{\mathbf{q}} \\ \dot{\mathbf{p}} \end{bmatrix} = \begin{bmatrix} \mathbf{O}_2 & \mathbf{M}^{-1}(\mathbf{q})\mathbf{M}_d(\mathbf{q}) \\ -\mathbf{M}_d(\mathbf{q})\mathbf{M}^{-1}(\mathbf{q}) & \mathbf{J}_2(\mathbf{q},\mathbf{p}) - \mathbf{R}_d(\mathbf{q}) \end{bmatrix} \nabla H_d(\mathbf{q},\mathbf{p}) + \begin{bmatrix} \mathbf{0}_2 \\ \mathbf{g}_i(\mathbf{q}) \end{bmatrix} (v + \delta_d), \tag{30}$$

where $\delta_d \in \mathbb{R}$ is the matched constant disturbance considered here and $v \in \mathbb{R}$ is a control input that will be used to compensate for the unknown disturbance. To obtain the dynamics (30), the control $u = u_{es} + u_{di} + v$ and the disturbance δ_d are used in (19). Notice that the disturbance shifts the equilibrium of the closed loop, defined by zero velocities (equivalently $\mathbf{p} = \mathbf{0}_2$), from the desired equilibrium $\mathbf{q}^\star$ to a new equilibrium $\bar{\mathbf{q}}$, which is the solution of

$$-\mathbf{M}_d(\bar{\mathbf{q}})\mathbf{M}(\bar{\mathbf{q}})^{-1}\nabla_q V_d(\bar{\mathbf{q}}) + \mathbf{g}_i(\bar{\mathbf{q}})\delta_d = \mathbf{0}_2.$$

This shows that the controller does not achieve the control objective in the presence of constant disturbances since $\mathbf{q}$ will not reach the desired value at the steady-state as desired. This motivates the implementation of an outer-loop controller to reject constant unknown disturbances.

In this section, the methodology to design integral-based controllers and enhance the robustness of the energy shaping controller reported in [1] will be introduced. The fundamental idea proposed in [1] is to find a dynamic control law $v(\mathbf{q}, \mathbf{p}, \zeta)$, where $\zeta \in \mathbb{R}$ is the state of the controller, and a change of coordinates such that the closed-loop dynamics that include the controller state expressed in the new coordinates can be written as a port-Hamiltonian system, thus stability is ensured. The proposed target port-Hamiltonian dynamics in new coordinates $\mathbf{z} \in \mathbb{R}^5$, where the state vector has been augmented by adding the controller state. The target port-Hamiltonian system is

$$\begin{bmatrix} \dot{\mathbf{z}}_1 \\ \dot{\mathbf{z}}_2 \\ \dot{z}_3 \end{bmatrix} = \begin{bmatrix} -\mathbf{\Gamma}_1 & \mathbf{M}^{-1}\mathbf{M}_d & -\mathbf{\Gamma}_2 \\ -\mathbf{M}_d\mathbf{M}^{-1} & -K_v\mathbf{g}_i\mathbf{g}_i^T & -\mathbf{g}_i K_3 \\ \mathbf{\Gamma}_2^T & K_3\mathbf{g}_i^T & -\Gamma_3 \end{bmatrix} \begin{bmatrix} \nabla_{\mathbf{z}_1} H_z \\ \nabla_{\mathbf{z}_2} H_z \\ \nabla_{z_3} H_z \end{bmatrix}, \tag{31}$$

with Hamiltonian

$$H_z(\mathbf{z}) = \frac{1}{2}\mathbf{z}_2^T\mathbf{M}_d^{-1}\mathbf{z}_2 + V_z(\mathbf{z}_1) + \frac{1}{2}K_I(z_3 - z_3^\star)^2, \tag{32}$$

where

$$V_z(\mathbf{z}_1) = V_d(\mathbf{q})\Big|_{\mathbf{q}=\mathbf{z}_1}, \tag{33}$$

and the gains are equal to

$$\begin{aligned}\boldsymbol{\Gamma}_1 &\triangleq K_1\mathbf{M}^{-1}\mathbf{g}_i\mathbf{g}_i^T\mathbf{M}^{-1},\\ \boldsymbol{\Gamma}_2 &\triangleq K_2\mathbf{M}^{-1}\mathbf{g}_i,\\ \Gamma_3 &\triangleq K_2K_3\mathbf{g}_i^T\mathbf{M}_d^{-1}\mathbf{g}_i,\\ z_3^\star &\triangleq K_I^{-1}\left(K_vK_2\mathbf{g}_i^T\mathbf{M}_d^{-1}\mathbf{g}_i+K_3\right)^{-1}\delta_d,\end{aligned}$$

where the new coordinates $\mathbf{z}=\psi(\mathbf{q},\mathbf{p},\zeta)$ are obtained by the state transformation

$$\mathbf{z}_1=\mathbf{q}, \tag{34}$$

$$\mathbf{z}_2=\mathbf{p}+K_1\mathbf{g}_i\mathbf{g}_i^T\mathbf{M}^{-1}\nabla V_d(\mathbf{q})+K_2K_I(\zeta-z_3^\star)\mathbf{g}_i, \tag{35}$$

$$z_3=\zeta, \tag{36}$$

with $K_v>0$, $K_I>0$, $K_1>0$, $K_3>0$ and $K_2=\left(\mathbf{g}_i^T\mathbf{M}_d^{-1}\mathbf{g}_i\right)^{-1}$.

Notice that Hamiltonian H_z in (32) has a minimum at $\mathbf{z}^\star=(\mathbf{q}^\star,\mathbf{0}_n,z_3^\star)$, which is the desired equilibrium. Therefore, we look for a control law that render the extended closed-loop dynamics in the form (31) to ensure stability of the equilibrium $\mathbf{z}^\star$. As shown in [1], the control law is obtained by matching the dynamics (30) and (31), and using the change of coordinates (34)–(36). It is shown that, under a few assumptions on the matrices $\mathbf{M}$, $\mathbf{M}_d$ and $\mathbf{g}_i^T$, the integral controller takes the form

$$\begin{aligned}v(\mathbf{q},\mathbf{p},\zeta)=&-\Big[K_vK_1\mathbf{g}_i^T\mathbf{M}_d^{-1}\mathbf{g}_i\mathbf{g}_i^T\mathbf{M}^{-1}+K_1\mathbf{g}_i^T\dot{\mathbf{M}}^{-1}\\ &+K_2K_I\Big(K_2+K_3\mathbf{g}_i^T\mathbf{M}_d^{-1}\mathbf{g}_iK_1\Big)\mathbf{g}_i^T\mathbf{M}^{-1}\Big]\nabla V_d\\ &-\Big[K_1\mathbf{g}_i^T\mathbf{M}^{-1}\nabla^2V_d\mathbf{M}^{-1}+(\mathbf{g}_i^T\mathbf{g}_i)^{-1}\mathbf{g}_i^T\mathbf{J}_2\mathbf{M}_d^{-1}\\ &+K_2K_IK_3\mathbf{g}_i^T\mathbf{M}_d^{-1}\Big]\mathbf{p}\\ &-\Big(K_vK_2\mathbf{g}_i^T\mathbf{M}_d^{-1}\mathbf{g}_i+K_3\Big)K_I\zeta,\end{aligned} \tag{37}$$

and

$$\begin{aligned}\dot{\zeta}=&\Big(K_2\mathbf{g}_i^T\mathbf{M}^{-1}+K_3K_1\mathbf{g}_i^T\mathbf{M}_d^{-1}\mathbf{g}_i\mathbf{g}_i^T\mathbf{M}^{-1}\Big)\nabla V_d\\ &+K_3\mathbf{g}_i^T\mathbf{M}_d^{-1}\mathbf{p}.\end{aligned} \tag{38}$$

The controller, composed by the control law (37) and the integrator (38), does not require the information of the constant disturbance δ_d, as desired. By construction, the dynamics (30) in a closed loop with the integral action controller (37)–(38) can be written in the form (31). The Hamiltonian form of the full closed-loop dynamics ensures the stability of the desired equilibrium. Indeed, the Hamiltonian in (32) has a minimum at the desired equilibrium $\mathbf{z}^\star$ and it qualifies as a Lyapunov function for the dynamics (31). The time derivative of H_z is

$$\begin{aligned}\dot{H}_z = & -K_1 \nabla^T V_d(\mathbf{z}_1)\mathbf{M}^{-1}\mathbf{g}_i\mathbf{g}_i^T\mathbf{M}^{-1}\nabla V_z - K_v \mathbf{z}_2^T\mathbf{M}_d^{-1}\mathbf{g}_i\mathbf{g}_i^T\mathbf{M}_d^{-1}\mathbf{z}_2 \\ & -K_I\Gamma_3 K_I(z_3 - z_3^\star)^2 \\ \leq & \ 0,\end{aligned}$$

which ensures stability. Asymptotic stability follows using LaSalle arguments and verifying that the maximum invariant set included in $\mathcal{S} = \{(\mathbf{z}_1, \mathbf{z}_2, z_3) | \mathbf{g}_i^T\mathbf{M}^{-1}\nabla V_z = 0, \mathbf{g}_i^T\mathbf{M}_d^{-1}\mathbf{z}_2 = 0, z_3 = z_3^\star\}$ is the desired equilibrium $\mathbf{z}^\star$.

4.2 *Control Design for Nonprehensile Systems*

In this section, the IDA-PBC method is used to design controllers for the DoD, the BnB, and the eccentric DoD. These examples show how a passivity-based framework can be used to solve the control problem of nonprehensile planar rolling manipulation.

4.3 *Case Studies*

4.3.1 Disk on Disk

Dynamic Model. The DoD is a rolling-balancing system shown in Fig. 2. The dynamics of the DoD can be described as displayed in Sect. 3.2.1. However, now we derive the dynamics in the pH form. Besides, we write the dynamics in terms of the angle of the hand θ_h and the deviation angle of object respect to the upright position $\varphi = \theta_h + \frac{s_h}{\rho_h}$. By overwriting the configuration vector $\mathbf{q}$ as $\mathbf{q} = \begin{bmatrix}\theta_h & \varphi\end{bmatrix}^T$, the DoD model takes the following pH form

$$\begin{bmatrix}\dot{\mathbf{q}} \\ \dot{\mathbf{p}}\end{bmatrix} = \begin{bmatrix}\mathbf{O}_2 & \mathbf{I}_2 \\ -\mathbf{I}_2 & \mathbf{O}_2\end{bmatrix}\begin{bmatrix}\nabla_q H \\ \nabla_p H\end{bmatrix} + \begin{bmatrix}\mathbf{0}_2 \\ \mathbf{g}_i\end{bmatrix} u, \tag{39}$$

where $\mathbf{g}_i = \begin{bmatrix}1 & 0\end{bmatrix}^T$. The Hamiltonian function is

$$H(\mathbf{q}, \mathbf{p}) = \frac{1}{2}\mathbf{p}^T M^{-1}\mathbf{p} + V(\mathbf{q}). \tag{40}$$

The elements of the mass matrix $\mathbf{M}(\mathbf{q})$ differ from those in Sect. 3.2.1 because of the new definition of $\mathbf{q}$. Therefore, the elements of $\mathbf{M}(\mathbf{q})$ are given by $m_{11} = \rho_h^2 m_o + I_h$, $m_{12} = -m_o\rho_h(\rho_o + \rho_h)$ and $m_{22} = 2m_o(\rho_o + \rho_h)^2$. The expression of the potential energy is instead $V(q) = V_0\cos(\varphi)$, with $V_0 = m_o g(\rho_o + \rho_h)$.

IDA-PBC Controller. The objective is to design a IDA-PBC controller for the DoD system that stabilises the point $q^* = (\theta_h^*,\ 0)$, where θ_h^* is the desired equilibrium for angle of the hand. This problem is solved by using energy shaping and damping injection as described in Sect. 4.1.2. That is, the control design aims to find a function V_d and matrices $\mathbf{M}_d$ and $\mathbf{J}_2$ that solve the KE-ME and PE-ME, (24) and (25) respectively. Thus, the energy shaping control is obtained from (26) and the damping injection control from (27).

Since the mass matrix $\mathbf{M}(\mathbf{q})$ of the DoD is constant, i.e., it does not depend on the coordinates $\mathbf{q}$, the desired mass matrix $\mathbf{M}_d$ is selected as a constant matrix as follows

$$\mathbf{M}_d = \begin{bmatrix} a_{11} & a_{12} \\ a_{12} & a_{22} \end{bmatrix}$$

where a_{11}, a_{12} and a_{22} are free constant parameters. To simplify the notation, we note

$$\mathbf{M}_d\mathbf{M}^{-1} = -\begin{bmatrix} \gamma & \delta \\ \alpha & \beta \end{bmatrix}.$$

Then, the PE-ME (25) is as follows

$$\begin{aligned} [0\ 1]\left\{\begin{bmatrix} 0 \\ -V_0\sin(\varphi) \end{bmatrix} + \begin{bmatrix} \gamma & \delta \\ \alpha & \beta \end{bmatrix}\begin{bmatrix} \nabla_{\theta_h} V_d \\ \nabla_{\varphi} V_d \end{bmatrix}\right\} &= 0 \\ -V_0\sin(\varphi) + \alpha\ \nabla_{\theta_h} V_d + \beta\ \nabla_{\varphi} V_d &= 0. \end{aligned} \tag{41}$$

A solution of the partial differential equation (41) for V_d, obtained by using a symbolic software (e.g. Mathematica, Maple), is

$$V_d(q) = -\frac{1}{\beta}V_0\cos(\varphi) + \frac{k_2}{2}\left(\theta_h - \frac{\alpha}{\beta}\varphi - k_1\right)^2 \tag{42}$$

where $k_1, k_2 \in \mathbb{R}$ are free constant parameters to be selected such that the potential function has a minimum at the desired equilibrium.

Since $\mathbf{M}_d$ was chosen as a constant matrix, it is clear that the KE-ME (24) is satisfied by selecting $\mathbf{J}_2(\mathbf{q}, \mathbf{p}) = \mathbf{O}_2$. The energy shaping design also requires that $\mathbf{M}_d$ is positive definite and that V_d has an isolated minimum at the desired equilibrium $\mathbf{q}^\star$.

The minimum of V_d is assigned by ensuring that the Jaccobian zero when it is evaluated at $\mathbf{q}^\star$ and Hessian is positive when it is evaluated at $\mathbf{q}^\star$. These conditions are verified through

R-I) $$\nabla_q V_d(q)\big|_{q=q^*} = 0 \Leftrightarrow \left.\begin{bmatrix} k_2\left(\theta_h - \frac{\alpha}{\beta}\varphi - k_1\right) \\ \frac{V_0}{\beta}\sin(\varphi) - \frac{k_2\alpha}{\beta}\left(\theta_h - \frac{\alpha}{\beta}\varphi - k_1\right) \end{bmatrix}\right|_{q=q^*} = 0$$

which is satisfied if $k_1 = \theta_h^*$.

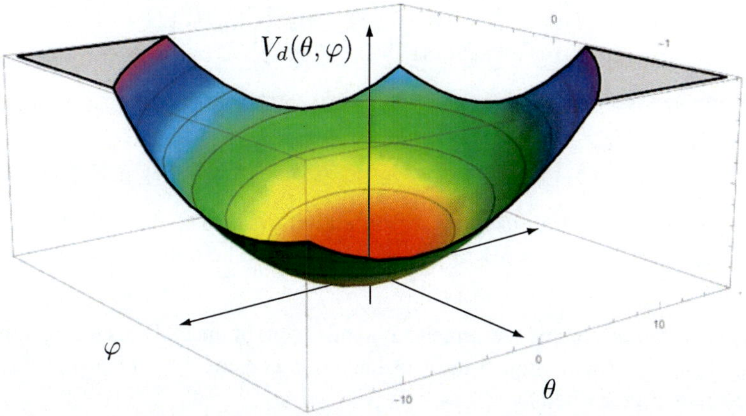

Fig. 4 Desired potential energy for the DoD case study with the pH formalism. In this picture $\theta_h = \theta$

R-II) $$\nabla_q^2 V_d(q)\big|_{q=q^*} > 0 \Leftrightarrow \left.\begin{bmatrix} k_2 & -k_2\dfrac{\alpha}{\beta} \\ -k_2\dfrac{\alpha}{\beta} & \dfrac{V_0}{\beta}\cos(\varphi) + k_2\left(\dfrac{\alpha}{\beta}\right)^2 \end{bmatrix}\right|_{q=q^*} > 0$$

which is satisfied provided that $k_2 > 0$ and $\beta > 0$ (equivalently $a_{12}m_{12} - a_{22}m_{11} > 0$).

The positive definiteness of M_d is ensured if $a_{11} > 0$ and $a_{11}a_{22} - a_{12}^2 > 0$.

Figure 4 shows that effectively the potential energy has a minimum at the desired equilibrium $(\theta_h^\star, \varphi^\star) = (0, 0)$ (the parameter values used to create the figure satisfy all the requirements R-I and R-II above).

Finally, the control law is computed from (26) to (27) as follows

$$u = u_{es} + u_{di} = \\ -\frac{\delta}{\beta}\nabla_\varphi V + k_2\left(\frac{\gamma\beta - \delta\alpha}{\beta}\right)\left(\theta_h - \frac{\alpha}{\beta}\varphi - \theta_h^\star\right) + K_v\,\beta\,\sigma\left(\dot{\theta}_h - \frac{\alpha}{\beta}\dot{\varphi}\right) \tag{43}$$

where $\sigma = \dfrac{m_{11}m_{22} - m_{12}^2}{a_{11}a_{22} - a_{12}^2}$ and the free parameters a_{11}, a_{12}, a_{22}, k_2 and K_v should satisfy

$$a_{11} > 0, \quad k_2 > 0, \quad K_v > 0,$$
$$a_{11}a_{22} - a_{12}^2 > 0,$$
$$a_{12}m_{12} - a_{22}m_{11} > 0.$$

Thus, the dynamics of the DoD system (39) in closed loop with the controller (43) can be written in the pH form

$$\begin{bmatrix} \dot{\mathbf{q}} \\ \dot{\mathbf{p}} \end{bmatrix} = \begin{bmatrix} \mathbf{O}_2 & \mathbf{M}^{-1}\mathbf{M}_d \\ -\mathbf{M}_d\mathbf{M}^{-1} & -K_v \mathbf{g}_i \mathbf{g}_i^T \end{bmatrix} \begin{bmatrix} \nabla_q H_d \\ \nabla_p H_d \end{bmatrix} \tag{44}$$

with

$$H_d = \frac{1}{2}\mathbf{p}^T \mathbf{M}_d^{-1}\mathbf{p} + V_d(\mathbf{q}).$$

Asymptotic stability of the desired equilibrium follows directly from the fact that the closed-loop dynamics has a pH form and the fact that the output $y_d = K_v \mathbf{g}_i^T \mathbf{M}_d^{-1}\mathbf{p}$ is dectable [17].

Integral Controller. In this section, the dynamics (44) of the DoD in closed loop with the IDA-PBC controller subject to disturbances is considered. Under this scenario, the dynamics is described as

$$\begin{bmatrix} \dot{\mathbf{q}} \\ \dot{\mathbf{p}} \end{bmatrix} = \begin{bmatrix} \mathbf{O}_2 & \mathbf{M}^{-1}\mathbf{M}_d \\ -\mathbf{M}_d\mathbf{M}^{-1} & -K_v \mathbf{g}_i \mathbf{g}_i^T \end{bmatrix} \begin{bmatrix} \nabla_q H_d \\ \nabla_p H_d \end{bmatrix} + \begin{bmatrix} \mathbf{0}_2 \\ \mathbf{g}_i \end{bmatrix} (v + \delta_d), \tag{45}$$

where v is the input used for integral control and δ_d is the disturbance. As shown in Sect. 4.1.3 the control law (37)–(38), specialised for the DoD case, compensates for the action of the disturbance and it preserves the stability of the desired equilibrium $\mathbf{q}^\star$. The IDA-PBC controller and the integral action controller were tested in both simulation and experimental setup. The results can be found in [1]. The results showed satisfactory performance of the closed loop and smooth time history of the control inputs and variables of the system.

4.3.2 Ball and Beam

Dynamic Model. The BnB system shown in Fig. 3 is another rolling-balancing benchmark addressed to test the controller addressed in this section. Now, we come back to the original definition of the configuration vector $\mathbf{q}$ as in Sect. 3.2.2, that is $\mathbf{q} = \begin{bmatrix}\theta_h & s_h\end{bmatrix}^T$. In these coordinates, the BnB model takes the pH form (39) with $\mathbf{g}_i = \begin{bmatrix}1 & 0\end{bmatrix}^T$, the potential function given by

$$V(\mathbf{q}) = m_o g[(d_h + \rho_o)\cos(\theta_h) - s_h \sin(\theta_h)], \tag{46}$$

and the mass matrix elements as in Sect. 3.2.2.

IDA-PBC Controller. The control objective for the BnB system is to stabilize the equilibrium $\mathbf{q}^\star = (0, s_h^\star)$, where $s_h^\star \in \mathbb{R}$ is the desired position of the ball on the beam. The control design follows the procedure proposed in [3] to compute energy shaping controllers for planar rolling manipulations, that is for systems in the form (19).

The procedure in [3] can be summarised as follows. Consider the vector of coordinates $\mathbf{q} = \begin{bmatrix}q_1 & q_2\end{bmatrix}^T$ and the desired mass matrix parametrized as follows

$$\mathbf{M}_d(\mathbf{q}) = \Delta \begin{bmatrix} a_{11}(\mathbf{q}) & a_{12}(\mathbf{q}) \\ a_{12}(\mathbf{q}) & a_{22}(\mathbf{q}) \end{bmatrix}, \tag{47}$$

where $\Delta = m_{11}(\mathbf{q})m_{22}(\mathbf{q}) - m_{12}^2(\mathbf{q}) > 0$, which allows to write the PE-ME (25) as

$$\mathbf{e}_2^T \left(\nabla_q V(\mathbf{q}) - \mathbf{\Gamma}(\mathbf{q})\nabla_q V_d(\mathbf{q})\right) = 0, \tag{48}$$

with

$$\mathbf{\Gamma}(\mathbf{q}) = \begin{bmatrix} a_{11}m_{22} - a_{12}m_{12} & a_{12}m_{11} - a_{11}m_{12} \\ a_{12}m_{22} - a_{22}m_{12} & a_{22}m_{11} - a_{12}m_{12} \end{bmatrix}. \tag{49}$$

The PDE (48) can be equivalently written as

$$\nabla_{q_2} V(\mathbf{q}) + \alpha(\mathbf{q})\nabla_{q_1} V_d(\mathbf{q}) + \beta(\mathbf{q})\nabla_{q_2} V_d(\mathbf{q}) = 0. \tag{50}$$

with

$$\alpha(\mathbf{q}) = a_{22}(\mathbf{q},)m_{12}(\mathbf{q}) - a_{12}(\mathbf{q})m_{22}(\mathbf{q}), \tag{51}$$

$$\beta(\mathbf{q}) = a_{12}(\mathbf{q},)m_{12}(\mathbf{q}) - a_{22}(\mathbf{q})m_{11}(\mathbf{q}). \tag{52}$$

Then, the scalar functions $\alpha(\mathbf{q}, \mathbf{c}_1)$ and $\beta(\mathbf{q}, \mathbf{c}_1)$ can be chosen to obtain a suitable solution of (50) such that $V_d(\mathbf{q})$ satisfies **C.2**. Once $\alpha(\mathbf{q})$ and $\beta(\mathbf{q})$ are selected, the entries $a_{12}(\mathbf{q})$ and $a_{22}(\mathbf{q})$ of the desired mass matrix are computed as

$$a_{12}(\mathbf{q}) = -\frac{\alpha(\mathbf{q})m_{11}(\mathbf{q}) + \beta(\mathbf{q})m_{12}(\mathbf{q})}{\Delta}, \tag{53}$$

$$a_{22}(\mathbf{q}) = -\frac{\alpha(\mathbf{q})m_{12}(\mathbf{q}) + \beta(\mathbf{q})m_{22}(\mathbf{q})}{\Delta}. \tag{54}$$

By this construction, the desired mass matrix is symmetric, and thus the condition **C.1** is fulfilled if and only if $a_{11}(\mathbf{q}) > 0$ and $a_{11}(\mathbf{q})a_{22}(\mathbf{q}) - a_{12}^2(\mathbf{q}) > 0$. Therefore, by selecting a_{11} as

$$a_{11}(\mathbf{q}) = \frac{k_a a_{12}^2(\mathbf{q})}{a_{22}(\mathbf{q}, \mathbf{c}_1)} > 0, \tag{55}$$

where $k_a > 1$ is a constant, the positiveness of $\mathbf{M}_d$ is satisfied if

$$\alpha(\mathbf{q})m_{12}(\mathbf{q}) + \beta(\mathbf{q})m_{22}(\mathbf{q}) < 0. \tag{56}$$

If this condition cannot be satisfied, then it is necessary to re-design $\alpha(\mathbf{q})$ and $\boldsymbol{\beta}(\mathbf{q})$ and find another solution for (50). Finally, the desired mass matrix takes the form

$$\mathbf{M}_d(\mathbf{q}) = \begin{bmatrix} -\dfrac{k_a(\alpha m_{11} + \beta m_{12})^2}{(\alpha m_{12} + \beta m_{22})} & -(\alpha m_{11} + \beta m_{12}) \\ -(\alpha m_{11} + \beta m_{12}) & -(\alpha m_{12} + \beta m_{22}) \end{bmatrix}. \tag{57}$$

In addition, the degree of freedom given by the matrix $\mathbf{J}_2(\mathbf{q}, \mathbf{p})$ is used to satisfy the KE-ME (24). The interconnection matrix $\mathbf{J}_2$ has the following structure

$$\mathbf{J}_2(\mathbf{q}, \mathbf{p}) = \begin{bmatrix} 0 & j_2(\mathbf{q}, \mathbf{p}) \\ -j_2(\mathbf{q}, \mathbf{p}) & 0 \end{bmatrix}, \tag{58}$$

with $j_2(\mathbf{q}, \mathbf{p})$ a scalar function. Since $\mathbf{e}_2^T \mathbf{J}_2(\mathbf{q}, \mathbf{p}) = -j_2(\mathbf{q}, \mathbf{p})\mathbf{e}_1^T$, then the KE-ME (24) can be expressed as

$$\mathbf{e}_2^T \nabla_q (\mathbf{p}^T \mathbf{M}^{-1}(\mathbf{q})\mathbf{p}) - \mathbf{e}_2^T \mathbf{M}_d(\mathbf{q})\mathbf{M}^{-1}(\mathbf{q})\nabla_q (\mathbf{p}^T \mathbf{M}_d^{-1}(\mathbf{q})\mathbf{p}) - 2 j_2(\mathbf{q}, \mathbf{p})\mathbf{e}_1^T \mathbf{M}_d^{-1}(\mathbf{q})\mathbf{p} = 0. \tag{59}$$

The scalar function $j_2(\mathbf{q}, \mathbf{p})$ can be from (59) as follows

$$j_2(\mathbf{q}, \mathbf{p}) = \left(2\mathbf{e}_1^T \mathbf{M}_d^{-1}(\mathbf{q})\mathbf{p}\right)^{-1} \left(\mathbf{e}_2^T \nabla_q (\mathbf{p}^T \mathbf{M}^{-1}(\mathbf{q})\mathbf{p}) - \mathbf{e}_2^T \mathbf{M}_d(\mathbf{q})\mathbf{M}^{-1}(\mathbf{q})\nabla_q (\mathbf{p}^T \mathbf{M}_d^{-1}(\mathbf{q})\mathbf{p})\right), \tag{60}$$

and the IDA-PBC law can be finally computed from (26).

Notice that the method used to satisfy the KE-ME, inspired by [18], provides a solution that is not always well-defined. Close to the equilibrium, the numerator of (60), which has a quadratic dependence on $\mathbf{p}$, tends towards zero faster than the denominator, which depends linearly on $\mathbf{p}$, which would avoid singularities. Despite this, a study about the denominator of the relation (60) reveals that, far from the equilibrium, it might be nullified if the equality $(\alpha(\mathbf{q})m_{12}(\mathbf{q}) + \beta(\mathbf{q})m_{22}(\mathbf{q}))p_1 = (\alpha(\mathbf{q})m_{11}(\mathbf{q}) + \beta(\mathbf{q})m_{12}(\mathbf{q}))p_2$ holds. This situation is addressed in practice by saturating the denominator of (60) when its absolute value is smaller than a suitable threshold. The simplification of the design proposed here is at the expense of the presence of possible singular solutions of (60), but these can always be numerically managed in the controller implementation. Recently, this problem has been overcome in [19].

For the BnB case study, the functions $\alpha(\mathbf{q})$ and $\beta(\mathbf{q})$ are selected as $\alpha(\theta_h) = k\text{sinc}(\theta_h)$ and $\beta(\theta_h) = -\text{sinc}(\theta_h)$, where $k \in \mathbb{R}$ is a constant parameter. Notice that the sinc$(\cdot)$ function is analytic everywhere. Assuming the domain of interest as $-\pi < \theta_h < \pi$, then $0 < \text{sinc}(\theta_h) < 1$. Using these functions in (50), the PE-ME becomes

$$-m_o g \sin(\theta_h) + k\text{sinc}(\theta_h)\nabla_{\theta_h} V_d(\mathbf{q}) - \text{sinc}(\theta_h)\nabla_{s_h} V_d(\mathbf{q}) = 0. \tag{61}$$

A solution of (61) is given by

$$V_d(\mathbf{q}) = \frac{m_o g \theta_h^2}{2k} + f\left(\frac{\theta_h + k s_h}{k}\right), \tag{62}$$

where $f(\cdot)$ is a function to be selected to satisfy **C.2**. Then, $f(\cdot)$ is chosen such that the desired potential function (62) results as follows

$$V_d(\mathbf{q}) = \frac{m_o g \theta_h^2}{2k} - \cos\left(\frac{k_f}{k}\left[\theta_h + k\left(s_h - s_h^{\star}\right)\right]\right), \tag{63}$$

with $k_f \in \mathbb{R}$ a constant parameter. The Jacobian of $V_d(\mathbf{q})$ is computed to verify that $\mathbf{q}^{\star}$ is a minimum of the desired potential function (63), which yields

$$\nabla V_d(\mathbf{q}) = \begin{bmatrix} \frac{m_o g}{k}\theta_h + \frac{k_f}{k}\sin\left(\frac{k_f}{k}[\theta_h + k(s_h - s_h^{\star})]\right) \\ k_f \sin\left(\frac{k_f}{k}[\theta_h + k(s_h - s_h^{\star})]\right) \end{bmatrix}, \tag{64}$$

where it is possible to verify that $\nabla V_d(\mathbf{q})$ is zero at $\mathbf{q}^{\star}$. Then, the corresponding Hessian is given by

$$\nabla^2 V_d(\mathbf{q}) = \begin{bmatrix} \frac{m_o g}{k} + \frac{k_f^2}{k^2}\cos\phi & \frac{k_f^2}{k}\cos\phi \\ \frac{k_f^2}{k}\cos\phi & k_f^2\cos\phi \end{bmatrix}, \tag{65}$$

with $\phi = \frac{k_f}{k}\left[\theta_h + k\left(s_h - s_h^{\star}\right)\right]$. It is possible to verify that $\nabla^2 V_d(\mathbf{q})$ is positive definite at the desired equilibrium $\mathbf{q}^{\star}$ if $k > 0$ and $k_f \neq 0$. The conditions on the Jacobian and the Hessian of $V_d(\mathbf{q})$ ensure that the desired potential function $V_d(\mathbf{q})$ has a minimum at the desired equilibrium $\mathbf{q}^{\star}$.

In addition, the inequality (56) must be satisfied to ensure the positiveness of $\mathbf{M}_d$. Using the selected functions $\alpha(\mathbf{q})$ and $\boldsymbol{\beta}(\mathbf{q})$, the inequality (56) becomes

$$km_{12} - m_{22} < 0, \tag{66}$$

which has the straightforward solution $k < \frac{m_{22}}{m_{12}}$. Since $\frac{m_{22}}{m_{12}} > 0$, the parameter k has to be selected to satisfy $0 < k < \frac{m_{22}}{m_{12}}$.

Finally, the entries $a_{12}(\mathbf{q})$ and $a_{22}(\mathbf{q})$ of $\mathbf{M}_d(\mathbf{q})$ are computed from (53) to (54) as follows

$$\begin{aligned} a_{12}(\mathbf{q}) &= -\frac{\operatorname{sinc}(\theta_h)(km_{11}(s_h) - m_{12})}{\Delta}, \\ a_{22}(\mathbf{q}) &= -\frac{\operatorname{sinc}(\theta_h)(km_{12} - m_{22})}{\Delta}, \end{aligned} \tag{67}$$

while $a_{11}(\mathbf{q})$ is computed as in (55). Therefore, the desired mass matrix is

$$\mathbf{M}_d(\mathbf{q}) = \begin{bmatrix} -\frac{k_a\left(km_{11}(s_h) - m_{12}\right)^2}{(km_{12} - m_{22})} & -\operatorname{sinc}(\theta_h)\left(km_{11}(s_h) - m_{12}\right) \\ -\operatorname{sinc}(\theta_h)\left(km_{11}(s_h) - m_{12}\right) & -\operatorname{sinc}(\theta_h)(km_{12} - m_{22}) \end{bmatrix}. \tag{68}$$

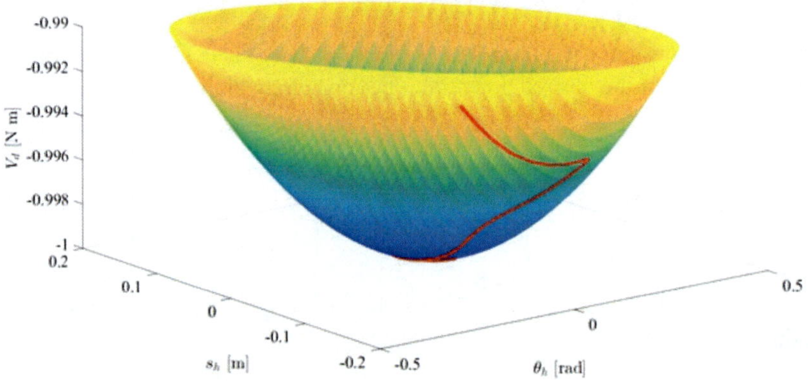

Fig. 5 Desired potential function and coordinate trajectory for the BnB case study with the pH formalism

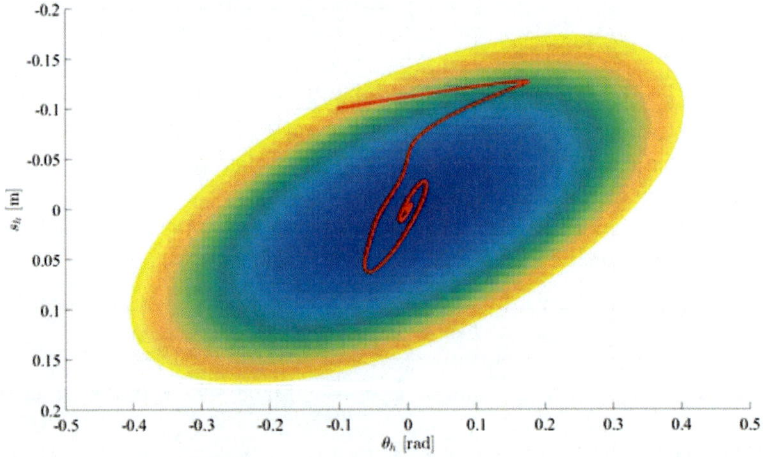

Fig. 6 Desired potential function and coordinate trajectory for the BnB case study with the pH formalism

The KE-ME (24) is satisfied using (60), and the IDA-PBC control law is computed from (28).

Figures 5 and 6 show the desired potential function (63) and a trajectory from a particular initial condition. As expected, the trajectory converges to the minimum, that is the desired equilibrium $\mathbf{q}^\star$. An exhaustive simulation study of the closed loop has been presented in [3].

4.3.3 Eccentric Disk-on-disk

Dynamic Model. The eccentric DoD system is represented in Fig. 7 [3]. In this system, the hand is the actuated bottom disk and the object is the non-actuated disk that sits on top. With respect to the classic DoD seen before, the hand's CoM and the actuation point are at a distance $\lambda > 0$. It can be shown that the mass matrix elements are $m_{11} = c_{b1} + c_{b2} \cos\left(\frac{s_h}{\rho_h}\right)$, $m_{12} = c_{b3} + c_{b4} \cos\left(\frac{s_h}{\rho_h}\right)$, and $m_{22} = I_o \kappa_r^2 + m_o \rho_o^2 \kappa_r^2$, where $c_{b1} = I_h + I_o + \lambda^2(m_h + m_o) + m_o(\rho_h + \rho_o)^2$, $c_{b2} = 2\lambda m_o(\rho_h + \rho_o)$, $c_{b3} = I_o \kappa_r + m_o \frac{(\rho_h+\rho_o)^2}{\rho_h}$, and $c_{b4} = m_o \lambda \rho_o \kappa_r$. The potential energy (4) for the eccentric DoD is

$$V(\mathbf{q}) = g(m_o(\rho_h + \rho_o) \cos\left(\theta_h + \frac{s_h}{\rho_h}\right) + (m_o + m_h)\lambda \cos(\theta_h)). \tag{69}$$

A detailed derivation of this model can be found in [3].

IDA-PBC Controller. The control objective is to balance the object on top of the hand. In this configuration, the CoM of the hand can be placed above or below its center of actuation. In both cases it is possible, under a suitable change of coordinates, to express the desired equilibrium point as $\mathbf{q}^\star = (0, 0)$ in both cases.

The controller for the eccentric DoD is computed using the same procedure used for the BnB in Sect. 4.3.2. Thus, the functions $\alpha(\mathbf{q})$ and $\beta(\mathbf{q})$ are selected as $\alpha(\theta_h, s_h) = \mathrm{sinc}\left(\theta_h + \frac{s_h}{\rho_h}\right)$ and $\beta(\theta_h, s_h, k) = k\mathrm{sinc}\left(\theta_h + \frac{s_h}{\rho_h}\right)$, where $k \in \mathbb{R}$ is a constant parameter. The function $\mathrm{sinc}(\cdot)$ satisfies $0 < \mathrm{sinc}\left(\theta_h + \frac{s_h}{\rho_h}\right) < 1$ in the domain of interest $-\pi < \left(\theta_h + \frac{s_h}{\rho_h}\right) < \pi$. Using the selected function $\alpha(\mathbf{q})$ and $\beta(\mathbf{q})$ in (50) yields

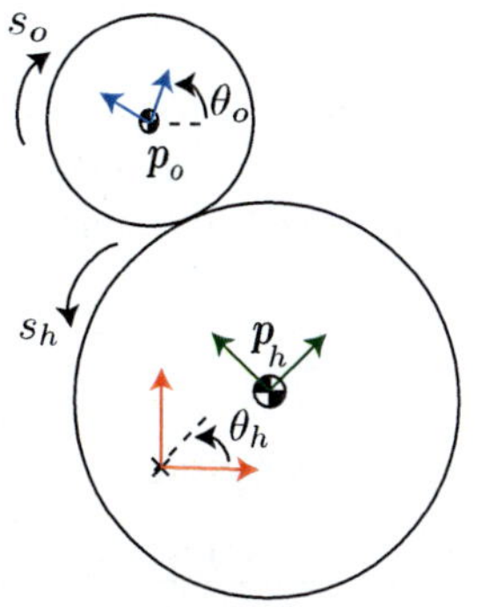

Fig. 7 A schematic of the eccentric DoD system

$$- c_v \sin\left(\theta_h + \frac{s_h}{\rho_h}\right) + \operatorname{sinc}\left(\theta_h + \frac{s_h}{\rho_h}\right) \nabla_{\theta_h} V_d(\mathbf{q}) + k\operatorname{sinc}\left(\theta_h + \frac{s_h}{\rho_h}\right) \nabla_{s_h} V_d(\mathbf{q}) = 0, \tag{70}$$

where $c_v = m_o g \dfrac{\rho_h + \rho_o}{\rho_h}$ is a positive constant parameter. A solution for (70) is given by

$$V_d(\mathbf{q}) = \frac{c_v \theta_h^2(\rho_h - k) + 2c_v\theta_h s_h}{2\rho_h} + f(s_h - k\theta_h), \tag{71}$$

where $f(\cdot)$ is a function to be chosen. To satisfy **C**.2, $f(\cdot)$ is selected such that the desired potential function (71) becomes

$$V_d(\mathbf{q}) = \frac{c_v \theta_h^2(\rho_h - k) + 2c_v\theta_h s_h}{2\rho_h} + k_f(s_h - k\theta_h)^2, \tag{72}$$

where $k_f \in \mathbb{R}$ is a controller gain.

To verify that $\mathbf{q}^\star$ is a minimum for (72), the Jacobian and the Hessian of the potential function are computed. The Jacobian is

$$\nabla V_d(\mathbf{q}) = \begin{bmatrix} \dfrac{c_v(-k\theta_h + \theta_h\rho_h + s_h)}{\rho_h} + 2kk_f(k\theta_h - s_h) \\ \dfrac{c_v\theta_h}{\rho_h} - 2kk_f\theta_h + 2k_f s_h \end{bmatrix}, \tag{73}$$

where it is possible to verify that $\nabla V_d(\mathbf{q})$ is zero at $\mathbf{q}^\star$. Also, the Hessian of $V_d(\mathbf{q})$ is

$$\nabla^2 V_d(\mathbf{q}) = \begin{bmatrix} c_v + 2k^2k_f - \dfrac{c_v k}{\rho_h} & -2kk_f + \dfrac{c_v}{\rho_h} \\ -2kk_f + \dfrac{c_v}{\rho_h} & 2k_f \end{bmatrix}. \tag{74}$$

It is possible to verify that $\nabla^2 V_d(\mathbf{q})$ is positive definite at the desired equilibrium $\mathbf{q}^\star$ if $k > -\rho_h$ and $k_f > \dfrac{c_v}{2\rho_h(k + \rho_h)}$. These conditions on the Jacobian and Hessian of the desired potential function ensure that $V_d(\mathbf{q})$ has a minimum at the desired equilibrium $\mathbf{q}^\star$.

Following the design procedure sketched out in Sect. 4.3.2, the inequality (56) must be solved. Using the functions $\alpha(\mathbf{q})$ and $\beta(\mathbf{q})$ selected above, equation (56) yields

$$m_{12}(s_h) + km_{22} < 0, \tag{75}$$

which has the straightforward solution $k < -\dfrac{m_{12}(s_h)}{m_{22}}$. Since it is possible to verify that $\rho_h > \dfrac{(c_{b3} - c_{b4})}{m_{22}}$ and together with the previous condition $k > -\rho_h$, then the gain k must satisfy $-\rho_h < k < -\dfrac{c_{b3} - c_{b4}}{m_{22}}$.

Finally, the entries of $\mathbf{M}_d(\mathbf{q})$ are computed as in (53) and (54)

$$a_{12}(\mathbf{q}) = -\frac{\text{sinc}\left(\theta_h + \dfrac{s_h}{\rho_h}\right)(m_{11}(s_h) + km_{12}(s_h))}{\Delta}, \quad (76)$$

$$a_{22}(\mathbf{q}) = -\frac{\text{sinc}\left(\theta_h + \dfrac{s_h}{\rho_h}\right)(m_{12}(s_h) + km_{22})}{\Delta},$$

while $a_{11}(\mathbf{q})$ is taken as in (55). The desired mass matrix is thus positive definite and it can be written as

$$\mathbf{M}_d(\mathbf{q}) = \begin{bmatrix} \Delta a_{11} & -(m_{11} + km_{12})\text{sinc}\left(\theta_h + \dfrac{s_h}{\rho_h}\right) \\ -(m_{11} + km_{12})\text{sinc}\left(\theta_h + \dfrac{s_h}{\rho_h}\right) & -(m_{12} + km_{22}\text{sinc}\left(\theta_h + \dfrac{s_h}{\rho_h}\right)) \end{bmatrix}. \quad (77)$$

The KE-ME (24) is satisfied using (60), while the IDA-PBC control law is computed from (28). The controller has been implemented in an real setup and the experiments show a satisfactory performance of the closed loop. The results can be found in [3].

5 Discussion and Conclusion

This chapter investigated the control design for nonprehensile planar rolling manipulation using FLin techniques, PBC methods, and pH theory. The dynamic model of nonprehensile holonomic rolling manipulation systems was presented in its general form and then used for control design purposes.

The first class of controllers presented in this chapter was obtained using FLin, thus requiring the cancellation of all system nonlinearities. Upon certain conditions given by the shapes of the hand the object in contact with it, it has been possible to found a diffeomorphism rendering the original system in a normal form (i.e., a chain of integrators). Therefore, any linear technique can be, in principle, employed. In this chapter, the EFL technique was employed on two benchmark systems: the DoD and the BnB.

The second class of controllers developed in this chapter was designed using IDA-PBC methods and pH dynamics. Two different designs were followed within this framework. First, the classical IDA-PBC procedure to stabilise mechanical systems was used to obtain a controller for the DoD. Also, an integral action controller was added in the loop to robustify the control system against disturbances. It was shown that the closed loop dynamics preserve the pH structure and thus its intrinsic passivity properties. This design requires solving a set of PDEs, which results from the so-called matching equation. To simplify solving PDEs, an alternative procedure was used to design the second class of controllers. This procedure was successfully applied to the BnB and the eccentric DoD systems.

The effectiveness of the controllers presented in this chapter was verified by simulations and experiments on real physical set-ups, and the results reported in [1–3, 5]. These positive results show that the set of methods presented in this chapter is suitable for controlling nonprehensile planar rolling manipulation systems and provides a potential framework for the control design of general dynamic robotic manipulation tasks. Future research will aim at the development of a framework for a general class of dynamic manipulations.

References

1. A. Donaire, J.G. Romero, R. Ortega, B. Siciliano, M. Crespo, Robust IDA-PBC for underactuated mechanical systems subject to matched disturbances. Int. J. Robust Nonlinear Control **27**(6), 1000–1016 (2017)
2. V. Lippiello, F. Ruggiero, B. Siciliano, The effects of shapes in input-state linearization for stabilization of nonprehensile planar rolling dynamic manipulation. IEEE Robot. Autom. Lett. **1**(1), 492–499 (2016)
3. D. Serra, F. Ruggiero, A. Donaire, L.R. Buonocore, V. Lippiello, B. Siciliano, Control of nonprehensile planar rolling manipulation: A passivity-based approach. IEEE Trans. Rob. **35**(2), 317–329 (2019)
4. D. Prattichizzo, J.C. Trinkle, Grasping, in *Springer Handbook of Robotics*, ed. by B. Siciliano, O. Khatib, 2nd edn. (Springer Science & Business Media, second edition, 2016), pp. 955–988
5. A. Donaire, F. Ruggiero, L.R. Buonocore, V. Lippiello, B. Siciliano, Passivity-based control for a rolling-balancing system: The nonprehensile disk-on-disk. IEEE Trans. Control Syst. Technol. **25**(6), 2135–2142 (2017)
6. J.-C. Ryu, F. Ruggiero, K.M. Lynch, Control of nonprehensile rolling manipulation: Balancing a disk on a disk. IEEE Trans. Rob. **29**(5), 1152–1161 (2013)
7. J. Hauser, S. Sastry, P. Kokotovic, Nonlinear control via approximate input-output linearization: The ball and beam example. IEEE Trans. Autom. Control **37**(3), 392–398 (1992)
8. R. Ortega, M.W. Spong, F. Gómez-Estern, G. Blankenstein, Stabilization of a class of underactuated mechanical systems via interconnection and damping assignment. IEEE Trans. Autom. Control **47**(8), 1218–1233 (2002)
9. M. Cefalo, L. Lanari, G. Oriolo, Energy-based control of the butterfly robot, in *International IFAC Symposium on Robot Control*, pp. 1–6 (2006)
10. K.M. Lynch, N. Shiroma, H. Arai, K. Tanie, The roles of shape and motion in dynamic manipulation: The butterfly example, in *1998 IEEE International Conference on Robotics and Automation*, pp. 1958–1963 (1998)
11. M. Surov, A. Shiriaev, L. Freidovich, S. Gusev, L. Paramonov, Case study in non-prehensile manipulation: Planning perpetual rotations for "Butterfly" robot, in *2015 IEEE International Conference on Robotics and Automation*, pp. 1484–1489 (2015)
12. B. Siciliano, L. Sciavicco, L. Villani, G. Oriolo, *Robotics: Modelling, Planning and Control* (Springer, London, UK, 2009)
13. C. Lanczos, *The Variational Principles of Mechanics* (Univ. of Toronto Press, 1960)
14. V. Hagenmeyer, S. Streif, M. Zeitz, Flatness-based feedforward and feedback linearisation of the ball & plate lab experiment, in *IFAC Symposium on Nonlinear Control Systems* (2004)
15. R. Ortega, A. Donaire, J.G. Romero, Passivity-based control of mechanical systems, in *Feedback Stabilization of Controlled Dynamical Systems*, pp. 167–199 (Springer, 2017)
16. H.K. Khalil, *Nonlinear Systems* (Prentice Hall, 2002)

17. A.J. van der Schaft, *L2-gain and Passivity Techniques in Nonlinear Control* (Springer, 2000)
18. M. Ryalat, D.S. Laila, A simplified IDA-PBC design for underactuated mechanical systems with applications. Eur. J. Control. **27**, 1–16 (2016)
19. P. Arpenti, F. Ruggiero, V. Lippiello, A constructive methodology for the IDA-PBC of underactuated 2-DoF mechanical systems with explicit solution of PDEs. Int. J. Control Autom. Syst. In press (2021)

Nonholonomic Rolling Nonprehensile Manipulation Primitive

Alejandro Gutierrez-Giles, Aykut C. Satici, Alejandro Donaire, Fabio Ruggiero, Vincenzo Lippiello, and Bruno Siciliano

Abstract This chapter reviews the problem of nonholonomic rolling in nonprehensile manipulation tasks through two challenging and illustrative examples: the robotic hula-hoop and the ballbot system. The hula-hoop consists of an actuated stick and an unactuated hoop. First, the corresponding kinematic model is derived. Second, the dynamic model is derived through the Lagrange-D'Alembert equations. Then a control strategy is designed to rotate the hoop at some desired constant speed whereas positioning it over a desired point on the stick surface. A stability analysis, which guarantees ultimate boundedness of all signals of interest, is carried out. The ballbot is an underactuated and nonholonomic constrained mobile robot whose upward equilibrium point must be stabilised by active controls. Coordinate-invariant equations of motion are derived for the ballbot. The linearised equations of motion are then derived, followed by the detailed controllability analysis. Excluding the rotary degree of freedom of the ball in the inertial vertical direction, the linear system turns out to be controllable. It follows that the nonlinear system is locally controllable, and a proportional-derivative type controller is designed to locally exponentially stabilise

A. Gutierrez-Giles
CECAv-UNAM, Av. Universidad 3000, Ciudad Universitaria, CDMX 04510, Mexico
e-mail: alejandro_giles@cecav.unam.mx

A. C. Satici
Boise State University, 1910 University Drive, Boise 83709, USA
e-mail: aykutsatici@boisestate.edu

A. Donaire
The University of Newcastle, University Drive, Callaghan, NSW 2308, Australia
e-mail: alejandro.donaire@newcastle.edu.eu

F. Ruggiero (✉) · V. Lippiello · B. Siciliano
Department of Electrical Engineering and Information Technology, PRISMA Lab, CREATE Consortium & University of Naples Federico II, Via Claudio 21, 80125 Naples, Italy
e-mail: fabio.ruggiero@unina.it

V. Lippiello
e-mail: vincenzo.lippiello@unina.it

B. Siciliano
e-mail: bruno.siciliano@unina.it

B. Siciliano and F. Ruggiero (eds.), *Robot Dynamic Manipulation*,
Springer Tracts in Advanced Robotics 144,
https://doi.org/10.1007/978-3-030-93290-9_7

the upward equilibrium point and the translation of the ball. Numerical simulations for these two examples illustrate the effectiveness of the proposed methods. This chapter is based on the works presented in [1–4].

1 Brief Introduction

In the previous chapter, the manipulation problem of nonprehensile planar rolling systems has been addressed. In planar rolling systems, the kinematic constraint associated with the pure rolling motion (only rolling, no sliding) is integrable. Therefore, the system is holonomic. In this chapter, the focus is shifted towards 3D systems in which the pure rolling constraint is not integrable. A system with a set of kinematic constraints, expressed in the so-called Pfaffian form [5], in which at least one of these constraints is not integrable is said to be a nonholonomic system. A nonholonomic constraint implies that some motion directions are instantaneously not allowed.

Therefore, this chapter tackles controlling an object manipulated through nonholonomic rolling motions without form or force closure grasps. The ball-on-plate gives the most illustrative example of it [6]. The primary objective is to steer the free-rolling sphere toward the desired position and/or orientation or along a desired path. It is worth pointing out that most of the works addressing the ball-and-plate application consider the prehensile case obtained by caging the sphere between two plates [7–9]. In such a configuration, one plate is actuated while the other one is fixed. Dismissing the fixed plate, the ball-and-plate application is addressed as a nonprehensile rolling manipulation system in which the sole supporting moving plate controls the ball. Therefore, position control of a basketball on a plate is tackled in [10]. An analysis of the kinematics of rolling, based on a coordinate-free approach, considering the cases of either pure rolling or twist-rolling, is proposed in [11]. An extension of the DoD to the 3D case is given by the stabilization of a ball free to roll on an actuated sphere in full gravity [12, 13].

This chapter first presents a control design for the hula-hoop task, which belongs to the nonholonomic rolling nonprehensile manipulation primitive. From a robotic point of view, this can be schematized through a hoop freely rolling around an actuated pole. A first mathematical derivation was proposed in [14] without taking correctly into account the nonholonomic constraints. This issue is overcome in [1], in which a control approach without velocity measurement is proposed. A formal mathematical analysis that guarantees ultimate boundedness of all coordinates is developed in [2].

Afterwards, this chapter addresses the control design for the ballbot. The ballbot is an underactuated, nonholonomically constrained, mobile robot whose upward equilibrium point has to be stabilised by active controls [3, 4]. The ballbot is thus a spherical robot with a turret that is actively controlled on a plane. It is very similar to the ball-on-plate problem but, in this case, the actuation is in the sphere and not in the plate.

The outline of the chapter shows then the two examples depicted above with modelling, control design, and simulation results. Table 1 collects the main symbols used in the chapter.

2 The Hula-Hoop Problem

The task consists of spinning the hoop around a stick at some desired angular speed while simultaneously driving the hoop to an arbitrarily desired point over the pole surface. First, the kinematic model is derived. Second, the dynamic model is derived through the Lagrange–D'Alembert equations. Then a control strategy is designed to rotate the hoop at some desired constant speed whereas positioning it over a desired point on the stick surface. A stability analysis, which guarantees ultimate boundedness of all signals of interest, is carried out. Finally, numerical simulations illustrate the effectiveness of the proposed method.

2.1 Contact Kinematics

A draw of the hoop and pole system considered in this work is shown in Fig. 1, where the inertial, pole (hand), hoop (object), and contact Cartesian frames are displayed, which in the subsequent are denoted by $\mathcal{W}$, $\mathcal{H}$, $\mathcal{O}$, and $\mathcal{C}$, respectively. A given coordinate frame $\mathcal{X}$ consists of an origin vector $\mathbf{o}_x \in \mathbb{R}^3$, and three orthonormal unit vectors $\mathbf{x}_x, \mathbf{y}_x, \mathbf{z}_x \in \mathbb{R}^3$. The contact frame $\mathcal{C}$ is defined as follows: $\mathbf{o}_c$ is located at the contact point, $\mathbf{x}_c$ is in the line connecting the hoop centre with the contact point and pointing outwards the pole, $\mathbf{z}_c$ is normal to the hoop equatorial plane passing through $\mathbf{o}_c$, and $\mathbf{y}_c$ is chosen to form an orthonormal frame.

In this section, the following notation is adopted: $\mathbf{v}_i^j$ means that the vector $\mathbf{v}_i$ is referenced to the j frame. When a vector or matrix is referenced to the world coordinate frame the superscript $(\cdot)^w$ is obviated.

The next coordinates are defined to describe the pole surface:

- $\theta \in \mathbb{R}$: an angle from one arbitrarily defined point on the pole surface to the contact point, measured by taking $\mathbf{z}_h$ as the rotation axis.
- $z_o > 0$: a distance form the origin $\mathbf{o}_h$ to the contact point over the $\mathbf{z}_h$ axis.

Defining the vector $\boldsymbol{\alpha}_h = \begin{bmatrix}\theta \; z_0\end{bmatrix}^T \in \mathbb{R}^2$, the pole surface in local coordinates is described by

$$\mathbf{c}_h(\boldsymbol{\alpha}_h) = \begin{bmatrix} r_h c_\theta \; r_h s_\theta \; z_o \end{bmatrix}^T \in \mathbb{R}^3 , \tag{1}$$

where $c_\theta = \cos(\theta)$ and $s_\theta = \sin(\theta)$. The tangent vectors are computed as

Table 1 Main symbols used in this chapter (pole and hoop are referred to the hula hoop system, ball and top are referred to the ballbot system)

Definition	Symbol
Frame attached to the pole's and ball's CoM	$\mathcal{H}$
Frame attached to the hoop's and top's CoM	$\mathcal{O}$
Frame attached to the contact point between the hoop and the pole	$\mathcal{C}$
Radius of the pole and the ball	$r_h > 0$
Pole metric tensor	$\mathbf{M}_h \in \mathbb{R}^{2\times 2}$
Pole curvature tensor	$\mathbf{K}_h \in \mathbb{R}^{2\times 2}$
Pole torsion form	$\mathbf{T}_h \in \mathbb{R}^{1\times 2}$
Radius of the hoop	$l_o > 0$
Thickness of the hoop	$2r_o > 0$
Hoop metric tensor	$\mathbf{M}_o \in \mathbb{R}^{2\times 2}$
Hoop curvature tensor	$\mathbf{K}_o \in \mathbb{R}^{2\times 2}$
Hoop torsion form	$\mathbf{T}_o \in \mathbb{R}^{1\times 2}$
Relative curvature tensor	$\widetilde{\mathbf{K}}_h \in \mathbb{R}^{2\times 2}$
Relative angular velocities between the contact frames	$\omega_x, \omega_y\ in \mathbb{R}$
Pfaffian matrix	$\mathbf{A}_c \in \mathbb{R}^{3\times 5}$
Generalised coordinates of the pole	$\mathbf{q}_h \in \mathbb{R}^m$
Position of the hoop's centre in $\mathcal{H}$	$\mathbf{p}_o^h \in \mathbb{R}^3$
Mass of the hoop	$m_o > 0$
Mass of the pole	$m_h > 0$
Gravity acceleration	$g \simeq 9.81$ m/s^2
Hoop inertia tensor with respect to $\mathcal{H}$	$\mathbf{I}_o \in \mathbb{R}^{3\times 3}$
Rotation of $\mathcal{O}$ with respect to $\mathcal{W}$	$\mathbf{R}_{st} \in SO(3)$
Angular velocity of the hoop in $\mathcal{W}$	$\omega_o \in \mathbb{R}^3$
Standard unit vectors	$\mathbf{e}_2 = \begin{bmatrix} 0 & 1 & 0 \end{bmatrix}^T$, $\mathbf{e}_3 = \begin{bmatrix} 0 & 0 & 1 \end{bmatrix}^T$
Distance from the ball's CoM and the top's CoM	$l > 0$
Position of the ball with respect to $\mathcal{W}$	$\mathbf{p}_{sb} \in \mathbb{R}^3$
Angular velocity of the ball with respect to $\mathcal{W}$	$\omega_{sb} \in \mathbb{R}^3$
Rotation of $\mathcal{H}$ with respect to $\mathcal{W}$	$\mathbf{R}_{sb} \in SO(3)$
Angular velocity of the top with respect to $\mathcal{W}$	$\omega_{st} \in \mathbb{R}^3$
Linear velocity of the top with respect to $\mathcal{W}$	$\mathbf{v}_{st} \in \mathbb{R}^3$
Linear velocity of the ball with respect to $\mathcal{W}$	$\mathbf{v}_{sb} \in \mathbb{R}^3$
Angular velocity of the top with respect to the ball	$\omega_{bt} \in \mathbb{R}^3$
Position of the top with respect to $\mathcal{W}$	$\mathbf{p}_{st} \in \mathbb{R}^3$
Inertia tensor of the ball with respect to $\mathcal{W}$	$\mathbf{I}_b \in \mathbb{R}^{3\times 3}$
Inertia tensor of the top with respect to $\mathcal{W}$	$\mathbf{I}_t \in \mathbb{R}^{3\times 3}$
Mass of the ball	$m_b > 0$
Mass of the top	$m_t > 0$

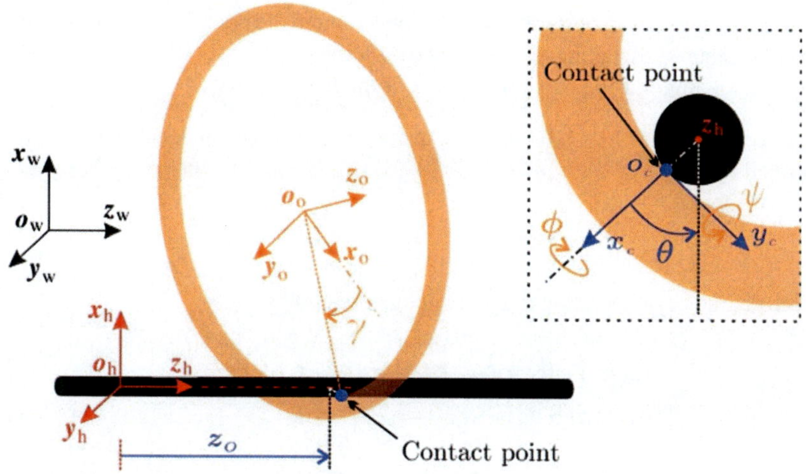

Fig. 1 Sketch of the pole and hoop system

$$\mathbf{c}_{h_u} = \begin{bmatrix} -r_h s_\theta \; r_h c_\theta \; 0 \end{bmatrix}^T \in \mathbb{R}^3 \tag{2}$$

$$\mathbf{c}_{h_v} = \begin{bmatrix} 0 \; 0 \; 1 \end{bmatrix}^T \in \mathbb{R}^3 . \tag{3}$$

It can be easily verified that $\|\mathbf{c}_{h_u}\| = r_h$ and $\|\mathbf{c}_{h_v}\| = 1$. The corresponding normal vector is given by

$$\mathbf{n}_h = \begin{bmatrix} c_\theta \; s_\theta \; 0 \end{bmatrix}^T \in \mathbb{R}^3 , \tag{4}$$

with partial derivatives

$$\mathbf{n}_{h_u} = \begin{bmatrix} -s_\theta \; c_\theta \; 0 \end{bmatrix}^T \in \mathbb{R}^3 \tag{5}$$

$$\mathbf{n}_{h_v} = \mathbf{0}_3 . \tag{6}$$

The second-order partial derivatives of $\mathbf{c}_{h_u}$, necessary to compute the torsion, are

$$\mathbf{c}_{h_{uu}} = \begin{bmatrix} -r_h c_\theta \; -r_h s_\theta \; 0 \end{bmatrix}^T \in \mathbb{R}^3 \tag{7}$$

$$\mathbf{c}_{h_{uv}} = \mathbf{0}_3 . \tag{8}$$

Therefore, following the definitions given in [15], the metric and curvature tensors and the torsion form of the pole surface are

$$\mathbf{M}_h = \begin{bmatrix} r_h & 0 \\ 0 & 1 \end{bmatrix} , \quad \mathbf{K}_h = \begin{bmatrix} 1/r_h & 0 \\ 0 & 0 \end{bmatrix} , \quad \mathbf{T}_h = \begin{bmatrix} 0 \; 0 \end{bmatrix} .$$

On the other hand, the following local coordinates for describing the hoop surface are defined:

- $\gamma \in \mathbb{R}$: an angle from one arbitrarily defined point in the hoop surface to the contact point measured by taking $\mathbf{z}_o$ as the rotation axis.
- $\psi \in \mathbb{R}$: an angle of the equatorial plane of the hoop over $\mathbf{y}_c$.

Collect $\boldsymbol{\alpha}_o = \begin{bmatrix}\gamma \ \psi\end{bmatrix}^T \in \mathbb{R}^2$ such that the hoop surface in local coordinates is described by

$$\mathbf{c}_o(\boldsymbol{\alpha}_o) = \begin{bmatrix}-(l_o - r_o c_\psi)c_\gamma \ (l_o - r_o c_\psi)s_\gamma \ -r_o s_\psi\end{bmatrix}^T \in \mathbb{R}^3. \tag{9}$$

Therefore, the corresponding tangent vectors are computed as

$$\mathbf{c}_{o_u} = \begin{bmatrix}-(l_o - r_o c_\psi)s_\gamma \ (l_o - r_o c_\psi)c_\gamma \ 0\end{bmatrix}^T \in \mathbb{R}^3 \tag{10}$$

$$\mathbf{c}_{f_v} = \begin{bmatrix}r_o c_\gamma s_\psi \ r_o s_\gamma s_\psi \ -r_o c_\psi\end{bmatrix}^T \in \mathbb{R}^3. \tag{11}$$

In such case, $\|\mathbf{c}_{o_u}\| = l_o - r_o c_\psi$ and $\|\mathbf{c}_{o_v}\| = r_o$. The normal vector is given by

$$\mathbf{n}_o = \begin{bmatrix}-c_\gamma c_\psi \ -s_\gamma c_\psi \ -s_\psi\end{bmatrix}^T \in \mathbb{R}^3, \tag{12}$$

with partial derivatives

$$\mathbf{n}_{o_u} = \begin{bmatrix}s_\gamma c_\psi \ -c_\gamma c_\psi \ 0\end{bmatrix}^T \in \mathbb{R}^3 \tag{13}$$

$$\mathbf{n}_{o_v} = \begin{bmatrix}c_\gamma s_\psi \ s_\gamma s_\psi \ -c_\psi\end{bmatrix}^T \in \mathbb{R}^3. \tag{14}$$

The second order partial derivatives of $\mathbf{c}_{o_u}$ are

$$\mathbf{c}_{o_{uu}} = \begin{bmatrix}-(l_o - r_o c_\psi)c_\gamma \ -(l_o - r_o c_\psi)s_\gamma \ 0\end{bmatrix}^T \in \mathbb{R}^3 \tag{15}$$

$$\mathbf{c}_{o_{uv}} = \begin{bmatrix}-r_o s_\gamma s_\psi \ r_o c_\gamma s_\psi \ 0\end{bmatrix}^T \in \mathbb{R}^3. \tag{16}$$

Regarding the hoop's surface, the metric and curvature tensors and the torsion form are given by

$$\mathbf{M}_o = \begin{bmatrix}l_o - r_o c_\psi & 0 \\ 0 & r_o\end{bmatrix}, \quad \mathbf{K}_o = \begin{bmatrix}-c_\psi / \left(l_o - r_o c_\psi\right) & 0 \\ 0 & 1/r_o\end{bmatrix},$$
$$\mathbf{T}_o = \begin{bmatrix}-s_\psi / \left(l_o - r_o c_\psi\right) \ 0\end{bmatrix}.$$

An additional coordinate, necessary to compute the contact kinematics, $\phi \in \mathbb{R}$ is the angle form $\mathbf{c}_{hu}$ to $\mathbf{c}_{ou}$, measured over the $\mathbf{x}_c$ axis. Thus, the relative curvature is computed to be

$$\widetilde{\mathbf{K}}_h = \frac{1}{r_h}\begin{bmatrix}c_\phi^2 & -s_\phi c_\phi \\ -s_\phi c_\phi & s_\phi^2\end{bmatrix}. \tag{17}$$

Now, by combining both surface geometric parameters, the Montana's equations in terms of the relative velocities between the contact frames, assuming a pure rolling, are given by

$$\begin{bmatrix} \dot{\gamma} \\ \dot{\psi} \end{bmatrix} = \frac{1}{l_o c_\phi^2 - (r_h + r_o)c_\psi} \begin{bmatrix} r_h + r_o s_\phi^2 & r_o s_\phi c_\phi \\ (l_o - r_o c_\psi) s_\phi c_\phi & (l_o - r_o c_\psi) c_\phi^2 - r_h c_\psi \end{bmatrix} \begin{bmatrix} -\omega_y \\ \omega_x \end{bmatrix} \tag{18}$$

$$\begin{bmatrix} \dot{\theta} \\ \dot{z}_o \end{bmatrix} = \frac{1}{l_o c_\phi^2 - (r_h + r_o)c_\psi} \begin{bmatrix} (l_o - r_o c_\psi) c_\phi & r_o s_\phi c_\psi \\ -(l_o - r_o c_\psi)(r_h + r_o) s_\phi & -r_o c_\phi \left(l_o - (r_h + r_o) c_\psi\right) \end{bmatrix} \begin{bmatrix} -\omega_y \\ \omega_x \end{bmatrix} \tag{19}$$

$$\dot{\phi} = \frac{(r_h + r_o s_\phi^2) s_\psi \omega_y - r_o s_\phi c_\phi s_\psi \omega_x}{l_o c_\phi^2 - (r_h + r_o) c_\psi} . \tag{20}$$

Let the contact coordinates vector be defined by

$$\mathbf{q}_c = \left[\gamma \ \psi \ \theta \ z_o \ \phi\right]^T \in \mathbb{R}^5 . \tag{21}$$

Thus, the kinematic equations (18)–(20) can be rewritten as

$$\dot{\mathbf{q}}_c = \mathbf{g}_1 \omega_x + \mathbf{g}_2 \omega_y , \tag{22}$$

where

$$\mathbf{g}_1 = \frac{1}{l_o c_\phi^2 - (r_o + r_o) c_\psi} \begin{bmatrix} r_o s_\phi c_\phi \\ (l_o - r_o c_\psi) c_\phi^2 - r_o c_\psi \\ r_o s_\phi c_\psi \\ -r_o c_\phi \left(l_o - (r_o + r_o) c_\psi\right) \\ -r_o s_\phi c_\phi s_\psi \end{bmatrix} \in \mathbb{R}^5 \tag{23}$$

$$\mathbf{g}_2 = \frac{1}{l_o c_\phi^2 - (r_o + r_o) c_\psi} \begin{bmatrix} -(r_o + r_o s_\phi^2) \\ -(l_o - r_o c_\psi) s_\phi c_\phi \\ -(l_o - r_o c_\psi) c_\phi \\ (l_o - r_o c_\psi)(r_o + r_o) s_\phi \\ (r_o + r_o s_\phi^2) s_\psi \end{bmatrix} \in \mathbb{R}^5 . \tag{24}$$

Assume that the hoop thickness can be neglected, i.e., $r_o = 0$. Therefore, by choosing a basis for the left null space of $\mathbf{G} = \left[\mathbf{g}_1 \ \mathbf{g}_2\right]$, a set of Pffafian constraints [16, p. 320] can be constructed as follows

$$\mathbf{A}_c(\mathbf{q}_c)\dot{\mathbf{q}}_c = \mathbf{0}_3 , \tag{25}$$

where

$$\mathbf{A}_c(\mathbf{q}_c) = \begin{bmatrix} -l_o c_\phi / r_o & 0 & 1 & 0 & 0 \\ s_\psi & 0 & 0 & 1 & 0 \\ l_o s_\phi & 0 & 0 & 0 & 1 \end{bmatrix} . \tag{26}$$

At this point, an arbitrary number of degrees of freedom for the pole is considered. The generalised coordinates of the pole express its pose (position and orientation) in the Cartesian space. As one of the goals of this work is to have the best controllability properties with the least actuated coordinates, the configuration space for the pole will be chosen after the controllability analysis given later in this section. Now, define the vectors $\mathbf{q}_r = \begin{bmatrix}\gamma \ \psi \ \mathbf{q}_h\end{bmatrix}^T \in \mathbb{R}^{m+2}$, $\mathbf{q}_s = \begin{bmatrix}\theta \ \phi \ z_o\end{bmatrix}^T \in \mathbb{R}^3$, and $\mathbf{q} = \begin{bmatrix}\mathbf{q}_r \ \mathbf{q}_s\end{bmatrix}^T \in \mathbb{R}^{m+5}$. Then, the constraints (26) can be written as

$$\mathbf{A}_r(\mathbf{q})\dot{\mathbf{q}}_r + \dot{\mathbf{q}}_s = \mathbf{0}\,, \tag{27}$$

where

$$\mathbf{A}_r(\mathbf{q}) = \begin{bmatrix} -l_o c_\phi / r_o & 0\ 0 \cdots 0 \\ s_\psi & 0\ 0 \cdots 0 \\ l_o s_\phi & 0\ 0 \cdots 0 \end{bmatrix}. \tag{28}$$

2.2 Dynamic Model

The vector $\mathbf{p}_o^h$ can be expressed in terms of the generalised coordinates is expressed as follows

$$\mathbf{p}_o^h = \begin{bmatrix} -l_o c_\theta c_\psi + r_o c_\theta + l_o s_\theta s_\phi s_\psi \\ r_o s_\theta - l_o c_\psi s_\theta - l_o c_\theta s_\phi s_\psi \\ l_o s_\psi + z_o \end{bmatrix}. \tag{29}$$

Given the expressions $\mathbf{I}_o = \text{diag}\left\{\frac{1}{2}m_o l_o^2, \frac{1}{2}m_o l_o^2, m_o l_o^2\right\}$ and

$$\mathbf{R}_o = \begin{bmatrix} c_\psi c_\theta - s_\phi s_\psi s_\theta & -c_\phi s_\theta & c_\theta s_\psi + c_\psi s_\phi s_\theta \\ c_\theta s_\phi s_\psi + c_\psi s_\theta & c_\phi c_\theta & -c_\psi c_\theta s_\phi + s_\psi s_\theta \\ -c_\phi s_\psi & s_\phi & c_\phi c_\psi \end{bmatrix}, \tag{30}$$

the hoop angular velocity can be obtained from $\mathbf{R}_{st}$ through $\mathbf{S}(\boldsymbol{\omega}_o) = \dot{\mathbf{R}}_o \mathbf{R}_o^T$, where $\mathbf{S}(\boldsymbol{\omega}_o)$ is a well–known skew symmetric matrix constructed from $\boldsymbol{\omega}_o$. The angular velocity can be computed from

$$\boldsymbol{\omega}_o = \begin{bmatrix} c_\theta \dot{\phi} - c_\phi s_\theta \dot{\psi} \\ s_\theta \dot{\phi} + c_\phi c_\theta \dot{\psi} \\ s_\phi \dot{\psi} + \dot{\theta} \end{bmatrix}. \tag{31}$$

The Lagrange–d'Alembert equations, subject to the Pfaffian constraints above, are given by

$$\left(\frac{\mathrm{d}}{\mathrm{d}t}\frac{\partial \mathcal{L}}{\partial \dot{\mathbf{q}}_r} - \frac{\partial \mathcal{L}}{\partial \mathbf{q}_r} - \boldsymbol{\Upsilon}_r\right) - \mathbf{A}_r^T(\mathbf{q})\left(\frac{\mathrm{d}}{\mathrm{d}t}\frac{\partial \mathcal{L}}{\partial \dot{\mathbf{q}}_s} - \frac{\partial \mathcal{L}}{\partial \mathbf{q}_s}\right) = \mathbf{0}_{\mathrm{m}+2}\,, \tag{32}$$

where

$$\boldsymbol{\Upsilon}_r = \begin{bmatrix} \mathbf{0}_2 \\ \mathbf{u} \end{bmatrix}, \tag{33}$$

with $\mathbf{u} \in \mathbb{R}^m$ the vector of generalised forces acting on the pole and $\mathcal{L} \in \mathbb{R}$ the system Lagrangian available in [1].

The vectors $\dot{\mathbf{q}}_s$ and $\ddot{\mathbf{q}}_s$ can be eliminated by following the procedure described in [16, Chap. 6], which results in the following dynamic model

$$\mathbf{B}_o(\mathbf{q})\ddot{\mathbf{q}}_o + \mathbf{c}_o(\mathbf{q}, \dot{\mathbf{q}}_o) + \mathbf{T}_o(\mathbf{q})\ddot{\mathbf{q}}_h = \mathbf{0}, \tag{34a}$$
$$\mathbf{B}_h(\mathbf{q})\ddot{\mathbf{q}}_h + \mathbf{c}_h(\mathbf{q}, \dot{\mathbf{q}}_o) + \mathbf{T}_o^T(\mathbf{q})\ddot{\mathbf{q}}_o = \mathbf{u}, \tag{34b}$$

where $\mathbf{q}_o = \begin{bmatrix}\gamma \ \psi\end{bmatrix}^T \in \mathbb{R}^2$, $\mathbf{B}_o(\mathbf{q}) \in \mathbb{R}^{2\times 2}$ is the inertia matrix of the hoop, $\mathbf{B}_h(\mathbf{q}) \in \mathbb{R}^{m\times m}$ is the inertia matrix of the pole, $\mathbf{c}_o(\mathbf{q}, \dot{\mathbf{q}}_o) \in \mathbb{R}^2$ is the Coriolis term related to hoop, $\mathbf{c}_h(\mathbf{q}, \dot{\mathbf{q}}_o) \in \mathbb{R}^m$ is the Coriols term related to the pole, and $\mathbf{T}_o(\mathbf{q}) \in \mathbb{R}^{2\times m}$ is the inertia coupling matrix whose effects are analysed below. Notice that $\mathbf{B}_o(\mathbf{q})$ is always invertible. Thus, Eq. (34b) can be solved for $\ddot{\mathbf{q}}_h$ and substituted into (34a) to obtain

$$\mathbf{M}_r(\mathbf{q})\ddot{\mathbf{q}}_o + \mathbf{c}_r(\mathbf{q}, \dot{\mathbf{q}}_o) = \mathbf{T}_r(\mathbf{q})\mathbf{u}, \tag{35}$$

where

$$\mathbf{M}_r(\mathbf{q}) = \mathbf{B}_o(\mathbf{q}) - \mathbf{T}_o(\mathbf{q})\mathbf{B}_h^{-1}(\mathbf{q})\mathbf{T}_o^T(\mathbf{q}) \tag{36}$$
$$\mathbf{c}_r(\mathbf{q}, \dot{\mathbf{q}}_o) = \mathbf{c}_o(\mathbf{q}, \dot{\mathbf{q}}_o) - \mathbf{T}_o(\mathbf{q})\mathbf{B}_h^{-1}(\mathbf{q})\mathbf{c}_h(\mathbf{q}, \dot{\mathbf{q}}_o) \tag{37}$$
$$\mathbf{T}_r(\mathbf{q}) = -\mathbf{T}_o(\mathbf{q})\mathbf{B}_h^{-1}(\mathbf{q}). \tag{38}$$

The dynamic model (35) can be further simplified to obtain

$$\ddot{\mathbf{q}}_o = \mathbf{f}(\mathbf{q}, \dot{\mathbf{q}}_o) + \mathbf{g}(\mathbf{q})\mathbf{u}, \tag{39}$$

with the definitions $\mathbf{f}(\mathbf{q}, \dot{\mathbf{q}}_o) = -\mathbf{M}_r^{-1}(\mathbf{q})\mathbf{c}_r(\mathbf{q}, \dot{\mathbf{q}}_o)$, and $\mathbf{g}(\mathbf{q}) = \mathbf{M}_r^{-1}(\mathbf{q})\mathbf{T}_r(\mathbf{q})$. The model (34) represents the dynamics of the hula-hoop system along with the nonholonomic constraints (27), which are equivalent to

$$\dot{\theta} = \frac{l_o c_\phi}{r_h}\dot{\gamma} \tag{40}$$
$$\dot{z}_o = -l_o s_\phi \dot{\gamma} \tag{41}$$
$$\dot{\phi} = -s_\psi \dot{\gamma}. \tag{42}$$

The inertial coupling matrix plays a crucial role for underactuated mechanical systems. In the present case, if $rank(\mathbf{T}_o(\mathbf{q})) = 2, \forall \mathbf{q}$, the underactuated system is said to be *strong inertially coupled* [17]. Whenever the mechanical system is strong inertially coupled, the Penrose's right pseudo-inverse matrix

$$\mathbf{T}_o^{+} = \mathbf{T}_o^T \left(\mathbf{T}_o \mathbf{T}_o^T\right)^{-1} \tag{43}$$

is well defined and the following orthogonal projection matrices can be constructed

$$\mathbf{P}_o = \mathbf{T}_o^{+}\mathbf{T}_o \in \mathbb{R}^{m\times m} \tag{44}$$

$$\mathbf{Q}_o = \mathbf{I}_m - \mathbf{P}_o \in \mathbb{R}^{m\times m}\,. \tag{45}$$

Notice that $\mathbf{P}_o$ projects every $\mathbb{R}^m$-vector onto the rank space of $\mathbf{T}_o$. Conversely, $\mathbf{Q}_o$ makes the projection into the null space of $\mathbf{T}_o$. It is straightforward to verify that the following relations hold: $\mathbf{P}_o\mathbf{T}_o^T = \mathbf{T}_o^T$, $\mathbf{T}_o\mathbf{P}_o = \mathbf{T}_o$, $\mathbf{Q}_o\mathbf{T}_o^T = \mathbf{O}_{2\times m}$, and $\mathbf{T}_o\mathbf{Q}_o = \mathbf{O}_2$.

2.2.1 Controllability Analysis

The controllability of the system (34a)–(34b) depends on the configuration of the pole. Several configurations of interest, namely (i) two rotations ($m = 2$), (ii) three translations ($m = 3$), (iii) two rotations plus two translations ($m = 4$), and (iv) three rotations plus three translations ($m = 6$), have been analysed in this work with the aid of a symbolic computing software (*Wolfram Mathematica*[1]). The corresponding dynamic model is strong inertially coupled for all the configurations mentioned above. Therefore, the main motivation for the comparison is to find the configuration with fewer degrees of freedom having the best controllability properties. The conclusions below are valid for all the cases mentioned.

For the model (34a)–(34b) the gravity torque of the underactuated part is not constant. The inertia matrix depends on the unactuated variables, so it never satisfies the structural necessary and sufficient conditions given in [18], and the nonholonomic constraints are of the *second-order* kind. As a consequence, the dynamic system is *strongly accessible* [19]. That means that, in principle, every possible configuration can be reached. However, this strong accessibility property "is far from being sufficient for the existence of a feedback control which asymptotically stabilises the underactuated system" [19]. For the system (34a)–(34b), it turns out from [20] that the Brockett's necessary condition for the existence of a continuous asymptotically stabilizing feedback control law is equivalent to check if the image of

$$\mathbf{B}_o^{-1}(\mathbf{q})\mathbf{c}_o(\mathbf{q}, \dot{\mathbf{q}}_r) \tag{46}$$

contains a neighbourhood of the origin in $\mathbb{R}^2$. This condition is satisfied by all the case studies, although it does not imply that there exists such a control law.

The so-called *small time local controllability* (STLC) is a stronger notion of controllability, which guarantees the existence of a piece-wise asymptotic stabilizing feedback control law [21]. The STLC property also guarantees the existence of an asymptotic stabilizing continuous time-periodic controller [22]. A sufficient condition to check the STLC property for mechanical systems is given in [19]. Unfortu-

[1] https://www.wolfram.com.

nately, this condition is not met by any of the case studies, and thus no conclusion can be made about the STLC property for the system (34a)–(34b). A necessary and sufficient condition for the STLC is given in [23]. Regrettably, this condition is much more challenging to check, even with the help of symbolic computing software.

Finally, the controller design and the dynamic analysis can be simplified by transforming the system into a *normal form* as proposed in [24]. Once again, this sufficient condition is not met by any of the cases under consideration.

2.3 Controller Design and Stability Analysis

The control objective is to spin the hoop at a desired angular velocity $\dot{\gamma}_d > 0$, while simultaneously driving it to the desired position $z_{od} \in \mathbb{R}$ over the pole surface, and maintaining it perpendicular to the pole. A design of a feedback model-based control for (34a)–(42) is a challenging problem from the control point of view. Some of the main difficulties are listed below.

- The kinematic constraints (40)–(42) are completely nonholonomic [16, p. 320]. In addition, the relative grow vector of the related control system is (2, 1, 2), and then it cannot be transformed into a chained form [25, p. 319].
- The model (34a)–(34b) is underactuated, and in the simplest case the shape coordinates are not actuated. Therefore, the result of [26] cannot be applied. In the remaining cases, the inertia matrices depend on both actuated and unactuated coordinates.
- The system trajectories must satisfy the nonholonomic constraints (40)–(42), hence it is not clear whether it is possible to induce a periodic motion for the unactuated coordinates satisfying the control objective stated above, which is a crucial step to apply the methodology of [27, 28].
- Because of the nonholonomic nature of the system, the control objective cannot be translated into a regulation problem, but it must be ensured tracking on the unactuated coordinates, for which the result of [29] does not apply.

Given the difficulties for designing a standard controller for the system under study, in the following development an *ad-hoc* strategy is employed to satisfy the control objective.

Let the input $\mathbf{u}$ be defined as

$$\mathbf{u} = \mathbf{B}_h \left(\mathbf{P}_o \mathbf{u}_P + \mathbf{Q}_o \mathbf{u}_Q \right) , \tag{47}$$

where $\mathbf{u}_P \in \mathbb{R}^m$ and $\mathbf{u}_Q \in \mathbb{R}^m$ are two independent inputs belonging to orthogonal subspaces. Taking into account (35), the noncollocated partial feedback linearisation (NPFL) [17] input can be defined as

$$\mathbf{u}_P = -\mathbf{T}_o^+ \left(\mathbf{c}_r + \mathbf{M}_r \mathbf{v}_P \right) , \tag{48}$$

with $\mathbf{v}_P \in \mathbb{R}^2$ a virtual control input to design. Because $\mathbf{M}_r$ is always full rank [30], it can be obtained

$$\ddot{\mathbf{q}}_o = \mathbf{v}_P\,, \tag{49}$$

with zero dynamics given in (70).

The control objective for the hoop coordinates consists in designing the input $\mathbf{v}_P$ to drive $\dot{\gamma} \to \dot{\gamma}_d$ and $(z_o, \psi, \phi) \to (z_{od}, 0, 0)$, while satisfying the nonholonomic constraints (40)–(42). For this purpose, first define

$$\boldsymbol{\xi} = \begin{bmatrix} \xi_1 \\ \xi_2 \\ \xi_3 \end{bmatrix} = \begin{bmatrix} z_o - z_{od} \\ -l_o s_\phi \\ l_o c_\phi s_\psi \end{bmatrix} \in \mathbb{R}^3, \tag{50}$$

whose time derivative are

$$\dot{\boldsymbol{\xi}} = \begin{bmatrix} \dot{\xi}_1 \\ \dot{\xi}_2 \\ \dot{\xi}_3 \end{bmatrix} = \begin{bmatrix} \xi_2 \dot{\gamma} \\ \xi_3 \dot{\gamma} \\ l_o s_\phi s_\psi^2 \dot{\gamma} + l_o c_\phi c_\psi \dot{\psi} \end{bmatrix}. \tag{51}$$

Next, consider the auxiliary definitions

$$\boldsymbol{\eta} = \begin{bmatrix} \eta_1 \\ \eta_2 \end{bmatrix} = \begin{bmatrix} \dot{\gamma} - \dot{\gamma}_d \\ \dot{\psi} - f_\psi \end{bmatrix} \in \mathbb{R}^2, \tag{52}$$

where

$$f_\psi = f_\psi(\psi, \phi, \dot{\gamma}, \boldsymbol{\xi}) = -\left(l_o s_\phi s_\psi^2 + \mathbf{k}_\xi^T \boldsymbol{\xi}\right) \frac{\dot{\gamma}}{l_o c_\phi c_\psi}\,, \tag{53}$$

defined for $-\pi/2 < \phi, \psi < \pi/2$, with $\mathbf{k}_\xi = \left[k_{\xi 1}\; k_{\xi 2}\; k_{\xi 3}\right]^T \in \mathbb{R}^3$ a vector of positive constant gains. Substituting (52) into (51) yields

$$\dot{\boldsymbol{\xi}} = \begin{bmatrix} \xi_2 \dot{\gamma}_d + \xi_2 \eta_1 \\ \xi_3 \dot{\gamma}_d + \xi_3 \eta_1 \\ -\mathbf{k}_\xi^T \boldsymbol{\xi} \dot{\gamma}_d - \mathbf{k}_\xi^T \boldsymbol{\xi} \eta_1 + l_o c_\phi c_\psi \eta_2 \end{bmatrix}. \tag{54}$$

In order to carry out the dynamic analysis, define the state

$$\boldsymbol{\zeta} = \left[\boldsymbol{\xi}^T\; \boldsymbol{\eta}^T\right]^T \in \mathbb{R}^5. \tag{55}$$

Proposition 2.1 ([2]) *Define the region $B_r = \{\boldsymbol{\zeta} : \|\boldsymbol{\zeta}\| \le l_o\}$ and let the control law be given by*

$$\mathbf{v}_P = \begin{bmatrix} v_{P1} \\ v_{P2} \end{bmatrix} = \begin{bmatrix} -k_{\eta 1} \eta_1 \\ \frac{\mathrm{d}}{\mathrm{d}t} f_\psi - k_{\eta 2} \eta_2 \end{bmatrix}, \tag{56}$$

where $k_{\eta 1}, k_{\eta 2} > 0$. There exists a bounded region $B_\delta \subset B_r$, and a combination of gains $\mathbf{k}_\xi$, $k_{\eta 1}$, k_{η_2} in (53) and (56), such that if the initial condition satisfies $\boldsymbol{\zeta}(t_0) \in B_\delta$, then $\boldsymbol{\zeta}(t) \in B_r$, $\forall t \geq t_0$. Furthermore, the system trajectories are ultimately bounded within an arbitrarily small region $B_\mu \subset B_r$, centred at the origin. □

Proof If $\boldsymbol{\eta} = \mathbf{0}_2$ in (54) one gets

$$\dot{\boldsymbol{\xi}} = \begin{bmatrix} 0 & 1 & 0 \\ 0 & 0 & 1 \\ -k_{\xi 1} & -k_{\xi 2} & -k_{\xi 3} \end{bmatrix} \boldsymbol{\xi} = \mathbf{A}_\xi \boldsymbol{\xi}, \tag{57}$$

that is a linear time-invariant system with $\mathbf{A}_\xi \in \mathbb{R}^{3\times 3}$ a Hurwitz matrix. A well-established result of linear control [31, Theorem 4.6] states that there exist two symmetric positive definite matrices $\mathbf{P}_\xi \in \mathbb{R}^{3\times 3}$ and $\mathbf{Q}_\xi \in \mathbb{R}^{3\times 3}$ satisfying

$$\mathbf{A}_\xi^T \mathbf{P}_\xi + \mathbf{P}_\xi \mathbf{A}_\xi = -\mathbf{Q}_\xi . \tag{58}$$

These matrices satisfy the bounds $\lambda_{Pm}\|\mathbf{x}\|^2 \leq \mathbf{x}^T\mathbf{P}_\xi\mathbf{x} \leq \lambda_{PM}\|\mathbf{x}\|^2$ and $\lambda_{Qm}\|\mathbf{x}\|^2 \leq \mathbf{x}^T\mathbf{Q}_\xi\mathbf{x} \leq \lambda_{QM}\|\mathbf{x}\|^2$ for every vector $\mathbf{x} \in \mathbb{R}^3$, with $0 < \lambda_{Pm} \leq \lambda_{PM}$, and $0 < \lambda_{Qm} \leq \lambda_{QM}$, where we denote by λ_{Hm} and λ_{HM} the minimum and the maximum eigenvalue, respectively, of a generic matrix $\mathbf{H} \in \mathbb{R}^{3\times 3}$. Next, let the scalar function

$$V = \boldsymbol{\xi}^T \mathbf{P}_\xi \boldsymbol{\xi} + \frac{1}{2}\boldsymbol{\eta}^T\boldsymbol{\eta} > 0, \tag{59}$$

satisfy the bounds

$$\lambda_{Vm}\|\boldsymbol{\zeta}\|^2 \leq V(\boldsymbol{\zeta}) \leq \lambda_{VM}\|\boldsymbol{\zeta}\|^2, \tag{60}$$

where $\lambda_{Vm} = \min\{1, \lambda_{Pm}\}$ and $\lambda_{VM} = \max\{1, \lambda_{PM}\}$. Define a region $B_\delta = \left\{\boldsymbol{\zeta} : \|\boldsymbol{\zeta}\| < \sqrt{\frac{\lambda_{Vm}}{\lambda_{VM}}} l_o\right\}$ and suppose that the initial condition satisfies $\boldsymbol{\zeta}(t_0) \in B_\delta$. Since $\frac{\lambda_{Vm}}{\lambda_{VM}} \leq 1$, B_δ is a subset of B_r.

By taking the time derivative of V along the system trajectories, one obtains

$$\dot{V} = -(\dot{\gamma}_d + \eta_1)\boldsymbol{\xi}^T\mathbf{Q}_\xi\boldsymbol{\xi} + 2\boldsymbol{\xi}^T\mathbf{P}_\xi\mathbf{b}\eta_2 + \eta_1\dot{\eta}_1 + \eta_2\dot{\eta}_2, \tag{61}$$

where $\mathbf{b} = \left[0\ 0\ l_o c_\phi c_\psi\right]^T \in \mathbb{R}^3$. Taking into account (49) and (52) and the control law (56) yields

$$\dot{V} = -\dot{\gamma}_d\boldsymbol{\xi}^T\mathbf{Q}_\xi\boldsymbol{\xi} - \eta_1\boldsymbol{\xi}^T\mathbf{Q}_\xi\boldsymbol{\xi} - k_{\eta 1}\eta_1^2 + 2\eta_2\boldsymbol{\xi}^T\mathbf{P}_\xi\mathbf{b} - k_{\eta 2}\eta_2^2 . \tag{62}$$

Within the set B_r, this function can be upper bounded by

$$\begin{aligned}\dot{V} \leq & -\dot{\gamma}_d \boldsymbol{\xi}^T \mathbf{Q}_\xi \boldsymbol{\xi} + \lambda_{QM} l_o^2 |\eta_1| - k_{\eta 1} |\eta_1|^2 \\ & + 2\lambda_{PM} l_o^2 |\eta_2| - k_{\eta 2} |\eta_2|^2 \\ \leq & -\dot{\gamma}_d \lambda_{Qm} \|\boldsymbol{\xi}\|^2 - |\eta_1| \left(k_{\eta 1}|\eta_1| - \lambda_{QM} l_o^2\right) \\ & - |\eta_2| \left(k_{\eta 2}|\eta_2| - 2\lambda_{PM} l_o^2\right) \\ = & -\dot{\gamma}_d \lambda_{Qm} \|\boldsymbol{\zeta}\|^2 - |\eta_1| \left(\left(k_{\eta 1} - \dot{\gamma}_d \lambda_{Qm}\right)|\eta_1| - \lambda_{QM} l_o^2\right) \\ & - |\eta_2| \left(\left(k_{\eta 2} - \dot{\gamma}_d \lambda_{Qm}\right)|\eta_2| - 2\lambda_{PM} l_o^2\right) , \end{aligned} \tag{63}$$

since $\|\boldsymbol{\zeta}\|^2 = \|\boldsymbol{\xi}\|^2 + |\eta_1|^2 + |\eta_2|^2$. It can be noticed that the term

$$-|\eta_1| \left(\left(k_{\eta 1} - \dot{\gamma}_d \lambda_{Qm}\right)|\eta_1| - \lambda_{QM} l_o^2\right) \tag{64}$$

is zero for $|\eta_1| = 0$ and negative for $|\eta_1| > \lambda_{QM} \dfrac{l_o^2}{k_{\eta 1} - \dot{\gamma}_d \lambda_{Qm}}$. Therefore, by continuity, a maximum for $|\eta_1|$ must exist. This maximum can be easily verified to be at $|\eta_1|_{max} = \lambda_{QM} \dfrac{l_o^2}{2\left(k_{\eta 1} - \dot{\gamma}_d \lambda_{Qm}\right)}$. Similar arguments can be used for the last term of (63). Overall, one has

$$\dot{V} \leq -\dot{\gamma}_d \lambda_{QM} \|\boldsymbol{\zeta}\|^2 + c_{\eta 1} + c_{\eta 2} , \tag{65}$$

where $c_{\eta 1} = \lambda_{QM}^2 \dfrac{l_o^4}{2\left(k_{\eta 1} - \dot{\gamma}_d \lambda_{Qm}\right)}$ and $c_{\eta 2} = 2\lambda_{PM}^2 \dfrac{l_o^4}{\left(k_{\eta 2} - \dot{\gamma}_d \lambda_{Qm}\right)}$, with $k_{\eta 1}, k_{\eta 2} > \dot{\gamma}_d \lambda_{Qm}$. Thus, it can be ensured that $\dot{V} \leq 0$ for

$$\|\boldsymbol{\zeta}\| \geq \sqrt{\frac{c_{\eta 1} + c_{\eta 2}}{\dot{\gamma}_d \lambda_{Qm}}} = \mu , \tag{66}$$

and the system trajectories are ultimately bounded by a region $B_\mu = \{\boldsymbol{\zeta} : \|\boldsymbol{\zeta}\| \leq \mu\}$. Because $k_{\eta 1}$ and $k_{\eta 2}$ can be chosen freely, the ultimate bound radius μ can be driven arbitrarily small. Moreover, μ can be easily forced to satisfy

$$\mu < \sqrt{\frac{\lambda_{Vm}}{\lambda_{VM}}} l_o , \tag{67}$$

to guarantee $B_\mu \subset B_r$.

There is a circularity in the proof[2]: when obtaining (63), it is implicitly assumed that $\boldsymbol{\zeta} \in B_r, \forall t \geq t_0$. To show that this is indeed the case, first notice that $\|\boldsymbol{\zeta}(t_0)\| \in B_\delta \implies \|\boldsymbol{\zeta}(t_0)\| < \sqrt{\frac{\lambda_{Vm}}{\lambda_{VM}}} l_o \leq l_o$. Suppose that $\boldsymbol{\zeta}$ leaves B_r. By continuity, a time $T > t_0$ exists such that $\|\boldsymbol{\zeta}(T)\| = l_o$. Notice that, in order to leave B_r, the trajectories

[2] See [32], remarks on Theorem 5.3.1.

cannot enter in B_μ since this set is positively invariant because $\dot{V} \le 0$ in its frontier. Therefore, the trajectories must remain within $B_r \backslash B_\mu$ before leaving B_r. On one hand, since $\dot{V} \le 0$ for $t \in [t_0, T)$, and after (60)

$$V(\zeta(T)) \le V(\zeta(t_0)) < \lambda_{Vm} l_o^2 \,. \tag{68}$$

On the other hand, from the assumption $\zeta(T) = l_o$ and (60), one has

$$V(\zeta(T)) = V(l_o) \ge \lambda_{Vm} l_o^2 \,. \tag{69}$$

By noticing that (68) and (69) are in contradiction, we can conclude that the original assumption is incorrect, and thus ζ must remain in B_r. □

In the interior of B_r, taking into account (50) and (52), $\zeta \approx \mathbf{0}$ implies $(z_o, \phi, \psi) \approx (z_{od}, 0, 0)$ and $\dot{\gamma} \approx \dot{\gamma}_d$, fulfilling the control objective. For simplicity, the upper-bound for the state ζ, which defines the region B_r in Proposition 2.1, is chosen to be l_o. This choice makes the stability proof clearer, yet it is very conservative. However, since this bound is arbitrary, it can be modified to enlarge the controller's domain of attraction.

The next step is to design a control strategy to stabilise the pole dynamics. For this, it is assumed that the hoop has reached stationary state, such that $\ddot{\mathbf{q}}_o \equiv \mathbf{0}_2$. From (34a) to (34b) and (47) one has

$$\ddot{\mathbf{q}}_h = \mathbf{f}_h + \mathbf{f}_{uP} + \mathbf{Q}_o \mathbf{u}_Q \,, \tag{70}$$

where $\mathbf{f}_h = \mathbf{M}_h^{-1} \mathbf{c}_h \in \mathbb{R}^m$ and $\mathbf{f}_{uP} = \mathbf{P}_o \mathbf{u}_P \in \mathbb{R}^m$.

Investigating the controllability of the nonlinear system (70) is challenging, as discussed earlier. For this reason, only a local result is pursued, based on the linearisation of (70) around its nominal trajectory

$$\mathbf{q}^* = \left[\dot{\gamma}_d t \;\, 0 \;\, z_{od} \;\, (l_o/r_h)\dot{\gamma}_d t \;\, 0 \;\, \mathbf{q}_h^{*T}\right]^T , \tag{71}$$
$$\dot{\mathbf{q}}^* = \left[\dot{\gamma}_d \;\, 0 \;\, 0 \;\, (l_o/r_h)\dot{\gamma}_d \;\, 0 \;\, \dot{\mathbf{q}}_h^{*T}\right]^T , \tag{72}$$

where $\mathbf{q}_h^* \in \mathbb{R}^m$ is the vector representing the nominal trajectory for the pole. Only two of the study cases are analysed here: (i) $m = 3$, three Cartesian directions of movement along the x_w, y_w, and z_w axes of $\mathcal{W}$[3] and (ii) $m = 4$, two Cartesian degree-of-freedom along x_w and y_w of $\mathcal{W}$, and two rotations around the same axes. The configuration coordinates for the three Cartesian degree-of-freedom case are the pole centre of mass coordinates (o_{hx}, o_{hy}, o_{hz}). For the latter case, the rotation matrix of the pole with respect to the inertial frame is given by the composition of two basic rotation matrices, namely

$$\mathbf{R}_o = \mathbf{R}_x(\alpha_1)\mathbf{R}_y(\alpha_2), \tag{73}$$

[3] This is the configuration studied in [1].

with $\alpha_1, \alpha_2 \in \mathbb{R}$ the rotation angles around x_w and y_w, respectively, and $\mathbf{R}_x(\alpha_1), \mathbf{R}_y(\alpha_2) \in SO(3)$. Hence, the configuration coordinates for the pole are $(o_{hx}, o_{hy}, \alpha_1, \alpha_2)$ when $m = 4$. Therefore, the nominal trajectories for the pole in both cases are $\mathbf{q}_h^* = \dot{\mathbf{q}}_h^* = \mathbf{0}_m$.

Definined the state space coordinates $\mathbf{x} = \left[\mathbf{q}_h^T \; \dot{\mathbf{q}}_h^T\right]^T \in \mathbb{R}^{2m}$, the following linearised model can be obtained

$$\dot{\mathbf{x}} = \mathbf{A}(t)\mathbf{x} + \mathbf{B}(t)\mathbf{u}_Q\,, \tag{74}$$

where

$$\mathbf{A}(t) = \begin{bmatrix} \mathbf{O}_m & \mathbf{I}_m \\ \mathbf{A}_{21}(t) & \mathbf{A}_{22}(t) \end{bmatrix}, \tag{75}$$

$$\mathbf{B}(t) = \mathbf{Q}_o\Big|_{\mathbf{q}^*,\dot{\mathbf{q}}^*}\,, \tag{76}$$

$$\mathbf{a}_{21}(t) = \frac{\partial \mathbf{f}_h}{\partial \mathbf{q}_h}\Big|_{\mathbf{q}^*,\dot{\mathbf{q}}^*}\,, \tag{77}$$

$$\mathbf{a}_{22}(t) = \frac{\partial \mathbf{f}_h}{\partial \dot{\mathbf{q}}_h}\Big|_{\mathbf{q}^*,\dot{\mathbf{q}}^*}\,. \tag{78}$$

The term $\mathbf{f}_{uP}$ in (70) is considered as an external bounded input to carry out the linearisation. The periodic linear time-varying system (74) is *not controllable* [33, Theorem 6.11] when the three Cartesian inputrs are considered for the pole (i.e., $m = 3$), while it is *controllable* [33, Theorem 6.12] when the two translations and the two rotations inputs are considered for the pole (i.e., $m = 4$). For this last case, it is possible to find a stabilising controller of the form [34]

$$\mathbf{u}_Q = -\boldsymbol{\Gamma}^{-1}\mathbf{B}^T(t)\mathbf{R}(t)\mathbf{x}\,, \tag{79}$$

where $\mathbf{R}(t) \in \mathbb{R}^{8\times 8}$ is a symmetric positive definite time-varying matrix satisfying the Riccati equation

$$\dot{\mathbf{R}}(t) + \mathbf{A}^T(t)\mathbf{R}(t) + \mathbf{R}(t)\mathbf{A}(t) + \mathbf{G} = \mathbf{R}(t)\mathbf{B}(t)\boldsymbol{\Gamma}^{-1}\mathbf{B}^T(t)\mathbf{R}(t)\,, \tag{80}$$

with $\mathbf{G} \in \mathbb{R}^{8\times 8}$ and $\boldsymbol{\Gamma} \in \mathbb{R}^{4\times 4}$ two positive definite matrices of constant gains. By employing the *quasi–linearisation of the periodic Riccati equation* method [35, p. 137], an approximation for the solution of the Riccati equation (80) can be found.

2.3.1 Conditions for Maintaining Contact

Rolling without slipping between the surfaces of the objects has been assumed for the modelling and control design presented in the above sections. Intuitively, it can be argued that there must be a minimal spinning speed for the hoop not to lose contact with the pole. For obtaining an approximation of the magnitude of this minimum speed, assume that stationary state has reached, i.e., the hoop is spinning in the orthogonal plane to the main pole axis. The Lagrange multipliers from the Lagrange-d'Alembert formulation, which was used to obtain the model (34a)–(34b), can be employed to compute the internal forces [16, p. 279]. A direct calculation of the radial component gives the contact condition

$$m_o l_o \dot{\theta}^2 - m_o g > 0\,, \tag{81}$$

which represents the difference between the centripetal and gravity forces acting on the hoop. This minimum velocity is intended for the best scenario (the hoop perpendicular to the pole), thus the desired velocity and the initial conditions should satisfy $\dot{\theta} >> \sqrt{\frac{g}{l_o}}$. If the pole is in the vertical position, keeping the contact depends on the static friction between the surfaces. Hence, the desired and initial condition for the spinning speed must satisfy $\dot{\theta} >> \sqrt{\frac{g}{f_c l_o}}$, where $f_c > 0$ is the static friction coefficient. On the other hand, the upper limit for the spinning speed depends on the system's bandwidth, which is mainly limited by the signals acquisition and processing time and the actuators' maximum speed and acceleration. As an assumption for maintaining contact as well as to fulfil the requirements of Proposition 2.1, the desired hoop spinning velocity $\dot{\gamma}_d$ must be sufficiently close to the initial speed $\dot{\gamma}$, which in turn must be strictly greater than zero. The required swing controller to obtain this initial spinning is considered out of the scope of this paper, although some planar-motion open-loop controllers can induce it (*e.g.*, the one proposed in [36]).

2.4 *Numerical Simulation*

A numerical simulation is carried out to test the validity of the proposed approach. The parameters employed for the simulation are listed in Table 2. The matrices and vectors in (34a)–(34b) were obtained through a symbolic computing software (*Wolfram Mathematica*) and are omitted here due to space constraints. It is assumed that the pole's apparent inertia can be assigned by the manipulator e.g., as proposed in [37] for the pole not to be affected by the hoop motion. Notice that the pole's actual mass must not have to be large, but only its apparent inertia. The pole sample time is considered to be $T = 0.005$ s, while the hoop coordinates are measured by means of a vision system with sample time $T_v = 0.02$ s.

The desired references are $\dot{\gamma}_d = 4\frac{\pi r_h}{l_o} \approx 1.0472$ rad/s and $z_{od} = 0.3$ m. The gains for the LQR controller in (79)–(80) are chosen as $\Gamma = diag\{0.5, 0.5, 1, 1\}$ and $\mathbf{G} = diag\{200, 200, 40000, 40000, 10, 10, 4, 4\}$. The hoop controller gains are chosen as $k_{\eta 1} = 20$, $k_{\eta 2} = 10$, $k_{\xi 1} = 40$, $k_{\xi 2} = 40$, and $k_{\xi 3} = 4$. The boundary condition

Table 2 Simulation parameters

Parameter	Value
m_o	0.05 kg
m_h	10 kg
l_o	0.3 m
r_o	0.005 m
r_h	0.025 m
l_h	0.7 m

for approximating $\mathbf{R}(t)$ is chosen as $\mathbf{R}(T_s) = \mathbf{O}_{8\times 8}$, where T_s is the period of the linearised system (74), which is given by $T_s = \frac{l_o}{r_h}\dot{\gamma}_d$. The initial conditions for the generalised positions are set to $\gamma(t_0) = 0$ rad, $\psi(t_0) = 0.05$ rad, $z_o(t_0) = 0.05$ m, $\theta(t_0) = \pi$ rad, and $\phi(t_0) = -0.05$ rad, while the initial conditions for the velocities are set to $\dot{\gamma}(t_0) = 0.8$ rad/s, $\dot{\psi}(t_0) = 0$ rad/s, $\dot{z}_o(t_0) = -l_o \sin(\phi(t_0))\dot{\gamma}(t_0)$ m/s, $\dot{\theta}(t_0) = l_o \cos(\phi(t_0))\frac{\dot{\gamma}(t_0)}{r_h}$ rad/s, and $\dot{\phi}(t_0) = -\sin(\psi(t_0))\dot{\gamma}(t_0)$ rad/s.

The contact coordinates time evolution is shown in Fig. 2, where it can be seen that the control objective is satisfied. In Fig. 3, the graphs of the ψ and ϕ coordinates are displayed, showing their ultimate boundedness within a small region around the origin. The $\dot{\theta}$ coordinate time evolution is displayed in Fig. 4, along with the minimum speed required to maintain contact. This condition is satisfied during all the simulation time with a considerably large margin. The Cartesian coordinates of the pole centre of mass are shown in Fig. 5, while the time evolution of the two angles describing the pole orientation is shown in Fig. 6. It can be seen that the proposed controller stabilises all the pole coordinates. Finally, the control inputs, i.e., the forces and torques acting on the pole, for the first 10 seconds of the simulation are shown in Fig. 7.

3 Ballbot

Contemporary research on robotics has steered towards the incorporation of robots into the everyday lives of humans. Robots are expected to interact with humans both outdoors and in human environments safely. This motivation requires robots not only to be mobile and slim but also tall enough to facilitate interaction. On the other hand, conventional multi-wheeled statically-stable robots are typically built to have a low center of gravity to prevent them from easily tipping over. The satisfaction of these two conflicting requirements urges the mobile robots to have large, wide, and heavy bases. At the cost of designing a more complicated controller, a more efficient method to tackle the interaction problem is to utilize dynamically stable robots.

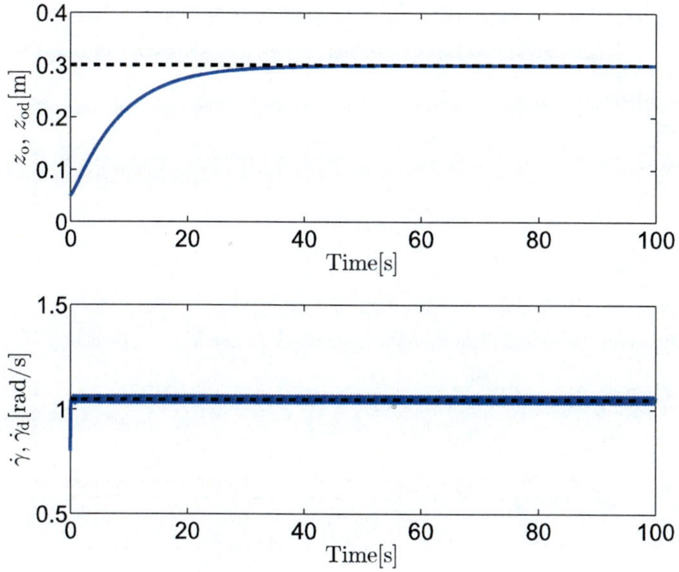

Fig. 2 Contact coordinates evolution: real (—), desired (- - -)

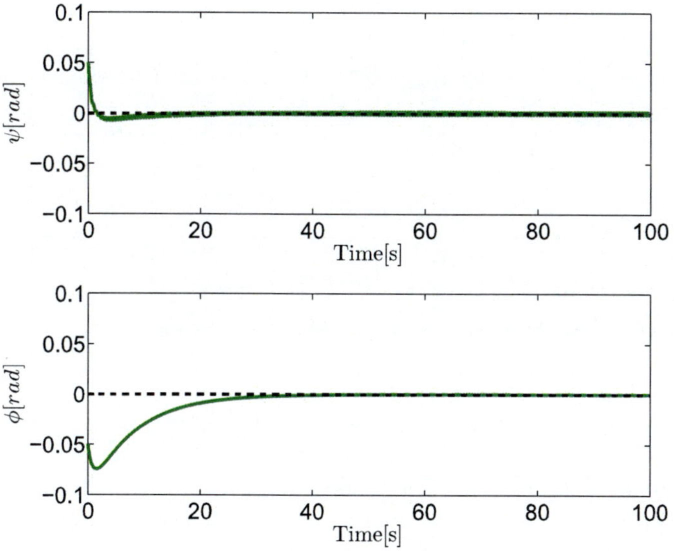

Fig. 3 Asymptotic stabilisation of the ψ and ϕ coordinates

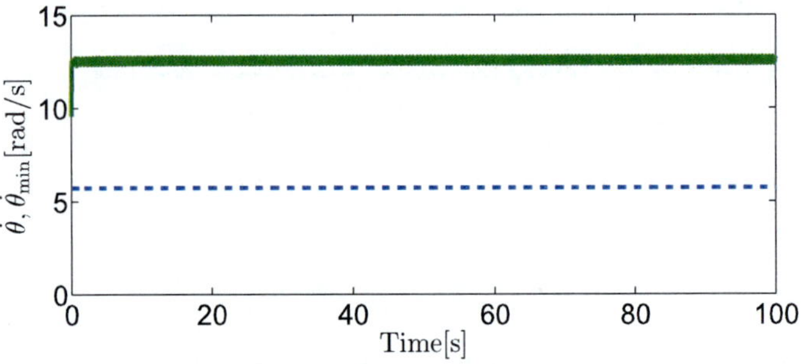

Fig. 4 Evolution of $\dot{\theta}$ (—) and the minimum value required to maintain contact (- - -)

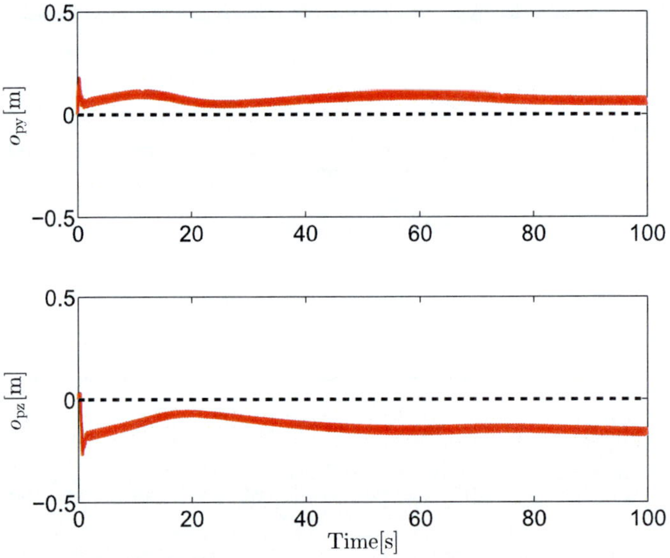

Fig. 5 Pole's Cartesian coordinates with respect to the initial position

One of the most popular dynamically balancing robots is the two-wheeled Segway [38]. The ballbot was introduced as a mobile robot moving on a single spherical omnidirectional wheel [39, 40]. The ballbot, whose design is detailed in [41, 42], is typically slim and as tall as an adult human, rendering it able to interact with humans while navigating constrained environments.

Even though a variant of this robot has been built by many laboratories [43, 44], its control framework has been restricted to the use of classic methods such as linearization about the desired equilibrium in coordinates and PID controllers [44, 45]. Derivation of the equations of motion of the ballbot with a 3-DoF manipulator mounted on top using both Lagrange's and Kane's methods have been performed in [46]. The authors have confirmed that the two approaches agree with each other with a numerical simulation. They have also designed two control laws for the planar

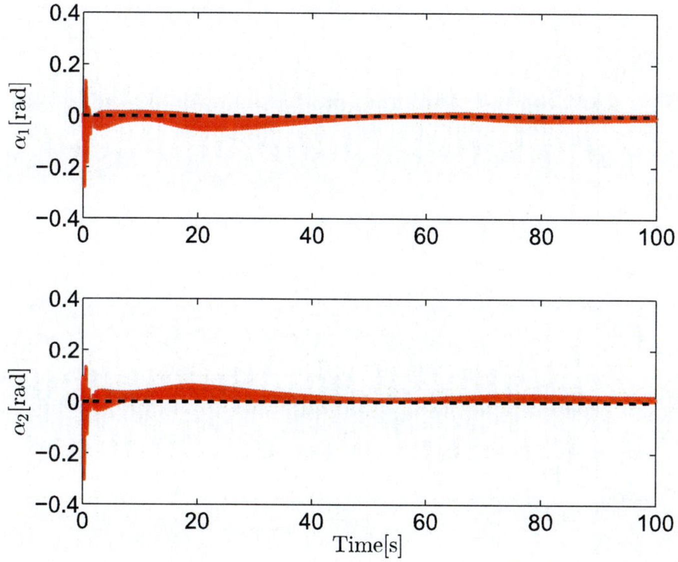

Fig. 6 Stabilisation of the pole's rotation angles

motion of the ballbot and manipulation, respectively, without explicitly addressing stability properties. Moreover, many controllers are typically developed by restricting the dynamics of the ballbot to a vertical 2D plane and applied to the 3D robot by an *ad-hoc* extension to two distinct vertical planes. This procedure inevitably ignores the energetic interaction of the full dynamics of the robot along these planes. Trajectory planning based on motion primitives has been presented in [47], while in [48], authors plan a trajectory for the ballbot equipped with right and left arms. A sliding-mode controller has also been designed for this system in [49]. For the most part, the equations of motion of the whole dynamics of the ballbot have been derived in coordinates, which injects a fair bit of unnecessary complexity into the problem formulation, requires the use of symbolic manipulation software and a decent amount of storage space in the computer [44]. The only exception to this trend has been provided in [50], where the authors derive a dynamic model of the ballbot, which additionally assumes that the body has no yaw motion relative to the ball using Newton's laws. This lengthy procedure, which was omitted from the chapter due to space considerations, leads to a dynamical model of the system, which is not particularly easy to work with for control synthesis. Lastly, in [51], a stabilization algorithm for a TWMM, which is a robot with similar characteristics to the ballbot, has been presented. The controller designed in this chapter utilizes the ZMP idea from the bipedal robotic literature to asymptotically stabilize the motion of the TWMM.

In this chapter, we derive the Euler-Lagrange equations of motion of the full dynamics of the ballbot without resort to any coordinate system. This yields a compact, yet explicit representation of the equations of motion, which recover the 2D

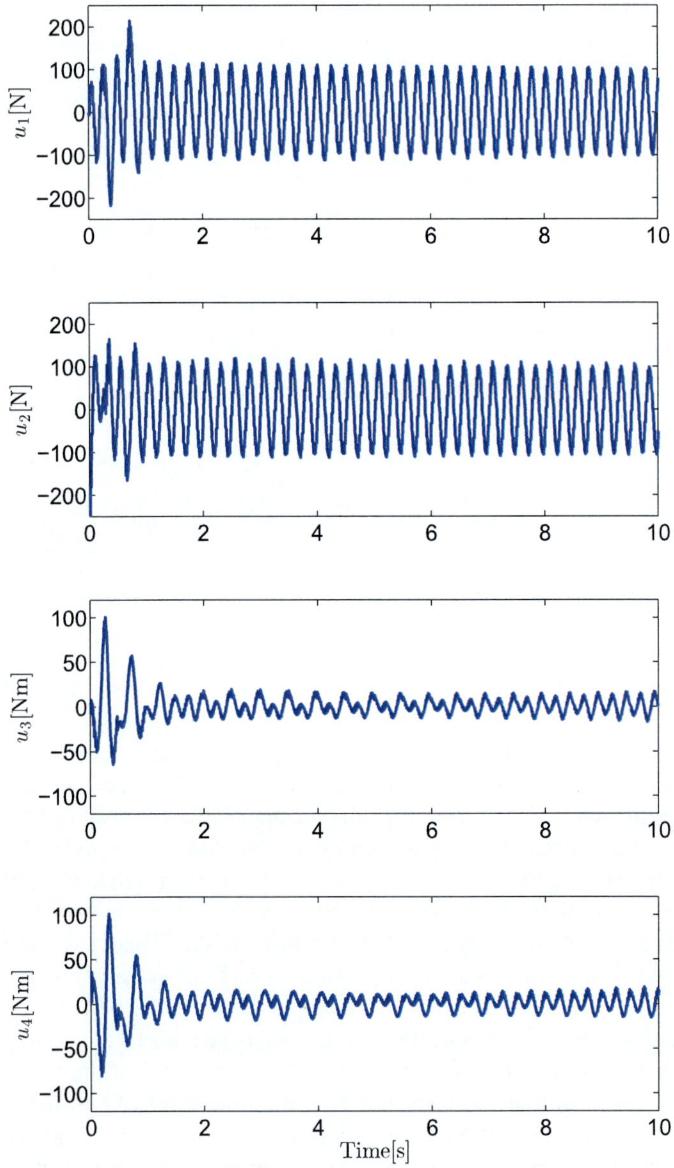

Fig. 7 Force and torque inputs on the pole

dynamics of the ballbot, restricted to a vertical plane, given in the literature [42]. Preliminary results on the derivation and linearization of the intrinsic dynamics of the ballbot are reported in [4]. The Euler-Poincaré dynamics developed in this chapter yield a reduced set of 10 first-order ODEs that govern the motion of the ballbot as opposed to the conventional Newton-Euler approach, which would yield 16 first-order ODEs. We develop energy-shaping control laws that use the available control inputs to make the system look like a new Lagrangian system with a desired asymptotically stable equilibrium point for both the 2D and the 3D dynamics of the ballbot. While in the 2D case we can follow the procedure outlined in [29] to shape the energy, this procedure needs to be extended for the 3D-case because the form of the Euler-Poincaré does not precisely match the form of the Euler-Lagrange equations as handled in that work. The derivation of the total energy shaping control law for the ballbot shows the approach's applicability to a system with considerably more states than the examples previously reported in the literature. The main contribution of this chapter is providing an intrinsic and unified framework to study dynamics and control of the balancing system consisting of a heavy top on a spherical wheel and providing nonlinear energy-shaping control law whose basin of attraction is almost global as long as the mechanism is judiciously designed.

3.1 *Lagrangian Dynamics of the Ballbot*

In this section, we present the background information to be used in the remainder of this chapter, including the kinematics and dynamics of the ballbot. We start by noting that every vector quantity in this paper is represented in the spatial world frame $\mathcal{W}$.

3.1.1 Background and Kinematics

The skeleton diagram of the ballbot is depicted in Fig. 8. It is constructed via the interconnection of a rigid spherical wheel and a rigid cylindrical body. The body is unable to translate with respect to the ball but is free to move otherwise. Therefore, the configuration manifold of the ballbot is $Q = \mathbb{R}^2 \times SO(3) \times SO(3)$. The world

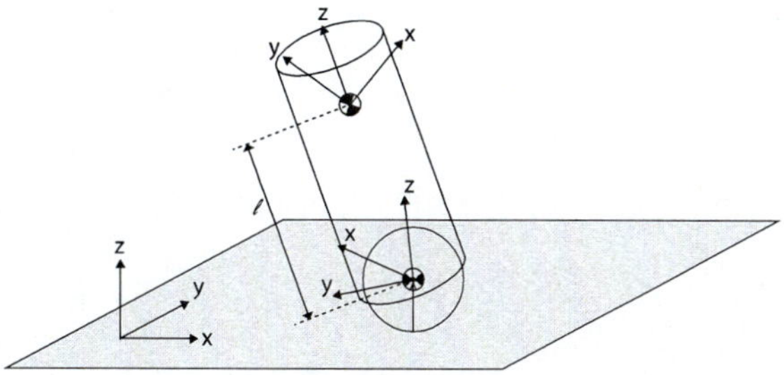

Fig. 8 Ballbot: bodies and frames

inertial frame $\mathcal{W}$ is fixed to a horizontal plane. The spherical wheel is represented by the frame $\mathcal{H}$ and is assumed to have its center of mass at its geometric center. As a result, the vector from the point of contact of the ball with the ground and its center of mass is given by $r_h\mathbf{e}_3$ in the inertial frame. The cylindrical rigid body situated on the wheel is referred to as the "top," and it is denoted by the reference frame $\mathcal{O}$. The center of mass of the top is assumed to lie on the central axis of its geometrical shape at a distance l from the center of the ball. The ball is assumed to roll without slipping, yielding the well-known nonholonomic constraint between the time derivative of its position vector $\mathbf{p}_{sb}$ and its spatial angular velocity $\boldsymbol{\omega}_{sb}$

$$\dot{\mathbf{p}}_{sb} = r_h\boldsymbol{\omega}_{sb} \times \mathbf{e}_3 = r_h\hat{\boldsymbol{\omega}}_{sb}\mathbf{e}_3, \tag{82}$$

where we introduced the *hat* $\wedge$ operator, which stands for the standard isomorphism between $\mathbb{R}^3$ and $\mathfrak{so}(3)$. Its inverse is denoted by the symbol $\vee$, known as the *vee* map [16].

The kinematics of the orientation of the ball and the top are given in the frame $\mathcal{W}$ by the familiar rigid body orientation kinematics

$$\dot{\mathbf{R}}_{sb} = \hat{\boldsymbol{\omega}}_{sb}\mathbf{R}_{sb}, \quad \dot{\mathbf{R}}_{st} = \hat{\boldsymbol{\omega}}_{st}\mathbf{R}_{st}. \tag{83}$$

Using notation and methods from [16], we express the velocity of the top with respect to the inertial frame $\mathbf{V}_{st} \in \mathbb{R}^6$ in terms of the velocity of the ball with respect to the inertial frame $\mathbf{V}_{sb} \in \mathbb{R}^6$ and the velocity of the top with respect to the ball $\mathbf{V}_{bt} \in \mathbb{R}^6$

$$\mathbf{V}_{st} = \begin{bmatrix} \mathbf{v}_{st} \\ \boldsymbol{\omega}_{st} \end{bmatrix} = \begin{bmatrix} \mathbf{v}_{sb} + \mathbf{p}_{sb} \times \mathbf{R}_{sb}\boldsymbol{\omega}_{bt} \\ \boldsymbol{\omega}_{sb} + \mathbf{R}_{sb}\boldsymbol{\omega}_{bt} \end{bmatrix}. \tag{84}$$

We can now compute the time derivative of $\mathbf{p}_{st}$ as a function of the time derivative of $\mathbf{p}_{sb}$, the angular velocity of the top with respect to $\mathcal{W}$ and the orientation of the top:

$$\dot{\mathbf{p}}_{st} = \dot{\mathbf{p}}_{sb} + l\boldsymbol{\omega}_{st} \times \mathbf{R}_{st}\mathbf{e}_3. \tag{85}$$

Throughout this chapter, some properties of the hat map that we freely use in the remainder are as follows

$$\begin{aligned} \hat{\mathbf{x}}\mathbf{y} &= \mathbf{x} \times \mathbf{y} = -\mathbf{y} \times \mathbf{x} = -\hat{\mathbf{y}}\mathbf{x}, \\ \mathbf{x}^T\hat{\mathbf{y}}\mathbf{z} &= \mathbf{y}^T\hat{\mathbf{z}}\mathbf{x} = \mathbf{z}^T\hat{\mathbf{x}}\mathbf{y}, \\ \hat{\mathbf{x}}\hat{\mathbf{y}}\hat{\mathbf{z}} &= \left(\mathbf{x}^T\mathbf{z}\right)\mathbf{y} - \left(\mathbf{x}^T\mathbf{y}\right)\mathbf{z}, \end{aligned}$$

for any $\mathbf{x}, \mathbf{y}, \mathbf{z} \in \mathbb{R}^3$.

3.1.2 Lagrangian

We write the Lagrangian of the ballbot in $\mathcal{W}$, that is, as seen by an observer stationary in the inertial frame. Note that it is imperative that the rolling constraint (82) not be inserted into the Lagrangian before its variation is taken. Suppose the variation of the Lagrangian is taken after the substitution of the nonholonomic constraints. In that case, this yields the vakonomic equations, which disagree with the dynamics of rigid bodies. Instead, one should take the variation before the imposition of the nonholonomic constraints, leading to the Lagrange-d'Alembert equations, the correct equations of motion [52, 53].

The kinetic energy of the ball, $K_b \in \mathbb{R}$, is given by the sum of its rotational and translational kinetic energies, while its potential energy, $V_b \in \mathbb{R}$, is zero, since its height with respect to the inertial frame remains a constant

$$\begin{aligned} K_b &= \frac{1}{2}\boldsymbol{\omega}_{sb}^T \mathbf{I}_b \boldsymbol{\omega}_{sb} + \frac{1}{2} m_b \dot{\mathbf{p}}_{sb}^T \dot{\mathbf{p}}_{sb}, \\ V_b &= 0. \end{aligned}$$

The potential energy of the top, $V_t \in \mathbb{R}$, is given by the height of its center of mass from the horizontal multiplied by its mass. The kinetic energy of the top, $K_t \in \mathbb{R}$, can be written in terms of the rotational velocity of the top and the translational velocity of the ball with respect to $\mathcal{W}$ by substituting from (85):

$$\begin{aligned} K_t &= \frac{1}{2}\boldsymbol{\omega}_{st}^T \mathbf{I}_t \boldsymbol{\omega}_{st} + \frac{1}{2} m_t \dot{\mathbf{p}}_{st}^T \dot{\mathbf{p}}_{st}^T \\ &= \frac{1}{2}\boldsymbol{\omega}_{st}^T \mathbf{I}_t \boldsymbol{\omega}_{st} + \frac{1}{2} m_t l^2 \boldsymbol{\omega}_{st}^T \boldsymbol{\omega}_{st} + \frac{1}{2} m_t \dot{\mathbf{p}}_{sb}^T \dot{\mathbf{p}}_{sb} \\ &\quad - \frac{1}{2} m_t l^2 \left(\boldsymbol{\omega}_{st}^T \mathbf{R}_{st} \mathbf{e}_3\right)^2 + m_t l \dot{\mathbf{p}}_{sb}^T \left(\boldsymbol{\omega}_{st} \times \mathbf{R}_{st} \mathbf{e}_3\right), \\ V_t &= m_t g l \mathbf{e}_3^T \mathbf{R}_{st} \mathbf{e}_3. \end{aligned}$$

Therefore, the Lagrangian $L = K - V = K_t + K_b - V_t \in \mathbb{R}$ is

$$\begin{aligned} L = &\frac{1}{2}\boldsymbol{\omega}_{st}^T \mathbf{I}_t \boldsymbol{\omega}_{st} + \frac{1}{2} m_t l^2 \boldsymbol{\omega}_{st}^T \boldsymbol{\omega}_{st} + \frac{1}{2}\boldsymbol{\omega}_{sb}^T \mathbf{I}_b \boldsymbol{\omega}_{sb} + \frac{1}{2}(m_b + m_t) \dot{\mathbf{p}}_{sb}^T \dot{\mathbf{p}}_{sb} \\ &- \frac{1}{2} m_t l^2 \left(\boldsymbol{\omega}_{st}^T \mathbf{R}_{st} \mathbf{e}_3\right)^2 + m_t l \dot{\mathbf{p}}_{sb}^T \left(\boldsymbol{\omega}_{st} \times \mathbf{R}_{st} \mathbf{e}_3\right) - m_t g l \mathbf{e}_3^T \mathbf{R}_{st} \mathbf{e}_3. \end{aligned} \tag{86}$$

Let us define an element of the unit 2-sphere $\boldsymbol{\gamma} := \mathbf{R}_{st} \mathbf{e}_3$. This quantity represents the direction of the center of mass of the top expressed in $\mathcal{W}$. Next, we write the Lagrangian in terms of $\boldsymbol{\gamma}$, the angular velocity of the top with respect to $\mathcal{W}$ and the angular velocity of the ball with respect to $\mathcal{O}$, all expressed in $\mathcal{W}$. We represent the latter quantity by $\bar{\omega}_{tb} \in \mathbb{R}^3$ and compute it by $\bar{\omega}_{tb} = \omega_{sb} - \omega_{st}$. When the Lagrangian (86) is expressed with these quantities, it takes the reduced form

$$\begin{aligned}\ell = &\frac{1}{2}\left\langle\omega_{st}, \left(\mathbf{I}_t + \mathbf{I}_b + m_t l^2 \mathbf{I}_3\right)\omega_{st}\right\rangle + \langle\omega_{st}, \mathbf{I}_b\bar{\omega}_{tb}\rangle \\ &+ \frac{1}{2}\langle\bar{\omega}_{tb}, \mathbf{I}_b\bar{\omega}_{tb}\rangle + \frac{1}{2}(m_b + m_t)\langle\dot{\mathbf{p}}_{sb}, \dot{\mathbf{p}}_{sb}\rangle \\ &- \frac{1}{2}m_t l^2\langle\omega_{st}, \gamma\rangle^2 + m_t l\langle\dot{\mathbf{p}}_{sb}, \omega_{st} \times \gamma\rangle - m_t g l\langle\mathbf{e}_3, \gamma\rangle. \end{aligned} \tag{87}$$

3.1.3 Euler-Poincaré equations of the ballbot

The equations of motion of the ballbot can be reduced from TQ to $\mathfrak{so}(3) \times \mathfrak{so}(3) \times \mathbb{S}^2 \times \mathbb{R}^2$ to yield the Euler-Poincaré equations [54] for the ballbot. We can derive the evolution of γ by differentiating its definition and using the kinematics of the rigid body

$$\dot{\gamma} = \dot{\mathbf{R}}_{st}\mathbf{e}_3 = \omega_{st} \times \mathbf{R}_{st}\mathbf{e}_3 = \omega_{st} \times \gamma, \tag{88a}$$

$$\dot{\gamma} + \gamma \times \omega_{st} = \mathbf{0}_3. \tag{88b}$$

We freely make use of the following identities when taking the variation of the reduced Lagrangian (87):

$$\delta\mathbf{R}^{-1} = -\mathbf{R}^{-1}\delta\mathbf{R}\mathbf{R}^{-1}, \tag{89a}$$

$$\delta\mathbf{I} = \delta\mathbf{R}\mathbf{R}^{-1}\mathbf{I} - \mathbf{I}\delta\mathbf{R}\mathbf{R}^{-1}, \tag{89b}$$

$$\delta\omega = \dot{\eta} + \eta \times \omega, \tag{89c}$$

$$\delta\dot{\mathbf{p}} = \frac{\mathrm{d}}{\mathrm{d}t}(\delta\mathbf{p}). \tag{89d}$$

where $\mathbf{R} \in SO(3), \omega \in \mathfrak{so}(3), \mathbf{p} \in \mathbb{R}^3$, and $\mathfrak{so}(3) \ni \hat{\eta} = \delta\mathbf{R}\mathbf{R}^{-1}$ are generic element.

The action integral is given by $s = \int \ell \,\mathrm{d}t$, whose variation, $\delta s = \int \delta\ell \,\mathrm{d}t$ is computed by

$$\delta s = \int\left(\frac{\delta\ell}{\delta\mathbf{R}_{st}}\delta\mathbf{R}_{st} + \frac{\delta\ell}{\delta\omega_{st}}\delta\omega_{st} + \frac{\delta\ell}{\delta\bar{\omega}_{tb}}\delta\bar{\omega}_{tb} + \frac{\delta\ell}{\delta\gamma}\delta\gamma + \frac{\delta\ell}{\delta\dot{\mathbf{p}}_{sb}}\delta\dot{\mathbf{p}}_{sb}\right)\mathrm{d}t.$$

Let us compute the individual terms of the above expression making use of the additional relation $\delta\gamma = \eta_{st} \times \gamma$

$$\int \frac{\delta \ell}{\delta \mathbf{R}_{st}} \delta \mathbf{R}_{st}\, \mathrm{d}t = \int \frac{1}{2} \left\langle \omega_{st}, \left(\hat{\eta}_{st} \mathbf{I}_t - \mathbf{I}_t \hat{\eta}_{st}\right) \omega_{st} \right\rangle \mathrm{d}t = \int \left\langle \mathbf{I}_t \omega_{st} \times \omega_{st}, \eta_{st} \right\rangle \mathrm{d}t,$$

$$\begin{aligned}
\int \frac{\delta \ell}{\delta \omega_{st}} \delta \omega_{st}\, \mathrm{d}t &= \int \left\langle \frac{\partial \ell}{\partial \omega_{st}}, \dot{\eta}_{st} + \eta_{st} \times \omega_{st} \right\rangle \mathrm{d}t \\
&= \int \left\{ \left\langle -\frac{\mathrm{d}}{\mathrm{d}t} \frac{\partial \ell}{\partial \omega_{st}} + \omega_{st} \times \frac{\partial \ell}{\partial \omega_{st}}, \eta_{st} \right\rangle \right\} \mathrm{d}t \\
&= \int \left\langle - \left(\hat{\omega}_{st} \mathbf{I}_t - \mathbf{I}_t \hat{\omega}_{st}\right) \omega_{st} - \left(\mathbf{I}_t + \mathbf{I}_b + m_t l^2 \mathbf{I}_3\right) \dot{\omega}_{st} - \mathbf{I}_b \dot{\bar{\omega}}_{tb} \right. \\
&\qquad + m_t l^2 \left(\langle \dot{\omega}_{st}, \gamma \rangle \gamma + \langle \omega_{st}, \gamma \rangle \omega_{st} \times \gamma \right) \\
&\qquad - m_t l \left((\omega_{st} \times \gamma) \times \dot{\mathbf{p}}_{sb} + \gamma \times \ddot{\mathbf{p}}_{sb} \right) + \omega_{st} \\
&\qquad \times \left(\mathbf{I}_t + \mathbf{I}_b + m_t l^2 \mathbf{I}_3\right) \omega_{st} + \omega_{st} \times \mathbf{I}_b \bar{\omega}_{tb} - m_t l^2 \langle \omega_{st}, \gamma \rangle \omega_{st} \\
&\qquad \left. \times \gamma + m_t l \left(\omega_{st} \times (\gamma \times \dot{\mathbf{p}}_{sb}) \right), \eta_{st} \right\rangle \mathrm{d}t,
\end{aligned}$$

$$\begin{aligned}
\int \frac{\delta \ell}{\delta \bar{\omega}_{tb}} \delta \bar{\omega}_{tb}\, \mathrm{d}t &= \int \langle -\frac{\mathrm{d}}{\mathrm{d}t} \frac{\partial \ell}{\partial \bar{\omega}_{tb}} + \bar{\omega}_{tb} \times \frac{\partial \ell}{\partial \bar{\omega}_{tb}} \rangle \,\mathrm{d}t \\
&= \int \left\langle -\mathbf{I}_b \left(\dot{\omega}_{st} + \dot{\bar{\omega}}_{tb}\right) + \bar{\omega}_{tb} \times \mathbf{I}_b \left(\omega_{st} + \bar{\omega}_{tb}\right), \bar{\eta}_{tb} \right\rangle \mathrm{d}t,
\end{aligned}$$

$$\begin{aligned}
&\int \frac{\delta \ell}{\delta \gamma} \delta \gamma\, \mathrm{d}t = \int \left\langle \frac{\partial \ell}{\partial \gamma}, \eta_{st} \times \gamma \right\rangle \mathrm{d}t = \int \left\langle \gamma \times \frac{\partial \ell}{\partial \gamma}, \eta_{st} \right\rangle \mathrm{d}t \\
&= \int \left\langle -m_t l^2 \langle \omega_{st}, \gamma \rangle \gamma \times \omega_{st} + m_t l \left(\gamma \times (\dot{\mathbf{p}}_{sb} \times \omega_{st}) \right) - m_t g l \gamma \times \mathbf{e}_3, \eta_{st} \right\rangle \mathrm{d}t,
\end{aligned}$$

$$\begin{aligned}
\int \frac{\delta \ell}{\delta \dot{\mathbf{p}}_{sb}} \delta \dot{\mathbf{p}}_{sb}\, \mathrm{d}t &= \int \langle -\frac{\mathrm{d}}{\mathrm{d}t} \frac{\partial \ell}{\partial \dot{\mathbf{p}}_{sb}}, \delta \mathbf{p}_{sb} \rangle \,\mathrm{d}t \\
&= \int \langle -(m_b + m_t) \ddot{\mathbf{p}}_{sb} - m_t l \left(\dot{\omega}_{st} \times \gamma + \omega_{st} \times (\omega_{st} \times \gamma) \right), \delta \mathbf{p}_{sb} \rangle \,\mathrm{d}t \\
&= \int \langle -(m_b + m_t) r_h \ddot{\mathbf{p}}_{sb} \\
&\qquad - m_t l \left(\dot{\omega}_{st} \times \gamma + \omega_{st} \times (\omega_{st} \times \gamma) \right), \left(\eta_{st} + \bar{\eta}_{tb}\right) \times \mathbf{e}_3 \rangle \,\mathrm{d}t \\
&= \int \langle -(m_b + m_t) r_h \mathbf{e}_3 \times \ddot{\mathbf{p}}_{sb} - m_t r_h l \left(\mathbf{e}_3 \times (\dot{\omega}_{st} \times \gamma) + \mathbf{e}_3 \right. \\
&\qquad \left. \times \left(\omega_{st} \times (\omega_{st} \times \gamma) \right) \right), \left(\eta_{st} + \bar{\eta}_{tb} \right) \rangle \,\mathrm{d}t.
\end{aligned}$$

Keeping accordance with the literature, we assume that the rotation of the ball along the z-axis of $\mathcal{W}$ cannot be actuated and is always a constant during the motion of the ballbot. We consider the scenario where the relative orientation between the ball and the top is actuated as in [42, 44]. In other words, the control input belongs to the subbundle of the cotangent bundle of Q, characterized by the annihilator of the relative angular velocity ω_{tb}: $\boldsymbol{\tau}' \in \{\boldsymbol{\sigma} \in \mathfrak{so}^*(3) : \langle \boldsymbol{\sigma}, \omega_{tb} \rangle = 0\}$, after its identification with $\mathbb{R}^3$. We notice that $\hat{\omega}_{tb} = \mathrm{Ad}_{\mathbf{R}_{st}^T} \left(\hat{\omega}_{sb} - \hat{\omega}_{st} \right)$, where Ad is the adjoint action, and using the dual of this mapping, we find the forced Euler-Lagrange equations of motion of the ballbot. We add the variations computed above and insert the rolling constraint (82) expressed as $\dot{\mathbf{p}}_{sb} = r\,(\boldsymbol{\omega}_{st} + \bar{\boldsymbol{\omega}}_{tb}) \times \mathbf{e}_3$ to arrive at

$$
\begin{aligned}
&\big(\mathbf{I}_t + m_t l^2 \left(\mathbf{I}_3 - \boldsymbol{\gamma} \otimes \boldsymbol{\gamma}^T\right) + \mathbf{I}_b \\
&\quad - (m_b + m_t) r_h^2 \hat{\mathbf{e}}_3^2 - m_t r_h l \left(\hat{\boldsymbol{\gamma}}\hat{\mathbf{e}}_3 + \hat{\mathbf{e}}_3\hat{\boldsymbol{\gamma}}\right)\big)\, \dot{\boldsymbol{\omega}}_{st} \\
&\quad + \left(\mathbf{I}_b - (m_b + m_t) r_h^2 b \hat{f} e_3^2 - m_t r_h l \hat{\boldsymbol{\gamma}}\hat{\mathbf{e}}_3\right) \dot{\bar{\boldsymbol{\omega}}}_{tb} + m_t l^2 \langle \boldsymbol{\omega}_{st}, \boldsymbol{\gamma}\rangle \hat{\boldsymbol{\gamma}} \boldsymbol{\omega}_{st} \\
&\quad + m_t r_h l \hat{\mathbf{e}}_3 \hat{\boldsymbol{\omega}}_{st}^2 \boldsymbol{\gamma} + \boldsymbol{\omega}_{st} \times \mathbf{I}_t \boldsymbol{\omega}_{st} - \boldsymbol{\omega}_{st} \times \mathbf{I}_b \bar{\boldsymbol{\omega}}_{tb} - m_t g l \mathbf{e}_3 \times \boldsymbol{\gamma} = \mathbf{0}_3, \qquad \text{(90a)}\\
&\left(\mathbf{I}_b - (m_b + m_t) r_h^2 \hat{\mathbf{e}}_3^2 - m_t r_h l \hat{\mathbf{e}}_3 \hat{\boldsymbol{\gamma}}\right) \dot{\boldsymbol{\omega}}_{st} \\
&\quad + \left(\mathbf{I}_\mathbf{b} - (m_b + m_t) r_h^2 \hat{\mathbf{e}}_3^2\right) \dot{\bar{\boldsymbol{\omega}}}_{tb} \qquad \text{(90b)}\\
&\quad + m_t r_h l \hat{\mathbf{e}}_3 \hat{\boldsymbol{\omega}}_{st}^2 \boldsymbol{\gamma} - \bar{\boldsymbol{\omega}}_{tb} \times \mathbf{I}_b \boldsymbol{\omega}_{st} = \mathbf{R}_{st} \boldsymbol{\tau},
\end{aligned}
$$

with $\boldsymbol{\tau} \in \mathbb{R}^3$ the control input. Notice that this system is defined on $\mathfrak{so}(3) \times \mathfrak{so}(3) \times \mathbb{S}^2 \times \mathbb{R}^2$, which has dimension 10, as opposed to the original system, which is defined on TQ, with a dimension count of 16. In case the translational dynamics of the ball, which does not affect the stability of the system, is not considered, the reduced equations evolve on an 8-dimensional manifold, whereas the original equations of motion evolve on a 12-dimensional one.

We note the following definitions to be utilized as in the subsequent sections:

$$
\begin{aligned}
\mathbf{M}_{11} &= \mathbf{I}_t + m_t l^2 \left(\mathbf{I}_3 - \boldsymbol{\gamma} \otimes \boldsymbol{\gamma}^T\right) + \mathbf{I}_b - (m_b + m_t) r_h^2 \hat{\mathbf{e}}_3^2 - m_t r_h l \left(\hat{\boldsymbol{\gamma}}\hat{\mathbf{e}}_3 + \hat{\mathbf{e}}_3\hat{\boldsymbol{\gamma}}\right), \\
\mathbf{M}_{12} &= \mathbf{I}_b - (m_b + m_t) r_h^2 \hat{\mathbf{e}}_3^2 - m_t r_h l \hat{\boldsymbol{\gamma}}\hat{\mathbf{e}}_3, \\
\mathbf{M}_{22} &= \mathbf{I}_b - (m_b + m_t) r_h^2 \hat{\mathbf{e}}_3^2.
\end{aligned}
$$

A comparison of the Euler-Poincaré dynamics and the conventional Euler-Lagrange equations derived using coordinates is made in Fig. 9, where the errors in $\mathbf{R}_{st}$, $\mathbf{R}_{sb}$, and $\mathbf{p}_{sb}$ between the two approaches have been plotted when the ballbot is operated freely under its drift vector field. Since the numerical integration error margin to be tolerated has been selected to be 10^{-7}, these errors are well within the tolerance range.

3.1.4 2D Dynamics

We are interested in finding out how the equations of motion restrict to the plane spanned by the inertial x-z axes of $\mathcal{W}$. In particular, we are going to use coordinates $x \in \mathbb{R}$ and $\theta \in \mathbb{R}$ on the circle for the rotation of the top with respect to $\mathcal{W}$ and the rotation of the top with respect to the ball, respectively. With this choice, the relevant quantities take on the values

$$
\mathbf{R}_{st} = \mathbf{R}_{\mathbf{e}_2, x}, \ \boldsymbol{\omega}_{st} = \dot{x}\mathbf{e}_2; \quad \mathbf{R}_{tb} = \mathbf{R}_{\mathbf{e}_2, \theta}, \ \boldsymbol{\omega}_{sb} = \dot{\theta}\mathbf{e}_2,
$$

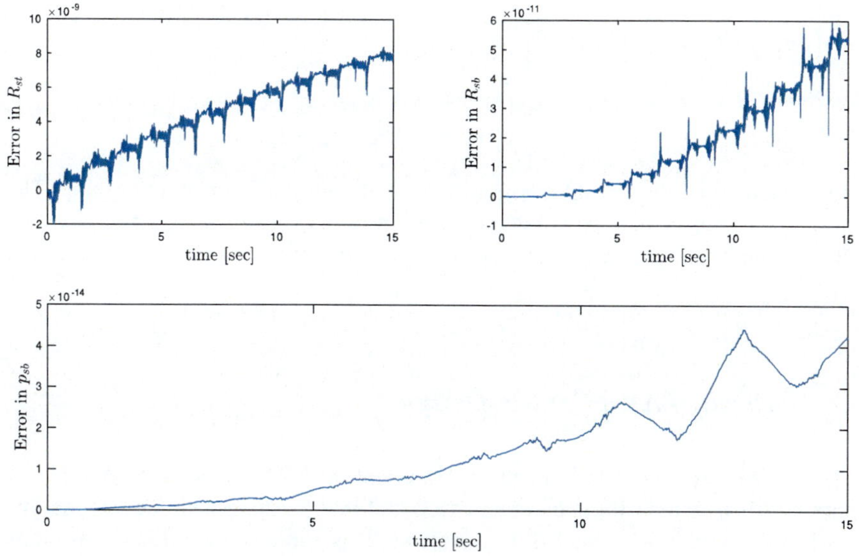

Fig. 9 Error between Euler-Poincaré equations and the conventional Lagrangian approach

where $R_{\mathbf{e}_2,\zeta}$ is the simple rotation matrix by $\zeta \in \mathbb{R}$ radians around the second standard basis vector $\mathbf{e}_2$. When restricted to the plane, the nonholonomic constraint becomes a holonomic one, and it is given by

$$\dot{\mathbf{p}}_{sb} = r_h \boldsymbol{\omega}_{sb} \times \mathbf{e}_3 = \left[r_h \left(\dot{x} + \dot{\theta} \right) 0 \; 0 \right]^T . \tag{91}$$

Using these quantities, the Lagrangian (86) restricted to the inertial x-z plane of $\mathcal{W}$ is computed to be

$$\begin{aligned} L = & \frac{1}{2} \left(I_t + m_t l^2 + m_t r_h l \cos(x) + I_b + (m_b + m_t) r_h^2 \right) \dot{x}^2 \\ & + \left(I_b + (m_b + m_t) r_h^2 + m_t r_h l \cos(x) \right) \dot{x} \dot{\theta} \\ & + \frac{1}{2} \left(I_b + (m_b + m_t) r_h^2 \right) \dot{\theta}^2 - m_t g l \cos x , \end{aligned}$$

where I_b and I_t are the components (2, 2) of the matrices $\mathbf{I}_b$ and $\mathbf{I}_t$, respectively. We can either use the conventional Euler-Lagrange equations with coordinates (x, θ) or directly the coordinate-invariant equations (90) derived in the previous section to compute the equations of motion of the ballbot restricted to the plane. It is readily checked that these two distinct methods yield exactly the same equations, which are given by

$$\mathbf{M}(q)\ddot{\mathbf{q}} + \mathbf{C}(\mathbf{q}, \dot{\mathbf{q}})\dot{\mathbf{q}} + \mathbf{g}(\mathbf{q}) = \begin{bmatrix} \alpha + \gamma + 2\beta \cos x & \alpha + \beta \cos x \\ \alpha + \beta \cos x & \alpha \end{bmatrix} \begin{bmatrix} \ddot{x} \\ \ddot{\theta} \end{bmatrix}$$
$$+ \begin{bmatrix} -\beta \sin x \dot{x} & 0 \\ -\beta \sin x \dot{x} & 0 \end{bmatrix} \begin{bmatrix} \dot{x} \\ \dot{\theta} \end{bmatrix} + \begin{bmatrix} -\mu \sin x \\ 0 \end{bmatrix} = \begin{bmatrix} 0 \\ 1 \end{bmatrix} \tau, \quad (92)$$

complemented by the rolling constraint (91) and with $\tau \in \mathbb{R}$ the control input. The various constants in these equations are given by $\alpha = I_b + (m_b + m_t)r_h^2$, $\beta = m_t r_h l$, $\gamma = I_t + m_t l^2$, and $\mu = m_t g l$. These equations correspond exactly to the ones given in [42].

3.2 Passivity Based Control Design

When $\boldsymbol{\tau} = \mathbf{0}_3$, we can determine the equilibria of the ballbot using the equations of motion (90) with the rolling contraints (82) and (88b). Along with the fact that the inertial z-axis rotation of the ball is assumed to be stationary, the rolling constraints yields $\dot{\mathbf{p}}_{sb} = \mathbf{0}_3 \iff \boldsymbol{\omega}_{\mathbf{sb}} = \boldsymbol{\omega}_{\mathbf{st}} + \bar{\boldsymbol{\omega}}_{\mathbf{tb}} = \mathbf{0}_3$. Inserting $\mathbf{p}_{sb}$ = constant and $\bar{\omega}_{tb} = \mathbf{0}_3$ into the equations of motion (90) along with $\boldsymbol{\omega}_{st} = \mathbf{0}_3$ yields $\mathbf{e}_3 \times \boldsymbol{\gamma} = \mathbf{0}_3$. In other words, the uncontrolled equilibria of the ballbot are given by

$$E_{\pm} = \{(\mathbf{p}_{sb}, \boldsymbol{\gamma}, \dot{\mathbf{p}}_{sb}, \boldsymbol{\omega}_{sb}, \boldsymbol{\omega}_{st}) \in TQ : \mathbf{p}_{sb} = \text{const}, \ \boldsymbol{\gamma} = \pm \mathbf{e}_3$$
$$\boldsymbol{\omega}_{st} = \bar{\boldsymbol{\omega}}_{tb} = \mathbf{0}_3, \dot{\mathbf{p}}_{sb} = \mathbf{0}_3\}.$$

Notice that E_+ corresponds to the upward equilibrium point, that is, the top points in the inertial positive z-direction and E_- corresponds to the downward equilibrium point. The control objective is to asymptotically stabilize the set E_+.

3.2.1 Passivity and Energy Considerations for the 2D Ballbot

Partial feedback linearization of (92) is achieved by the following feedback

$$\tau = \left(\alpha - \frac{(\alpha + \beta \cos(x))^2}{\alpha + \gamma + 2\beta \cos(x)}\right) u + \left(\frac{\alpha + \beta \cos(x)}{\alpha + \gamma + 2\beta \cos(x)} - 1\right) \beta \sin(x) \dot{x}^2$$
$$+ \frac{\mu(\alpha + \beta \cos(x))}{\alpha + \gamma + 2\beta \cos(x)} \sin(x),$$

which yields

$$(\alpha + \gamma + 2\beta \cos(x))\ddot{x} - \beta \sin(x)\dot{x}^2 - \mu \sin(x) =$$
$$-(\alpha + \beta \cos(x))\, u, \quad (93a)$$
$$\ddot{\theta} = u, \quad (93b)$$

with $u \in \mathbb{R}$ a virtual control input.

The following are two passive outputs

$$y_1 = \dot{\theta}, \tag{94a}$$

$$y_2 = -\left(\alpha + \beta\cos(x)\right)\dot{x}, \tag{94b}$$

with the corresponding storage functions

$$H_1 = \frac{1}{2}\dot{\theta}^2, \tag{95a}$$

$$H_2 = \frac{1}{2}\left(\alpha + \gamma + 2\beta\cos(x)\right)\dot{x}^2 + \mu\cos(x). \tag{95b}$$

3.2.2 2D Energy-Shaping Control

Let us consider the following Lyapunov function candidate

$$H_d = k_e\left(k_1H_1 + k_2H_2\right) + \frac{1}{2}k_k\left(k_1y_1 + k_2y_2\right)^2 + \frac{1}{2}k_I\left(k_1\theta - k_2\left(\alpha x + \beta\sin(x)\right)\right)^2, \tag{96}$$

with $k_e, k_1, k_2, k_k, k_I > 0$. Notice that this Lyapunov function candidate comes from the desired energy function that can be written as $H_d = \frac{1}{2}\left[\dot{\theta}\ \dot{x}\right]\mathbf{M}_d\begin{bmatrix}\dot{\theta}\\ \dot{x}\end{bmatrix} + V_d$, where

$$\mathbf{M}_d = \begin{bmatrix} k_ek_1 + k_1^2k_k & -k_1k_2k_k \\ -k_1k_2k_k & k_ek_2\left(\alpha+\gamma+2\beta\cos(x)\right) + k_2^2k_k\left(\alpha+\beta\cos(x)\right)^2 \end{bmatrix},$$

$$V_d = k_ek_2\mu\cos(x) + \frac{1}{2}k_I\left(k_1\theta - k_2\left(\alpha x + \beta\sin(x)\right)\right)^2.$$

The conditions under which $\mathbf{M}_d$ and V_d can be selected such that H_d is a Lyapunov function are developed in the next subsection for the 3D-dynamics of the ballbot. Taking the Lie derivative of (96) along the solutions of (93), we get

$$\begin{aligned}\dot{H}_d = \left(k_1y_1 + k_2y_2\right)\Bigg[&\left(k_e + k_1k_k + k_2k_k\frac{\left(\alpha+\beta\cos(x)\right)^2}{\alpha+\gamma+2\beta\cos(x)}\right)u \\ &+ k_2k_k\left(-(\alpha+\beta c_x)\left(\frac{\beta s_x}{\alpha+\gamma+2\beta c_x}\dot{x}^2 + \frac{\mu s_x}{\alpha+\gamma+2\beta c_x}\right) + \beta s_x\dot{x}^2\right) \\ &+ k_I\left(k_1\theta - k_2\left(\alpha x + \beta s_x\right)\right)\Bigg],\end{aligned}$$

where c_x and s_x are shortened notations for $\cos(x)$ and $\sin(x)$, respectively.

Once we select the control as follows

$$u = -\frac{1}{k}\left[k_2 k_k \left(-(\alpha+\beta c_x)\left(\frac{\beta s_x}{\alpha+\gamma+2\beta c_x}\dot{x}^2 + \frac{\mu s_x}{\alpha+\gamma+2\beta c_x}\right) + \beta s_x \dot{x}^2\right) \right.$$
$$\left. + k_I\left(k_1\theta - k_2\left(\alpha x + \beta s_x\right)\right) + k_p\left(k_1 y_1 + k_2 y_2\right)\right].$$

where $k = k_e + k_1 k_k + k_2 k_k \dfrac{(\alpha+\beta c_x)^2}{\alpha+\gamma+2\beta c_x}$. The time derivative of H_d becomes

$$\dot{H}_d = -k_p\left(k_1 y_1 + k_2 y_2\right)^2 \leq 0.$$

Once the detectability of the output $y = k_1 y_1 + k_2 y_2$ is proven, this implies that the desired equilibrium point is asymptotically stable. The detectability of this output is proven in the next section for the full dynamics of the ballbot. It is omitted in this section because that calculation can be applied to the 2D dynamics verbatim.

3.2.3 Passivity and Energy Considerations for the 3D Ballbot

Partial feedback linearization of the equations of motion (90) on the second factor yields

$$\left(\mathbf{I}_t + m_t l^2\left(\mathbf{I}_3 - \gamma\otimes\gamma^T\right) + \mathbf{I}_b - (m_b+m_t)r_h^2\hat{\mathbf{e}}_3^2 - m_t r_h l\left(\hat{\gamma}\hat{\mathbf{e}}_3 + \hat{\mathbf{e}}_3\hat{\gamma}\right)\right)\dot{\omega}_{st}$$
$$+ m_t l^2\langle\omega_{st},\gamma\rangle\hat{\gamma}\omega_{st} + m_t r_h l\hat{\mathbf{e}}_3\hat{\omega}_{st}^2\gamma + \omega_{st}$$
$$\times\mathbf{I}_t\omega_{st} - \omega_{st}\times\mathbf{I}_b\bar{\omega}_{tb} - m_t g l\mathbf{e}_3\times\gamma = -\left(\mathbf{I}_b - (m_b+m_t)r_h^2\hat{\mathbf{e}}_3^2 - m_t r_h l\hat{\gamma}\hat{\mathbf{e}}_3\right)\mathbf{u}, \tag{97a}$$

$$\dot{\bar{\omega}}_{tb} = \mathbf{u}, \tag{97b}$$

with $\mathbf{u}\in\mathbb{R}^3$ a virtual control input and where the feedback linearizing torque is given by

$$\tau = \mathbf{R}_{st}^T\left(\mathbf{M}_{22} - \mathbf{M}_{12}^T\mathbf{M}_{11}^{-1}\mathbf{M}_{12}\right)\left(\mathbf{u} + m_t r_h l\left(\mathbf{I}_3 - \mathbf{m}_{12}^T\mathbf{M}_{11}^{-1}\right)\hat{\mathbf{e}}_3\hat{\omega}_{st}^2\gamma\right.$$
$$\left. - \mathbf{M}_{12}^T\mathbf{M}_{11}^{-1}m_t l^2\langle\omega_{st},\gamma\rangle\hat{\gamma}\omega_{st} + \mathbf{M}_{12}^T\mathbf{M}_{11}^{-1}m_t g l\left(\mathbf{e}_3\times\gamma\right)\right).$$

The following are two passive outputs

$$\mathbf{y}_1 = \bar{\omega}_{tb}, \tag{98a}$$
$$\mathbf{y}_2 = -\left(\mathbf{I}_b - (m_b+m_t)r_h^2\hat{\mathbf{e}}_3^2 - \beta\hat{\mathbf{e}}_3\hat{\gamma}\right)\omega_{st}. \tag{98b}$$

with the corresponding storage functions

$$H_1 = \frac{1}{2}\bar{\omega}_{tb}^T\bar{\omega}_{tb}, \tag{99a}$$

$$H_2 = \frac{1}{2}\omega_{st}^T\left(\mathbf{I}_b - (m_b + m_t)r_h^2\hat{\mathbf{e}}_3^2 - \beta\left(\hat{\gamma}\hat{\mathbf{e}}_3 + \hat{\mathbf{e}}_3\hat{\gamma}\right) + \mathbf{I}_t + m_t l^2\mathbf{I}_3\right)\omega_{st} - \frac{1}{2}m_t l^2\left(\omega_{st}^T\gamma\right)^2 + \mu\mathbf{e}_3{}^T\gamma. \tag{99b}$$

The passivity of the pair (bfy_1, H_1) is readily seen

$$\frac{\mathrm{d}H_1}{\mathrm{d}t} = \langle\bar{\omega}_{tb}, \mathbf{u}\rangle = \langle\mathbf{y}_1, \mathbf{u}\rangle.$$

To prove the same statement for the pair (y_2, H_2), we calculate

$$\begin{aligned}\frac{\mathrm{d}H_2}{\mathrm{d}t} &= \omega_{st}^T\Big[\left(\mathbf{I}_b - (m_b + m_t)r_h^2\hat{\mathbf{e}}_3^2 - \beta\left(\hat{\gamma}\hat{\mathbf{e}}_3 + \hat{\mathbf{e}}_3\hat{\gamma}\right) + \mathbf{I}_t + m_t l^2\right)\dot{\omega}_{st} - \mu\mathbf{e}_3 \\ &\qquad \times\gamma + \frac{1}{2}\beta\left(\gamma\times(\omega_{st}\times(\mathbf{e}_3\times\omega_{st})) - \mathbf{e}_3\times(\omega_{st}\times(\gamma\times\omega_{st}))\right)\Big] \\ &= \omega_{st}^T\Big[-\left(\mathbf{I}_b - (m_b + m_t)r_h^2\hat{\mathbf{e}}_3^2 - \beta\hat{\gamma}\hat{\mathbf{e}}_3\right)\mathbf{u} - \beta\left(\mathbf{e}_3\times(\omega_{st}\times(\omega_{st}\times\gamma))\right) \\ &\qquad -\frac{1}{2}\beta\left(2\mathbf{e}_3\times(\omega_{st}\times(\gamma\times\omega_{st}))\right)\Big] \\ &= \left\langle-\left(\mathbf{I}_b - (m_b + m_t)r_h^2\hat{\mathbf{e}}_3^2 - \beta\hat{\mathbf{e}}_3\hat{\gamma}\right)\omega_{st}, \mathbf{u}\right\rangle = \langle\mathbf{y}_2, \mathbf{u}\rangle,\end{aligned}$$

where the second to the last step follows by noticing that the first two terms in the final expression below are orthogonal to ω_{st}

$$\begin{aligned}\gamma\times(\omega_{st}\times(\mathbf{e}_3\times\omega_{st})) &= -\omega_{st}\times((\mathbf{e}_3\times\omega_{st})\times\gamma) - (\mathbf{e}_3\times\omega_{st})\times(\gamma\times\omega_{st}) \\ &= -\omega_{st}\times((\mathbf{e}_3\times\omega_{st})\times\gamma) - \omega_{st} \\ &\quad \times((\gamma\times\omega_{st})\times\mathbf{e}_3) - \mathbf{e}_3\times(\omega_{st}\times(\gamma\times\omega_{st})).\end{aligned}$$

Lemma 3.1 *The integrals of the passive outputs can be computed to be*

$$\dot{\bar{\theta}}_{tb} = \mathbf{y}_1, \tag{100a}$$

$$\frac{\mathrm{d}}{\mathrm{d}t}\{-\mathbf{M}_{22}\theta_{st} - \beta\mathbf{e}_3\times\gamma\} = \mathbf{y}_2, \tag{100b}$$

with $\bar{\theta}_{tb}, \theta_{st} \in \mathbb{R}^3$.

Proof We first compute the integral of part of the second output, $-\beta\mathbf{e}_3\times(\gamma\times\omega_{st})$. Let $\phi(\gamma, \omega_{st}) = \beta\mathbf{e}_3\times\gamma$, then

$$\frac{\mathrm{d}\Phi}{\mathrm{d}t} = \beta\left(\mathbf{e}_3\times\dot{\gamma}\right) = -\beta\left(\mathbf{e}_3\times(\gamma\times\omega_{st})\right). \tag{101}$$

To locally express the integral of the remaining terms, we use the exponential mapping from $\mathfrak{so}(3)$ to $SO(3)$ [55] to express the rotation so that

$$\mathbf{R}_{st} = \mathrm{e}^{\widehat{\boldsymbol{\theta}}_{st}}, \quad \mathbf{R}_{tb} = \mathrm{e}^{\widehat{\boldsymbol{\theta}}_{tb}} = \mathrm{e}^{-\widehat{\boldsymbol{\theta}}_{st}}\mathrm{e}^{\widehat{\boldsymbol{\theta}}_{sb}}.$$

Upon differentiation and utilization of the rigid body kinematics (83), we have

$$\begin{aligned}
\hat{\omega}_{st} &= \dot{\mathbf{R}}_{st}\mathbf{R}_{st}^{T} = \dot{\hat{\boldsymbol{\theta}}}_{st},\\
\hat{\bar{\omega}}_{tb} &= \mathrm{Ad}_{\mathbf{R}_{st}}\hat{\omega}_{tb} = \mathrm{Ad}_{\mathbf{R}_{st}}\left(\dot{\mathbf{R}}_{tb}\mathbf{R}_{tb}^{T}\right) = -\dot{\hat{\boldsymbol{\theta}}}_{st} + \dot{\hat{\boldsymbol{\theta}}}_{sb} =: \dot{\hat{\bar{\boldsymbol{\theta}}}}_{tb}.
\end{aligned}$$

Combining these with (101) yields the assertions of the lemma. □

3.2.4 3D Energy-Shaping Control

Let us consider the following Lyapunov function candidate

$$\begin{aligned}
H_d = k_e\left(k_1 H_1 + k_2 H_2\right) &+ \frac{1}{2}\|k_1\mathbf{y}_1 + k_2\mathbf{y}_2\|_{\mathbf{K}_k}^2\\
&+ \frac{1}{2}\|k_1\bar{\boldsymbol{\theta}}_{tb} + k_2\left(-\mathbf{M}_{22}\boldsymbol{\theta}_{st} - \beta\mathbf{e}_3 \times \gamma\right)\|_{\mathbf{K}_I}^2, \qquad (102)
\end{aligned}$$

with $\mathbf{K}_k, \mathbf{K}_I \in \mathbb{R}^{3\times 3}$ constant matrices. Notice that this Lyapunov function candidate comes from a desired energy function that can be written as $H_d = \frac{1}{2}\left[\omega_{tb}\ \omega_{st}\right]\mathbf{M}_d\begin{bmatrix}\omega_{tb}\\ \omega_{st}\end{bmatrix} + V_d$, where

$$\mathbf{M}_d = \begin{bmatrix} k_e k_1\mathbf{I}_3 + k_1^2\mathbf{K}_k & -k_1k_2\mathbf{K}_k\mathbf{M}_{12}^T\\ -k_1k_2\mathbf{M}_{12}\mathbf{K}_k & k_e k_2\mathbf{M}_{11} + k_2^2\mathbf{K}_k\mathbf{M}_{12}\mathbf{M}_{12}^T\end{bmatrix}, \qquad (103a)$$

$$V_d = k_e k_2 V + \frac{1}{2}\|k_1\bar{\boldsymbol{\theta}}_{tb} + k_2\left(-\mathbf{M}_{22}\boldsymbol{\theta}_{st} - \beta\mathbf{e}_3\times\gamma\right)\|_{\mathbf{K}_I}^2. \qquad (103b)$$

Let $\mathbf{q}^* = (\mathbf{R}_{st}, \mathbf{R}_{tb}, \omega_{st}, \bar{\omega}_{tb}) = \left(\mathrm{e}^{\rho\hat{\mathbf{e}}_3}, \mathrm{e}^{\sigma\hat{\mathbf{e}}_3}, \mathbf{0}_3, \mathbf{0}_3\right)$, for some constants $\rho, \sigma \in \mathbb{R}$. In order to qualify H_d as a Lyapunov function, we need to make sure that $\mathbf{M}_d(\mathbf{q}^*)$ is posititve definite, $\delta V_d(\mathbf{q}^*) = 0$, $\delta^2 V_d(\mathbf{q}^*)$ is full rank along the directions orthogonal to the combined, but functionally related rotation of the ball and the top along the inertial vertical axis, and $\dot{H}_d \leq 0$. As long as the yaw rotation of the ball is restricted by its friction with the ground, by Lagrange-Dirichlet stability criterion, these conditions will ensure that both the ball and the top will converge to the desired orientation.

Theorem 3.1 *At* $\mathbf{q}^*$, V_d *has a global minimum* V_d^*, *which is shared by a line of points characterized by a combined rotation of the ball and the top along the inertial vertical axis.* □

Proof We observe from the expression of V_d that it achieves a minimum only if each term individually achieves a minimum. While the minimum of the second term is zero, the minimum $k_e k_2 \mu$ of the first term $k_e k_2 \mu \mathbf{e}_3^T \gamma$ is attained when $\gamma = \mathbf{e}_3$, provided that $k_2 < 0$ and $k_e > 0$. Computing the first variation of V_d yields

$$\begin{aligned}\delta V_d &= k_e k_2 \mu \boldsymbol{\eta}_{st}^T (\gamma \times \mathbf{e}_3)\\ &\quad + \left(k_1 \bar{\boldsymbol{\theta}}_{tb} - k_2 (\mathbf{M}_{22}\boldsymbol{\theta}_{st} + \beta \mathbf{e}_3 \times \gamma)\right)^T \mathbf{K}_I \left(k_1 \bar{\boldsymbol{\eta}}_{tb} - k_2 \left(\mathbf{M}_{22} - \beta \hat{\mathbf{e}}_3 \hat{\gamma}\right) \boldsymbol{\eta}_{st}\right),\end{aligned}$$

which vanishes at $\mathbf{q}^*$. Note that when $\boldsymbol{\theta}_{st} = \mathbf{0}_3$, it follows that $\gamma = \mathbf{R}_{st}\mathbf{e}_3 = \mathrm{e}^{\hat{\boldsymbol{\theta}}_{st}}\mathbf{e}_3 = \mathrm{e}^{\hat{\mathbf{0}}_3}\mathbf{e}_3 = \mathbf{e}_3$.

Computing the second variation $\delta^2 V_d$ of V_d at $\mathbf{q}^* = 0$ yields

$$\delta^2 V_d = k_e k_2 \boldsymbol{\eta}_{st}^T \hat{\mathbf{e}}_3^2 \boldsymbol{\eta}_{st} + \| k_1 \bar{\boldsymbol{\eta}}_{tb} - k_2 \left(\mathbf{M}_{22} - \beta \hat{\mathbf{e}}_3^2\right) \boldsymbol{\eta}_{st} \|_{\mathbf{K}_I}^2 .$$

This expression shows that $\delta^2 V_d$ positive semidefinite and is degenerate only on the subspace spanned by $\boldsymbol{\eta}_{st} = \mathbf{e}_3$ and $\bar{\boldsymbol{\eta}}_{tb} = \dfrac{k_2}{k_1}\mathbf{M}_{22}\mathbf{e}_3$. Since $\mathbf{M}_{22}$ is a diagonal matrix, $\bar{\boldsymbol{\eta}}_{tb}$ is a multiple of $\mathbf{e}_3$ by a negative constant. □

The assumption that the yaw rotation of the ball is constrained by frictional forces implies that $\boldsymbol{\eta}_{sb}^T \mathbf{e}_3 = 0$. Since $\bar{\boldsymbol{\eta}}_{tb} = -\boldsymbol{\eta}_{st} + \boldsymbol{\eta}_{sb}$, it follows that $\bar{\boldsymbol{\eta}}_{tb}^T \mathbf{e}_3 = -\boldsymbol{\eta}_{st}^T \mathbf{e}_3$. Notice that this subspace and the nullspace of $\delta^2 V_d$ intersect only at the zero section of the tangent bundle (state space).

Theorem 3.2 *If the ballbot is strongly inertially coupled [17], i.e., the rank of* $\mathbf{M}_{12}(\mathbf{q})$ *is three, then appropriate gains* k_e, k_1, k_2, $\mathbf{K}_k$ *can be chosen such that* $\mathbf{M}_d(\mathbf{q})$ *is positive definite.* □

Proof $\mathbf{M}_d$ is positive definite if $k_e k_1 \mathbf{I}_3 + k_1^2 \mathbf{K}_k$ is positive definite and $\boldsymbol{\Delta}_{11}$ is positive definite, where $\boldsymbol{\Delta}_{11}$ is the Schur-complement of the $(1, 1)$ block of $\mathbf{M}_d$, that is,

$$\boldsymbol{\Delta}_{11} = k_2 \left(k_e \mathbf{M}_{11} + k_2 \mathbf{M}_{12}\mathbf{K}_k \mathbf{M}_{12}^T\right) - k_1 k_2^2 \mathbf{M}_{12}\mathbf{K}_k \left(k_e \mathbf{I}_3 + k_1 \mathbf{K}_k\right)^{-1} \mathbf{K}_k \mathbf{M}_{12}^T.$$

The condition that $k_e k_1 \mathbf{I}_3 + k_1^2 \mathbf{K}_k$ is positive definite holds if $k_e, k_1 > 0$ and $\mathbf{K}_k$ is positive definite. These constraints on the gains will be in force in the sequel. Let $\underline{\lambda}_k \in \mathbb{R}$ be the smallest eigenvalue of $\mathbf{K}_k$, $\bar{\lambda}_{11} \in \mathbb{R}$ denote the maximum of eigenvalue of $\mathbf{M}_{11}$, and $\underline{\lambda}_{12} \in \mathbb{R}$ denote the minimum eigenvalue of $\mathbf{M}_{12}\mathbf{M}_{12}^T$ as γ varies over $\mathbb{S}^2$. Note that when the hypothesis of the theorem holds, then $\underline{\lambda}_{12}$ is bounded away from zero. Since $k_2 < 0$ and $\mathbf{M}_{11}$ is positive definite, we have that $\boldsymbol{\Delta}_{11}$ is positive definite if and only if the matrix

$$k_e \mathbf{M}_{11} + k_2 \mathbf{M}_{12}\mathbf{K}_k \left(\mathbf{I}_3 - k_1 \left(k_e \mathbf{I}_3 + k_1 \mathbf{K}_k\right)^{-1} \mathbf{K}_k\right) \mathbf{M}_{12}^T$$

is negative definite. This implies that also the matrix

$$k_e\bar{\lambda}_{11}\mathbf{I}_3 + k_2\underline{\lambda}_k\left(1 - \frac{k_1\underline{\lambda}_k}{k_e + k_1\underline{\lambda}_k}\right)\mathbf{M}_{12}\mathbf{M}_{12}^T$$

is negative definite. Finally, this implies that

$$k_e\bar{\lambda}_{11} + k_2\underline{\lambda}_k\left(1 - \frac{k_1\underline{\lambda}_k}{k_e + k_1\underline{\lambda}_k}\right)\underline{\lambda}_{12} < 0.$$

This last implication shows that choosing the quantity $\frac{|k_2|\,\underline{\lambda}_k}{k_e}$ large enough ensures that the desired mass matrix $\mathbf{M}_d$ is positive definite at all points $\mathbf{q}$ where the system is strongly inertially coupled. □

Whether or not the system is strongly inertially coupled depends on the design of the mechanism. If the mass and inertia of the ball are large enough with respect to the top's mass, then this property holds everywhere in the configuration space. Therefore, the following control law achieves global asymptotic stability for those mechanism designs.

Proposition 3.1 *Consider the partially feedback linearized dynamics of the ball-bot (97) in close loop with the control law*

$$\mathbf{u} = -\mathbf{K}^{-1}\left[\mathbf{s} + \mathbf{K}_p\left(k_1\mathbf{y}_1 + k_2\mathbf{y}_2\right)\right], \tag{104}$$

with the expressions for the matrices $\mathbf{K} \in \mathbb{R}^{3\times3}$ and $\mathbf{s} \in \mathbb{R}^3$ that are as given in the proof, $\mathbf{K}_p \in \mathbb{R}^{3\times3}$ a positive definite matrix, and k_1, k_2 satisfy the conditions that render $\mathbf{M}_d$ a positive definite matrix, $\delta V_d(\mathbf{q}^*) = 0$, and $\delta^2 V_d(\mathbf{q}^*)$ a positive definite matrix.

Then, $\mathbf{q}^*$ is an asymptotically stable equilibrium of the closed-loop system. □

Proof Taking the Lie derivative of (102) along the solutions of (97) yields

$$\begin{aligned}\dot{H}_d = \big\langle k_1\mathbf{y}_1 + k_2\mathbf{y}_2, \big[\left(k_e\mathbf{I}_3 + k_1\mathbf{K}_k + k_2\mathbf{K}_k\mathbf{M}_{12}^T\mathbf{M}_{11}^{-1}\mathbf{M}_{12}\right)\mathbf{u}ght. \\ + k_2\mathbf{K}_k\left(-\dot{\mathbf{M}}_{12}^T\omega_{st} + \mathbf{M}_{12}^T\mathbf{M}_{11}^{-1}\left(\mathbf{c}_1 + \mathbf{g}_1\right)\right) \\ + \mathbf{K}_I\left(k_1\bar{\boldsymbol{\theta}}_{tb} + k_2\left(-\mathbf{M}_{22}\boldsymbol{\theta}_{st} - \beta\mathbf{e}_3 \times \gamma\right)\right)\big]\big\rangle \\ = \langle k_1\mathbf{y}_1 + k_2\mathbf{y}_2, \mathbf{K}\mathbf{u} + \mathbf{s}\rangle,\end{aligned}$$

where $\mathbf{c}_1 = m_t l^2\langle\omega_{st}, \gamma\rangle\hat{\gamma}\omega_{st} + \beta\hat{\mathbf{e}}_3\hat{\omega}_{st}^2\gamma + \omega_{st} \times \mathbf{I}_t\omega_{st} - \omega_{st} \times \mathbf{I}_b\hat{\omega}_{tb}$, $\mathbf{g}_1 = -\mu\mathbf{e}_3 \times \gamma$. We thus select

$$\mathbf{u} = -\mathbf{K}^{-1}\left[\mathbf{s} + \mathbf{K}_p\left(k_1\mathbf{y}_1 + k_2\mathbf{y}_2\right)\right], \tag{105}$$

where $\mathbf{K} = \left(k_e\mathbf{I}_3 + k_1\mathbf{K}_k + k_2\mathbf{K}_k\mathbf{M}_{12}^T\mathbf{M}_{11}^{-1}\mathbf{M}_{12}\right)$, $\mathbf{s}$ consists of all the terms not multiplied by $\mathbf{u}$ in the second factor of the natural pairing and $\mathbf{K}_p$ being positive definite, yielding

$$\dot{H}_d = -\|k_1\mathbf{y}_1 + k_2\mathbf{y}_2\|^2_{\mathbf{K}_p}.$$

This implies that $k_1\mathbf{y}_1 + k_2\mathbf{y}_2 \xrightarrow[t\to\infty]{} \mathbf{0}_3$. Let us analyze the smallest invariant set within $\mathcal{E} = \{k_1\mathbf{y}_1 + k_2\mathbf{y}_2 = \mathbf{0}_3\}$. We have

$$\mathbf{0}_3 = \mathbf{k}_1\mathbf{K}_\mathbf{k}\dot{\mathbf{P}}_1 + \mathbf{k}_2\mathbf{K}_\mathbf{k}\dot{\mathbf{P}}_2 = \left(\mathbf{k}_1\mathbf{K}_\mathbf{k} + \mathbf{k}_2\mathbf{K}_\mathbf{k}\mathbf{M}_{12}^\mathbf{T}\mathbf{M}_{11}^{-1}\mathbf{M}_{12}\right)\mathbf{u} + \boldsymbol{\delta}_1,$$

where $\boldsymbol{\delta}_1 = k_2\mathbf{K}_k\left(-\dot{\mathbf{M}}_{12}^T\boldsymbol{\omega}_{st} + \mathbf{M}_{12}^T\mathbf{m}_{11}^{-1}\left(\mathbf{c}_1 + \mathbf{g}_1\right)\right)$. Thus, we can write the above equation as $\left(\mathbf{K} - k_e\mathbf{I}_3\right)\mathbf{u} = -\boldsymbol{\delta}_1$. From the control law (104), we have $\mathbf{K}\mathbf{u} = -\left(\boldsymbol{\delta}_1 + \boldsymbol{\delta}_2\right)$, where $\boldsymbol{\delta}_2 = \mathbf{K}_I\left(k_1\bar{\boldsymbol{\theta}}_{tb} + k_2\left(-\mathbf{M}_{22}\boldsymbol{\theta}_{st} - \beta\mathbf{e}_3 \times \gamma\right)\right)$. Therefore, from the following two equations

$$\begin{aligned}\left(\mathbf{K} - k_e\mathbf{I}_3\right)\mathbf{u} &= -\boldsymbol{\delta}_1,\\ \mathbf{K}\mathbf{u} &= -\left(\boldsymbol{\delta}_1 + \boldsymbol{\delta}_2\right),\end{aligned}$$

we deduce that $\mathbf{u} = -\frac{1}{k_e}\boldsymbol{\delta}_2$ on $\mathcal{E}$. Plugging this into (97), shows that the first of these equations is unstable unless $\mathbf{u} = \boldsymbol{\delta}_2 = \mathbf{0}_3$. This implies $\bar{\omega}_{tb} = \text{constant} = \mathbf{0}_3$, which, in turn, implies $\boldsymbol{\omega}_{st} = \mathbf{0}_3$ because otherwise its dynamics would again be unstable. This discussion shows that the system asymptotically converges to an equilibrium point. Since the closed-loop system comes from a Lagrangian system, it is readily shown that, the only stable equilibrium point is the upward equilibrium point, with $\boldsymbol{\theta}_{st} = \left[0\ 0\ \text{constant}\right]^T$ and $\boldsymbol{\theta}_{tb} = \left[0\ 0\ \text{constant}\right]^T$, where the inertial z-axis rotation of the system goes to a certain constant because this motion is uncontrollable. □

3.2.5 Controlling the Ball Position

We can use a similar technique as in Sect. 3.2.4 to control the position of the ball as well as the upward equilibrium point of the top. In order to do this, we start from (97) and switch to the error system dynamics, with the error defined by $\bar{\mathbf{e}}_{tb} = \bar{\omega}_{tb} - \bar{\omega}_{tb}^d$, where $\bar{\omega}_{tb}^d$ is the desired angular velocity of the ball with respect to the top expressed in $\mathcal{W}$. We compute this quantity from the desired angular velocity of the ball with respect to the spatial frame, $\boldsymbol{\omega}_{sb}^d \in \mathbb{R}^3$, which is, in turn, computed from the rolling constraint (82). In order to get $\mathbf{p}_{sb} \to \mathbf{p}_{sb}^d$, where $\mathbf{p}_{sb}{}^d \in \mathbb{R}^3$ is the desired position of the ball on the x-y plane (the z component is a constant), we would like the rolling constraint Eq. (82) to look like

$$\dot{\mathbf{p}}_{sb} = r_h\boldsymbol{\omega}_{sb}^d \times \mathbf{e}_3 = k_t\hat{\mathbf{e}}_3^2\left(\mathbf{p}_{sb} - \mathbf{p}_{sb}^d\right).$$

As a result, we set $\boldsymbol{\omega}_{sb}^d = -\frac{k_t}{r_h}\hat{\mathbf{e}}_3\left(\mathbf{p}_{sb} - \mathbf{p}_{sb}^d\right)$, where $k_t > 0$. This implies that the desired angular velocity of the ball with respect to the top expressed in $\mathcal{W}$ is, $\bar{\omega}_{tb}^d = \boldsymbol{\omega}_{sb}^d - \boldsymbol{\omega}_{st}$. We use (82) in the time derivative of this relation to get

$$\dot{\bar{\omega}}_{tb}^d = \dot{\omega}_{sb}^d - \dot{\omega}_{st} = k_t \hat{\mathbf{e}}_3^2 (\bar{\omega}_{tb} + \omega_{st}) - \dot{\omega}_{st}. \tag{106}$$

We set the control $\mathbf{u}$ in Eq. (97b) to $\mathbf{u} = \dot{\bar{\omega}}_{tb}^d + \mathbf{v}$, with $\mathbf{v} \in \mathbb{R}^3$ a virtual input, which yields the error system dynamics

$$(\mathbf{M}_{11} - \mathbf{M}_{12}) \dot{\omega}_{st} + \mathbf{c}_2 + \mathbf{g}_1 = -\mathbf{M}_{12}\mathbf{v}, \tag{107a}$$

$$\dot{\bar{\mathbf{e}}}_{tb} = \mathbf{v}, \tag{107b}$$

where $\mathbf{c}_2 = \mathbf{c}_1 + k_t \mathbf{M}_{12} \hat{\mathbf{e}}_3^2 (\bar{\omega}_{tb} + \omega_{st})$. Solving (107b) for $\dot{\omega}_{st}$ and substituting into (106), we derive the form of the control $\mathbf{u}$ as

$$\mathbf{u} = \left(\mathbf{I}_3 + (\mathbf{M}_{11} - \mathbf{M}_{12})^{-1} \mathbf{M}_{12}\right) \mathbf{v} + (\mathbf{M}_{11} - \mathbf{M}_{12})^{-1} (\mathbf{c}_2 + \mathbf{g}_1) + k_t \hat{\mathbf{e}}_3^2 (\bar{\omega}_{tb} + \omega_{st}).$$

We identify a passive output $\mathbf{y}_1 = \bar{\mathbf{e}}_{tb}$ with the storage function $H_1 = \frac{1}{2}\bar{\mathbf{e}}_{tb}^T \bar{\mathbf{e}}_{tb}$. We keep the second output $\mathbf{y}_2 = -\mathbf{M}_{12}^T \omega_{st}$ and modify its storage to $H_2' = H_2 - \omega_{st}^T \mathbf{M}_{12} \omega_{st}$. Note that this output with the storage function H_2' is not passive any more due to the additional terms appearing in $\mathbf{c}_2$. It is also important to note that, we do not need to asymptotically stabilize the orientation of the ball to a desired value. We would rather asymptotically stabilize $\bar{\mathbf{e}}_{tb}$ to zero, which implies $\omega_{sb} \to \omega_{sb}^d$, which, in turn, asymptotically stabilizes $\mathbf{p}_{sb}$ to $\mathbf{p}_{sb}^d$. This observation implies that, we only need to devise a desired potential energy which has $\gamma = \mathbf{e}_3$ a point of minimum. This is easily achieved by setting $V_d = k_e k_2 V$. Recall that the product $k_e k_2$ is negative and since the original potential energy V has a minimum at $\gamma = -\mathbf{e}_3$, this desired potential energy is good enough.

To shape the kinetic energy of the system, we use the same desired mass matrix as the one (103b) used in the previous subsection. We use the following desired Hamiltonian (energy) for the closed loop system

$$H_d = k_e \left(k_1 H_1 + k_2 H_2'\right) + \frac{1}{2} \|k_1 \mathbf{y}_1 + k_2 \mathbf{y}_2\|_{\mathbf{K}_k}^2 + \frac{1}{2} \left\langle \mathbf{p}_{sb} - \mathbf{p}_{sb}^d, \mathbf{p}_{sb} - \mathbf{p}_{sb}^d \right\rangle.$$

Proceeding analogously to Proposition 3.1, we compute the time derivative of H_d to be

$$\begin{aligned} \dot{H}_d = &\Big\langle k_1 \mathbf{y}_1 + k_2 \mathbf{y}_2, \Big[\left(k_e \mathbf{I}_3 + k_1 \mathbf{K}_k + k_2 \mathbf{K}_k \mathbf{M}_{12}^T (\mathbf{M}_{11} - \mathbf{M}_{12})^{-1} \mathbf{M}_{12}\right) \mathbf{v} \\ &+ k_2 \mathbf{K}_k \left(-\dot{\mathbf{M}}_{12}^T \omega_{st} + \mathbf{M}_{12}^T (\mathbf{M}_{11} - \mathbf{M}_{12})^{-1} (\mathbf{c}_2 + \mathbf{g}_1)\right)\Big]\Big\rangle \\ &- \frac{\beta}{2} \left\langle \omega_{st}, \hat{\mathbf{e}}_3^2 \hat{\omega}_{st}^2 \gamma \right\rangle + \left\langle -k_e k_t \hat{\mathbf{e}}_3^2 \left(\mathbf{y}_1 - \frac{k_t}{r_h} \hat{\mathbf{e}}_3 (\mathbf{p}_{sb} - \mathbf{p}_{sb}^d)\right), k_2 \mathbf{y}_2 \right\rangle \\ &+ \left\langle k_t \hat{\mathbf{e}}_3^2 (\mathbf{p}_{sb} - \mathbf{p}_{sb}^d), \mathbf{p}_{sb} - \mathbf{p}_{sb}^d \right\rangle + \left\langle r_h \mathbf{y}_1, \hat{\mathbf{e}}_3 (\mathbf{p}_{sb} - \mathbf{p}_{sb}^d) \right\rangle, \end{aligned}$$

where the second and third natural pairings arise because of the extra terms in the new Coriolis term, $\mathbf{c}_2$. We select the virtual input term $\mathbf{v}$ as

$$\mathbf{v} = -\mathbf{K}_b^{-1}\left[\mathbf{s}_b + \mathbf{K}_p\left(k_1\mathbf{y}_1 + k_2\mathbf{y}_2\right) + \frac{k_e k_t^2}{r_h}\hat{\mathbf{e}}_3^2\left(\mathbf{p}_{sb} - \mathbf{p}_{sb}^d\right)\right], \tag{108}$$

where $\mathbf{K}_b = \left(k_e\mathbf{I}_3 + k_1\mathbf{K}_k + k_2\mathbf{K}_k\mathbf{M}_{12}^T\left(\mathbf{M}_{11} - \mathbf{M}_{12}\right)^{-1}\mathbf{M}_{12}\right)$, $\mathbf{s}_b \in \mathbb{R}^3$ consists of all the terms not multiplied by $\mathbf{v}$ in the second factor of the first natural pairing, and $\mathbf{K}_p \in \mathbb{R}^{3\times3}$ is a positive definite constant matrix. This selection yields to the following expression for the time derivative of H_d along the solutions of the system

$$\begin{aligned}\dot{H}_d = &-\|k_1\mathbf{y}_1 + k_2\mathbf{y}_2\|_{\mathbf{K}_p}^2 + k_e k_2 k_t\left\langle\hat{\mathbf{e}}_3^2\mathbf{y}_1, \mathbf{y}_2\right\rangle - \frac{\beta}{2}\left\langle\omega_{st}, \hat{\mathbf{e}}_3^2\hat{\omega}_{st}^2\gamma\right\rangle \\ &+ k_t\left\langle\hat{\mathbf{e}}_3^2\left(\mathbf{p}_{sb} - \mathbf{p}_{sb}^d\right), \mathbf{p}_{sb} - \mathbf{p}_{sb}^d\right\rangle + \left\langle\left(r_h - \frac{k_1 k_e k_t^2}{r_h}\right)\mathbf{y}_1, \hat{\mathbf{e}}_3\left(\mathbf{p}_{sb} - \mathbf{p}_{sb}^d\right)\right\rangle.\end{aligned}$$

Selecting $k_t = \dfrac{r_h}{\sqrt{k_1 k_e}}$ ensures that the last natural pairing in this expression vanishes. While the first natural pairing may be absorbed into the very first term, using the Cauchy-Schwarz inequality, by a proper selection of the gain $\mathbf{K}_p$, the second natural pairing satisfies the linear growth condition as long as strong inertial coupling condition is satisfied. As a result, this pairing may also be dominated by the first term semi-globally by increasing the magnitude of the gain k_2. These arguments prove that $\dot{H}_d \leq 0$. Since the detectability of the output $\mathbf{y} = k_1\mathbf{y}_1 + k_2\mathbf{y}_2$ is proven in exactly the same way as in the proof of Proposition 3.1, we can summarize the result in the following proposition

Proposition 3.2 *Consider the partially feedback linearized dynamics of the ballbot (97b) in closed loop with the control law*

$$\mathbf{v} = -\mathbf{K}_b^{-1}\left[\mathbf{s}_b + \mathbf{K}_p\left(k_1\mathbf{y}_1 + k_2\mathbf{y}_2\right) + \frac{k_e k_t^2}{r_h}\hat{e}_3^2\left(\mathbf{p}_{sb} - \mathbf{p}_{sb}^d\right)\right], \tag{109a}$$

$$\begin{aligned}\mathbf{u} = &\left(\mathbf{I}_3 + \left(\mathbf{M}_{11} - \mathbf{M}_{12}\right)^{-1}\mathbf{M}_{12}\right)\mathbf{v} \\ &+ \left(\mathbf{M}_{11} - \mathbf{M}_{12}\right)^{-1}\left(\mathbf{c}_2 + \mathbf{g}_1\right) + k_t\hat{\mathbf{e}}_3^2\left(\bar{\omega}_{tb} + \omega_{st}\right),\end{aligned} \tag{109b}$$

with the expressions for $\mathbf{K}_b$, $\mathbf{s}_b$ and k_t are as given above the proposition, $\mathbf{K}_p$ a positive definite matrix, and k_1, k_2 satisfy the conditions that render $\mathbf{M}_d$ positive definite, $\delta V_d(\gamma = \mathbf{e}_3) = 0$, and $\delta^2 V_d(\gamma = \mathbf{e}_3)$ a positive definite matrix.

Then, $\gamma = \mathbf{e}_3$, $\omega_{st} = \mathbf{0}_3$, $\bar{\omega}_{tb} = \mathbf{0}_3$, and $\mathbf{p}_{sb} = \mathbf{p}_{sb}^d$ is an asymptotically stable equilibrium of the closed-loop system. □

Remark 7.1 Although the control design in the section is inspired by the development in [29], the technique presented in that work cannot be applied verbatim. The fundamental reason for this shortcoming is that the dynamics (90b) is not derived directly from Euler-Lagrange equations, but are reduced to yield the Euler-Poincaré equations. Consequently, the construction of a second passive output in (98b), a corresponding storage function in (99b) and an integral of this passive output (100) has to be novelly performed. Once these quantities have been obtained, we can construct the Lyapunov function (102).

3.3 Numerical Examples

The inertial properties of the ballbot that are utilized in the simulations are given in Table 3.

First, simulation results for the ballbot whose dynamics is restricted to the 2D x-z plane are presented. The initial conditions and the control gains used in the simulation are given in the first row of Table 4. The corresponding simulation results are shown in Fig. 10. The top two plots illustrate the convergence of the orientation of the top and the ball, in other words, the fact that x and θ tend both to zero. The bottom left plot shows the evolution of the passive output $y = k_1 y_1 + k_2 y_2$, while the bottom right plot shows the evolution of the closed-loop energy functional H_d (96).

Second, simulation results are presented that show the response of the ballbot to a feedback control (104). In this simulation, the initial conditions and the control gains are given by the second row of Table 4. Notice that the initial conditions are quite far away from the desired upward equilibrium $\mathbf{R}_{st} = \mathrm{e}^{\rho \hat{\mathbf{e}}_3}$ and the desired ball orientation $\mathbf{R}_{sb} = \mathrm{e}^{\sigma \hat{\mathbf{e}}_3}$, where ρ and σ are constant real numbers. On the left of Fig. 11, we illustrate the asymptotic stabilization. The top two plots illustrates the fact that both the ball and the top move to an orientation such that $\mathbf{R}_{st} e_3 = \mathbf{e}_3$ and $\mathbf{R}_{sb} e_3 = \mathbf{e}_3$, which is another way to state that $\mathbf{R}_{st} = \mathrm{e}^{\rho \hat{\mathbf{e}}_3}$ and $\mathbf{R}_{sb} = \mathrm{e}^{\sigma \hat{\mathbf{e}}_3}$ for some constant numbers ρ and σ. The bottom left plot shows the evolution of the passive output $\mathbf{y} = k_1 \mathbf{y}_1 + k_2 \mathbf{y}_2$, while the bottom right plot shows the evolution of the closed loop energy functional H_d (102). The path tracked by the ballbot on the plane is shown on the right in Fig. 11.

Table 3 System parameters

Parameter	Symbol	Value
Ball radius	r_h	0.1058 m
Ball mass	m_b	244 kg
Ball inertia	I_b	1.821 kg·m^2
Top center of mass height	l	0.69 m
Roll moment of inertia	$I_{t,11}$	12.59 kg·m^2
Pitch moment of inertia	$I_{t,22}$	12.48 kg·m^2
Yaw moment of inertia	$I_{t,33}$	0.66 kg·m^2
Top mass	m_t	51.66 kg

Table 4 Initial conditions and control gains

	Initial conditions		Control gains		
Sim 1	$x(0) = \frac{2\pi}{3}$	$\theta(0) = 4\pi$	$k_1 = 5$	$k_2 = -70$	$k_e = 20$
	$\dot{x}(0) = 0$	$\dot{\theta} = 0$	$k_k = 11$	$k_p = 10$	$k_I = 1$
Sim 2	$\mathbf{R}_{st}(0) = \mathbf{R}_{y,\frac{\pi}{3}}\mathbf{R}_{x,-\frac{\pi}{2}}$	$\mathbf{R}_{sb}(0) = \mathbf{I}_3$	$k_1 = 5$	$k_2 = -70$	$k_e = 20$
	$\omega_{st}(0) = \omega_{sb} = \mathbf{0}_3$	$\mathbf{p}_{sb} = \mathbf{0}_3$	$\mathbf{K}_k = 11\mathbf{I}_3$	$\mathbf{K}_p = 10\mathbf{I}_3$	$\mathbf{K}_I = \mathbf{I}_3$
Sim 3	$\mathbf{R}_{st}(0) = \mathbf{R}_{z,\frac{4\pi}{3}}\mathbf{R}_{y,1.74}\mathbf{R}_{x,1}$	$\mathbf{R}_{sb}(0) = \mathbf{I}_3$	$k_1 = 5$	$k_2 = -70$	$k_e = 20$
	$\omega_{st}(0) = \omega_{sb} = \mathbf{0}_3$	$\tilde{\mathbf{p}}_{sb}(0) = (\frac{1}{4}, \frac{2}{y}, 0)$	$\mathbf{K}_k = 11\mathbf{I}_3$	$\mathbf{K}_p = 7.5\mathbf{I}_3$	$k_t = \frac{r}{\sqrt{k_1 k_e}}$

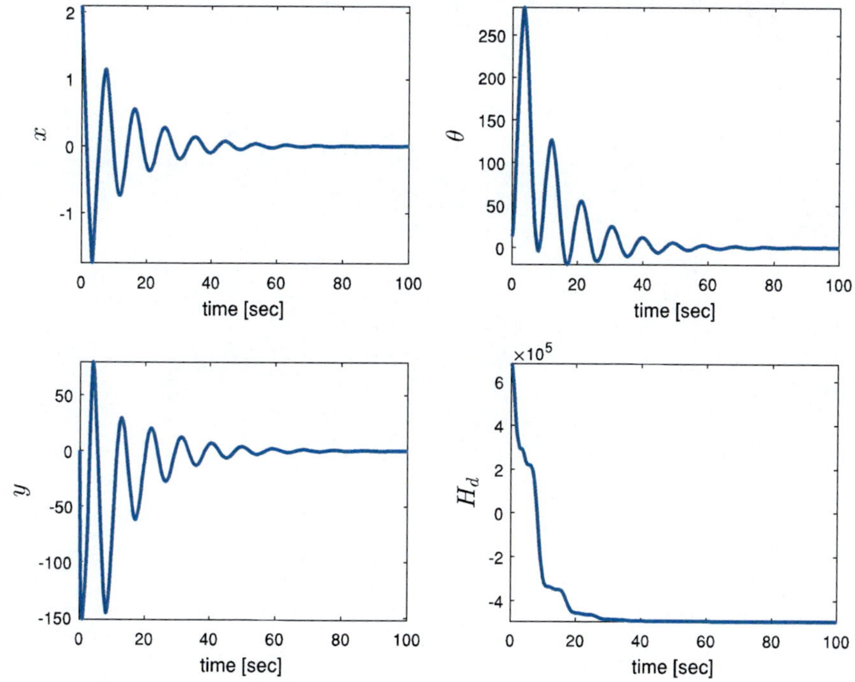

Fig. 10 Asymptotic stabilization of the 2D Ballbot

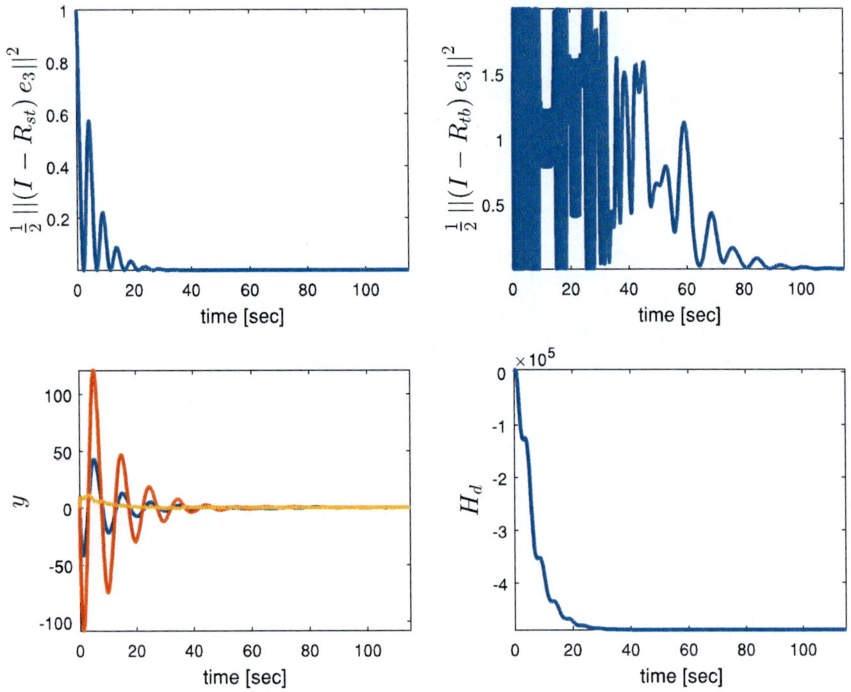

(a)

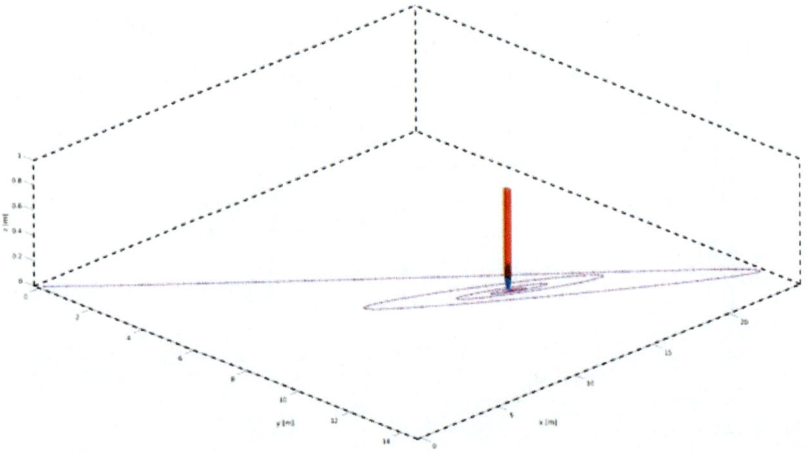

(b)

Fig. 11 **a** Time evolution of quantities of interest **b** The path tracked by the Ballbot

Finally, we present the simulation results that use the controller (109), designed to stabilize the position of the ball along with the upward equilibrium of the top. The initial conditions and the control gains are selected as the third row of Table 4. On the left of Fig. 12, the top left plot shows that the top asymptotically converges to the upward equilibrium point, where $\gamma = \mathbf{e}_3$. The top left figure depicts the evolution of the position error of the ball, which converges to zero, indicating that $\mathbf{p}_{sb} \rightarrow \mathbf{p}_{sb}^d$. Additionally, while the bottom left plot shows that the passive output is asymptotically driven to zero, the bottom right plot shows that the closed-loop energy function H_d converges to its minimum value, as expected due to the detectability of the passive output $\mathbf{y}$. Again, the path tracked by the ballbot on the plane is shown on the right of Fig. 12.

4 Discussion and Conclusion

This chapter investigated the control design for nonprehensile rolling manipulation dealing with nonholonomic constraints. Two systems were addressed, namely, the hula-hoop and the ballbot.

The considered robotic hula-hoop is an underactuated mechanical system subject to second-order nonholonomic constraints. We designed a locally stable controller scheme by exploiting the null space of the inertia coupling matrices, making it possible to simultaneously satisfy the control objective of (i) spinning the hoop at a desired angular velocity on the desired position over the pole surface and (ii) stabilizing the pole coordinates. A formal proof that guarantees locally ultimate boundedness of the hoop coordinates was presented, with arbitrary small ultimate bound on the tracking error and boundedness of the pole coordinates. To validate the proposed approach, we presented a numerical simulation that showed our solution's good performance. The main challenges for the experimental setup are the necessity of a high-speed reconstruction of the hoop position and orientation (in the simulation is was considered to be implemented at a 20 Hz rate) with good resolution and the high velocities and acceleration required for the actuator (e.g., for the humanoid-like robot used in the simulations).

The ballbot robotic system is made of an actuated ball with a top. A reduced set of dynamic equations, whose configuration space is $Q = \mathbb{R}^2 \times SO(3) \times SO(3)$, was derived. These 10 first-order ODEs can express the system's motion comprehensibly, given the kinematic relations that the system has to satisfy. We can analyze dynamic properties and derive control laws that achieve asymptotic stabilization for several purposes thanks to the compact form of these equations of motion. In particular, we identified two passive outputs for the restricted 2D dynamics and the full 3D dynamics, which were then used to devise energy-shaping control laws making the system behave as a new Lagrangian system whose desired equilibrium point is asymptotically stable. The basin of attraction was shown to be global as long as the mechanism

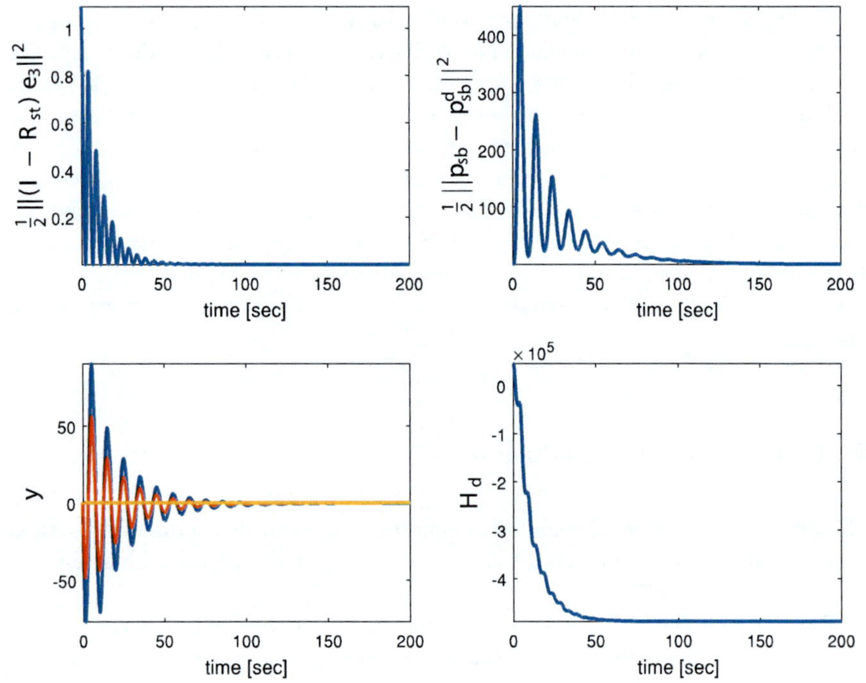

(a)

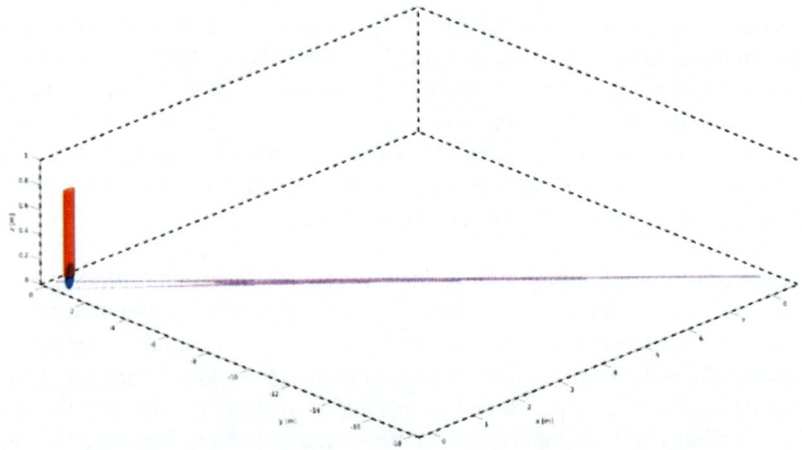

(b)

Fig. 12 **a** Time evolution of quantities of interest. **b** The path tracked by the Ballbot is designed to be strongly inertially coupled. We emphasize that modeling, analysis, and computations can be carried out directly in a geometric coordinate-free framework, as illustrated for the ballbot in this chapter. This fact facilitates the analysis of the dynamics and control synthesis for complex systems such as the ballbot.

References

1. A. Gutierrez-Giles, F. Ruggiero, V. Lippiello, B. Siciliano, Modelling and control of a robotic hula-hoop system without velocity measurements, in *20th World IFAC Congress*, pp. 9808–9814 (2017)
2. A. Gutierrez-Giles, F. Ruggiero, V. Lippiello, B. Siciliano, Nonprehensile manipulation of an underactuated mechanical system with second-order nonholonomic constraints: The robotic hula-hoop. IEEE Robot. Autom. Lett. **3**(2), 1136–1143 (2018)
3. A.C. Satici, A. Donaire, B. Siciliano, Intrinsic dynamics and total energy-shaping control of the ballbot systems. Int. J. Control **90**, 2734–2747 (2017)
4. A.C. Satici, F. Ruggiero, V. Lippiello, and B. Siciliano, Intrinsic Euler-Lagrange dynamics and control analysis of the ballbot, in *2016 American Control Conference*, pp. 5685–5690 (2016)
5. B. Siciliano, L. Sciavicco, L. Villani, G. Oriolo, *Robotics: Modelling, Planning and Control* (Springer, London, UK, 2009)
6. D. Serra, J. Ferguson, F. Ruggiero, A. Siniscalco, A. Petit, V. Lippiello, B. Siciliano. On the experiments about the nonprehensile reconfiguration of a rolling sphere on a plate, in *26th Mediterranean Conference on Control and Automation*, pp. 13–20, Zadar, HR (2018)
7. H. Date, M. Sampei, M. Ishikawa, M. Koga, Simultaneous control of position and orientation for ball plate manipulation problem based on time state control form. IEEE Trans. Robot. Autom. **20**(3), 465–480 (2004)
8. A. Marigo, A. Bicchi, Rolling bodies with regular surface: Controllability theory and applications. IEEE Trans. Autom. Control **45**(9), 1586–1599 (2000)
9. G. Oriolo, M. Vendittelli, A framework for the stabilization of general nonholonomic systems with an application to the plate-ball mechanism. IEEE Trans. Rob. **21**(2), 162–175 (2005)
10. K.-K. Lee, G. Bätz, and D. Wollherr. Basketball robot: Ball on plate with pure haptic information, in *2008 IEEE International Conference on Robotics and Automation*, Pasadena, CA, USA, pp. 2410–2415 (2008)
11. L. Cui, J.S. Dai, A coordinate-free approach to instantaneous kinematics of two rigid objects with rolling contact and its implications for trajectory planning, in *2009 IEEE International Conference on Robotics and Automation*, Kobe, J, pp. 612–617 (2009)
12. R. Gahleitner, Ball on ball: Modeling and control of a novel experiment set-up. IFAC-PapersOnLine **48**(1), 796–801 (2015)
13. S.-Y. Liu, Y. Rizal, and M.-T. Ho. Stabilization of a ball and sphere system using feedback linearization and sliding mode control, in *8th Asian Control Conference*, Kaohsiung, Taiwan, pp. 1334–1339 (2011)
14. J. Nishizaki, S. Nakamura, M. Sampei, Modeling and control of hula-hoop system, in *48th IEEE Conference on Decision and Control*, Shanghai, C (2009), pp. 4125–4130
15. D.J. Montana, The kinematics of contact and grasp. Int. J. Robot. Res. **7**(3), 17–32 (1988)
16. R.M. Murray, Z. Li, S.S. Sastry, *A Mathematical Introduction to Robotic Manipulation* (CRC press, 1994)
17. M.W. Spong, Partial feedback linearization of underactuated mechanical systems, in *1994 IEEE/RSJ/GI International Conference on Intelligent Robots and Systems*, pp. 314–321 (1994)
18. G. Oriolo, Y. Nakamura, Control of mechanical systems with second-order nonholonomic constraints: Underactuated manipulators, in *30th IEEE Conference on Decision and Control*, pp. 2398–2403 (1991)

19. M. Reyhanoglu, A. van der Schaft, N.H. McClamroch, I. Kolmanovsky, Dynamics and control of a class of underactuated mechanical systems. IEEE Trans. Autom. Control **44**(9), 1663–1671 (1999)
20. R.W. Brockett, Asymptotic stability and feedback stabilization, in *Differential Geometric Control Theory* (Birkhauser, 1983), pp. 181–191
21. H.J. Sussmann, Subanalytic sets and feedback control. J. Differ. Equ. **31**(1), 31–52 (1979)
22. J.-M. Coron, On the stabilization in finite time of locally controllable systems by means of continuous time-varying feedback law. SIAM J. Control. Optim. **33**(3), 804–833 (1995)
23. H.J. Sussmann, V. Jurdjevic, Controllability of nonlinear systems. J. Differ. Equ. **12**(1), 95–116 (1972)
24. R. Olfati-Saber, Cascade normal forms for underactuated mechanical systems. in *39th IEEE Conference on Decision and Control* (2000), pp. 2162–2167
25. A. De Luca, G. Oriolo, Modelling and control of nonholonomic mechanical systems, in *Kinematics and Dynamics of Multi-body Systems* (Springer, 1995), pp. 277–342
26. A. Donaire, J.G. Romero, R. Ortega, B. Siciliano, M. Crespo, Robust IDA-PBC for underactuated mechanical systems subject to matched disturbances. Int. J. Robust Nonlinear Control **27**(6), 1000–1016 (2017)
27. A.S. Shiriaev, L.B. Freidovich, S.V. Gusev, Transverse linearization for controlled mechanical systems with several passive degrees of freedom. IEEE Trans. Autom. Control **55**(4), 893–906 (2010)
28. A.S. Shiriaev, L.B. Freidovich, M.W. Spong, Controlled invariants and trajectory planning for underactuated mechanical systems. IEEE Trans. Autom. Control **59**(9), 2555–2561 (2014)
29. A. Donaire, R. Mehra, R. Ortega, S. Satpute, J.G. Romero, F. Kazi, N.M. Sing, Shaping the energy of mechanical systems without solving partial differential equations. IEEE Trans. Autom. Control **61**(4), 1051–1056 (2016)
30. Y.-L. Gu, Y. Xu, A normal form augmentation approach to adaptive control of space robot systems. Dyn. Control **5**(3), 275–294 (1995)
31. H.K. Khalil, *Nonlinear Systems* (Prentice Hall, 2002)
32. M. Vidyasagar, *Nonlinear Systems Analysis*, 2nd edn. (Society for Industrial and Applied Mathematics, Philadelphia, PA, 2002)
33. C.-T. Chen, *Linear System Theory and Design*, 3rd edn. (Oxford University Press, 1999)
34. A. Shiriaev, J.W. Perram, C. Canudas-de Wit, Constructive tool for orbital stabilization of underactuated nonlinear systems: Virtual constraints approach. IEEE Trans. Autom. Control **50**(8), 1164–1176 (2005)
35. S. Bittanti, A.J. Laub, J.C. Willems, *The Riccati Equation*, 1st edn. (Springer, 1991)
36. A.P. Seyranian, A.O. Belyakov, How to twirl a hula hoop. Am. J. Phys. **79**(7), 712–715 (2011)
37. F. Ficuciello, L. Villani, B. Siciliano, Variable impedance control of redundant manipulators for intuitive human-robot physical interaction. IEEE Trans. Rob. **31**(4), 850–863 (2015)
38. H.G. Nguyen, J. Morrell, K.D. Mullens, A.B. Burmeister, S. Miles, N. Farrington, K.M. Thomas, D.W. Gage, Segway robotic mobility platform, in *Mobile Robots XVII*, vol. 5609 (SPIE, 2004), pp. 207–220
39. T. Lauwers, G. Kantor, R. Hollis, One is enough! in *12th International Symposium for Robotics Research*, pp. 12–15 (2005)
40. T.B. Lauwers, G.A. Kantor, R.L. Hollis, A dynamically stable single-wheeled mobile robot with inverse mouse-ball drive, in *2006 IEEE International Conference on Robotics and Automation*, pp. 2884–2889 (2006)
41. U. Nagarajan, A. Mampetta, G.A. Kantor, R.L. Hollis, State transition, balancing, station keeping, and yaw control for a dynamically stable single spherical wheel mobile robot, in *2009 IEEE International Conference on Robotics and Automation*, pp. 998–1003 (2009)
42. U. Nagarajan, G. Kantor, R. Hollis, The ballbot: An omnidirectional balancing mobile robot. Int. J. Robot. Res. **33**(6), 917–930 (2013)
43. L. Hertig, D. Schindler, M. Bloesch, C.D. Remy, R. Siegwart, Unified state estimation for a ballbot, in *2013 IEEE International Conference on Robotics and Automation*, pp. 2471–2476 (2013)

44. S. Leutenegger, P. Fankhauser, Modeling and control of a ballbot. Bachelor thesis, Swiss Federal Institute of Technology (2010)
45. M. Kumaga, T. Ochiai. Development of a robot balanced on a ball: Application of passive motion to transport, in *2009 IEEE International Conference on Robotics and Automation*, pp. 4106–4111 (2009)
46. P. Asgari, P. Zarafshan, S.A.A. Moosavian, Manipulation control of an armed ballbot with stabilizer. Proc. Inst. Mech. Eng. Part I: J. Syst. Control Eng. **229**(5), 429–439 (2015)
47. U. Nagarajan, G. Kantor, R. Hollis, Integrated planning and control for graceful navigation of shape-accelerated underactuated balancing mobile robots, in *2012 IEEE International Conference on Robotics and Automation*, pp. 136–141 (2012)
48. U. Nagarajan, B. Kim, R. Hollis, Planning in high-dimensional shape space for a single-wheeled balancing mobile robot with arms, in *2012 IEEE International Conference on Robotics and Automation*, pp. 130–135 (2012)
49. C.-W. Liao, C.-C. Tsai, Y.Y. Li, C.-K. Chan, Dynamic modeling and sliding-mode control of a ball robot with inverse mouse-ball drive. In SICE Annual Conference **2008**, 2951–2955 (2008)
50. A.N. Inal, O. Morgul, U. Saranli, A 3D dynamic model of a spherical wheeled self-balancing robot, in *2012 IEEE/RSJ International Conference on Intelligent Robots and Systems*, pp. 5381–5386 (2012)
51. S.R. Larimi, P. Zarafshan, S.A.A. Moosavian, A new stabilization algorithm for a two-wheeled mobile robot aided by reaction wheel. J. Dyn. Syst. Meas. Control **137**(1), 011009–1–011009–8 (2015)
52. A.D. Lewis, R.M. Murray, Variational principles for constrained systems: theory and experiment. Int. J. Non-Linear Mech. **30**(6), 793–815 (1995)
53. J. Baillieul, A.M. Bloch, P. Crouch, J. Marsden, *Nonholonomic Mechanics and Control, Interdisciplinary Applied Mathematics* (Springer, New York, 2008)
54. D. Schneider, Non-holonomic Euler-Poincaré equations and stability in Chaplygin's sphere. Dyn. Syst.: Int. J. **17**(2), 87–130 (2002)
55. K.M. Lynch, F.C. Park, *Modern Robotics: Mechanics, Planning, and Control*, 1st edn. (Cambridge University Press, 2017)

A Coordinate-Free Framework for Robotic Pizza Tossing and Catching

Aykut C. Satici, Fabio Ruggiero, Vincenzo Lippiello, and Bruno Siciliano

Abstract This chapter presents a solution to the problem of autonomous pizza tossing and catching. Under the assumption that robotic fingers grasp the pizza dough with soft contact, the grasp constraints are formulated and used to derive the individual and combined Euler-Lagrange dynamic equations of motion of the robotic manipulator and the dough. In particular, the dynamics of the dough is a modified version of the rigid-body dynamics, taking into account the change of inertia due to its deformation. Through these mathematical models, the two control problems of tossing and catching are formulated. For the tossing phase, an exponentially convergent controller that stabilizes a desired velocity of the dough as it leaves the fingers, is derived. On the other hand, to catch the dough, an optimal trajectory for the end-effector of the robotic manipulator is generated. Finally, the control laws to make the optimal trajectory exponentially attractive are derived. The developed theory is demonstrated with an elaborate simulation of the tossing and catching phases. This chapter is based on the works presented in [1].

1 Introduction

The problem of tossing and catching a pizza dough is a procedure that is frequently dexterously performed by human pizza makers. There are at least three important reasons why tossing the dough during the preparation of the pizza is attractive: (i) the

A. C. Satici (✉)
Boise State University, 1910 University Drive, Boise, ID 83725, USA
e-mail: aykutsatici@boisestate.edu

F. Ruggiero · V. Lippiello · B. Siciliano
Department of Electrical Engineering and Information Technology, PRISMA Lab, CREATE Consortium & University of Naples Federico II, Via Claudio 21, 80125 Naples, Italy
e-mail: fabio.ruggiero@unina.it

V. Lippiello
e-mail: vincenzo.lippiello@unina.it

B. Siciliano
e-mail: bruno.siciliano@unina.it

B. Siciliano and F. Ruggiero (eds.), *Robot Dynamic Manipulation*,
Springer Tracts in Advanced Robotics 144,
https://doi.org/10.1007/978-3-030-93290-9_8

dough is stretched to the desired size, (ii) the dough naturally assumes a configuration that is thicker at the ends and thinner in the middle, and (iii) as the spinning dough freely falls, the outside of the dough dries, making it crunchy in the outside but light in the middle. The pizza maker is trained to perform a streamlined hand motion to toss and catch the dough, and we are aiming to achieve a similar feat with a humanoid robot.

There are several challenges to achieve this kind of manipulation. The fact that the object that we are trying to manipulate is deformable complicates the several previously well-studied problems, such as catching or grasping [2], motion estimation [3, 4] and manipulation of rigid objects [5]. Furthermore, such nonprehensile manipulation tasks typically require high-speed sensing and control action [6, 7]. While the deformation is beneficial for the grasping phase as it introduces more surface area where normal force may be applied, motion estimation is more problematic because the dynamic model of the rigid body is no longer valid. In this chapter, we estimate the equations of motion of the dough by those of a rigid body whose mass is constant but whose inertia varies due to its varying shape. We assume the ability to receive the shape information via a vision sensor, which enables us to estimate the motion of the dough while it is in the air. The specific literature for robotic tossing and catching of pizza dough is limited. In [8] the authors have studied how different forms of tossing might yield different desired behaviors.

With perfect knowledge of the motion of the dough, we can generate optimal trajectories to intercept it. The generation of optimal trajectories in $SE(3)$ has been studied in [9, 10]. Since we only estimate this motion, we repeat the optimal trajectory generation as we are fed with new sensor information. The optimal trajectories are generated to match the initial position, velocity, acceleration, and final velocity and accelerations. Therefore, it is at least thrice continuously differentiable. An optimal trajectory whose initial and final accelerations are desired to be prescribed has to satisfy a sixth-order boundary value problem (BVP). We generate such a BVP using the necessary conditions for a path to minimize a convex combination of the jerk and acceleration functionals. While minimizing the jerk functional reduces the vibrations in the structure of the robotic manipulator, minimizing the acceleration functional reduces the total amount of energy expended during the catching motion [11]. We can determine which aspect to emphasize by choosing the convex coefficients suitably. As soon as the first optimal trajectory is generated, the robot is instructed to move along this trajectory and update its path as new measurements come in.

We derive control laws for both the tossing and catching phases. The control law for tossing the object makes sure the object has the desired velocity as it leaves the fingers, while the control law for catching moves the manipulator's end-effector along the previously generated optimal trajectory. The robotic manipulator is chosen with a redundant, S-R-S topology to mimic the human arm with 7 degrees of freedom (DoF). All of the quantities and mathematical operators needed to analyze the combined robotic and dough system are presented in Table 1. The extra DoF is exploited to flow to the configuration with a maximum manipulability index along the nullspace of the manipulator Jacobian. Finally, we demonstrate the developed theory with an elaborate tossing and catching simulation.

Table 1 Main symbols used in this chapter

Definition	Symbol
Word (inertial) frame	W
Homogeneous transformation between frames A and B	T_{ab}
Line brackets of the vector fields v and w	[v, w]
Wedge and hat operators either between $\mathbb{R}^3$ and $SO(3)$ or $\mathbb{R}^3$ and $SO(3)$	$\wedge, \vee$
Frame at the ith contact point moving with the dough	$\mathcal{C}_i$
Frame at the ith contact point moving with the i-th finger	$\mathcal{F}_i$
Frame at the tool's center (hand frame)	$\mathcal{H}$
Frame at the dough's center of mass	$\mathcal{O}$
Standard unit vector	$\mathbf{e}_3 = \begin{bmatrix} 0 & 0 & 1 \end{bmatrix}^T$
Velocity (twist) of $\mathcal{C}_i$ with respect to $\mathcal{F}_i$, expressed in body coordinates	$\mathbf{v}^b_{f_i c_i} \in \mathbb{R}^6$
Homogeneous transformation between a generic frame $\mathcal{A}$ and a frame $\mathcal{B}$	$\mathbf{T}_{ab} \in SE(3)$
Velocity (twist) of $\mathcal{H}$ with respect to $\mathcal{W}$	$\mathbf{v}_{sh} \in \mathbb{R}^6$
Velocity (twist) of $\mathcal{O}$ with respect to $\mathcal{W}$	$\mathbf{v}_{so} \in \mathbb{R}^6$
Body manipulator Jacobian	$\mathbf{J}^b_{sh} = \mathbf{J} \in \mathbb{R}^{6\times n}$
Vector of joint angles	$\boldsymbol{\theta} = \begin{bmatrix} \theta_1 & \cdots & \theta_n \end{bmatrix}^T \in \mathbb{R}^n$
Radius of the pizza dough	$r > 0$
Height of the pizza dough's geometric center	$h > 0$
Orientation of $\mathcal{O}$ with respect to $\mathcal{W}$	$\mathbf{R}_{so} \in SO(3)$
Angular velocity of $\mathcal{O}$ with respect to $\mathcal{W}$ expressed in the body frame $\mathcal{O}$	$\boldsymbol{\omega}_o \in \mathbb{R}^3$
Inertia tensor of the pizza dough	$\mathbf{I}_o \in \mathbb{R}^{3\times 3}$
External torque applied to the pizza dough expressed in $\mathcal{O}$	$\boldsymbol{\upsilon} \in \mathbb{R}^3$.
Gravity acceleration	$\gamma \simeq 9.81$ m/s^2
Mass of the pizza dough	$m_o > 0$
Linear velocity of $\mathcal{O}$ with respect to $\mathcal{W}$ expressed in the body frame $\mathcal{O}$	$\mathbf{v} \in \mathbb{R}^3$
External force applied to the pizza dough expressed in $\mathcal{O}$	$\mathbf{f} \in \mathbb{R}^3$.
Mass matrix of the manipulator	$\mathbf{M}_r(\boldsymbol{\theta}) \in \mathbb{R}^{n\times n}$
Coriolis matrix of the manipulator	$\mathbf{C}_r(\boldsymbol{\theta}, \dot{\boldsymbol{\theta}}) \in \mathbb{R}^{n\times n}$
Gravitational terms of the manipulator	$\mathbf{n}(\boldsymbol{\theta}) \in \mathbb{R}^n$
Manipulator joint torques	$\boldsymbol{\tau} \in \mathbb{R}^n$
Mass matrix of the pizza dough	$\mathbf{M}_o \in \mathbb{R}^{6\times 6}$
Wrench acting on the pizza dough	$\mathbf{f}_o \in \mathbb{R}^6$
Coriolis and gravity term of the pizza dough	$\boldsymbol{\phi}_o \in \mathbb{R}^6$
Contact force at the ith finger	$\mathbf{f}_{c_i} \in \mathbb{R}^4$

2 Grasp Constraints

For the formulation of the contact constraint, we follow [5]. We assume that the fingers contact the dough softly and that the contact points are fixed. The contact frame $\mathcal{C}_i$ moves with the object, while the finger frame $\mathcal{F}_i$ moves with the hand. The system and the frames are schematically displayed in Fig. 1.

The soft contact condition yields the equation $\mathbf{B}_{c_i}^T \mathbf{v}_{f_i c_i}^b = \mathbf{0}_4$, for each finger enumerated by the index $i \in \mathfrak{I} := \{1, \ldots, k\}$, where $\mathbf{B}_{c_i} = \begin{bmatrix} \mathbf{I}_3 & \mathbf{0}_3 \\ \mathbf{O}_3 & \mathbf{e}_3 \end{bmatrix} \in \mathbb{R}^{6\times4}$. We adopt the convention that when the superscript over a quantity is omitted, it is expressed in the body coordinate frame, "b". After some algebra, we find that

$$\mathbf{v}_{f_i c_i} = -\mathrm{Ad}_{\mathbf{T}_{hc_i}}^{-1} \mathbf{v}_{sh} + \mathrm{Ad}_{\mathbf{T}_{oc_i}}^{-1} \mathbf{v}_{so}.$$

where $\mathrm{Ad} : SE(3) \to \mathrm{Aut}\,(\mathfrak{se}(3))$ is the adjoint operator on $SE(3)$ whose range is the set of isomorphisms of $\mathfrak{se}(3)$.Therefore, we have the contact constraint

$$\mathbf{B}_{c_i}^T \left(-\mathrm{Ad}_{\mathbf{T}_{hc_i}}^{-1} \mathbf{v}_{sh} + \mathrm{Ad}_{\mathbf{T}_{oc_i}}^{-1} \mathbf{v}_{so} \right) = \mathbf{0}_4,$$

where the $\vee$ and $\wedge$ operators denote the isomorphism from $\mathfrak{se}(3)$ to $\mathbb{R}^6$ and its inverse, respectively. Because $\mathbf{v}_{sh} = \mathbf{J}\dot{\boldsymbol{\theta}}$, we have

$$\mathbf{B}_{c_i}^T \mathbf{v}_{f_i c_i} = \mathbf{B}_{c_i}^T \left(-\mathrm{Ad}_{\mathbf{T}_{hc_i}}^{-1} J\dot{\theta} + \mathrm{Ad}_{\mathbf{T}_{oc_i}}^{-1} \mathbf{v}_{so} \right) = \mathbf{0}_4. \tag{1}$$

Defining the following constraint matrices

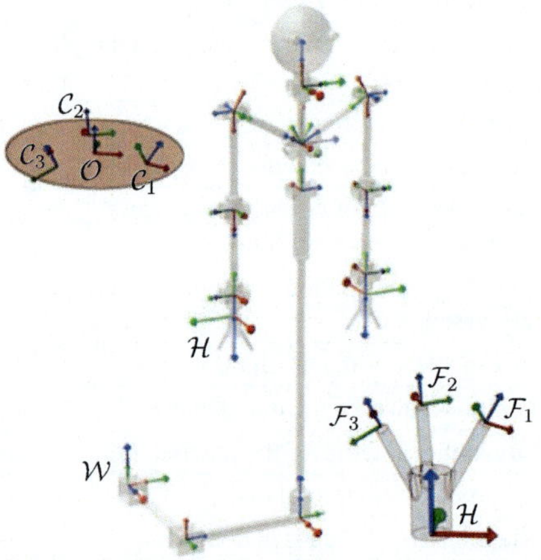

Fig. 1 RoDyMan Prototype: frames and joint axes

$$\mathbf{B}_c = \begin{bmatrix} \mathbf{B}_{c_1} & \mathbf{O}_{6\times4} & \cdots & \mathbf{O}_{6\times4} \\ \mathbf{O}_{6\times4} & \mathbf{B}_{c_2} & \cdots & \mathbf{O}_{6\times4} \\ \vdots & \vdots & \ddots & \vdots \\ \mathbf{O}_{6\times4} & \mathbf{O}_{6\times4} & \cdots & \mathbf{B}_{c_k} \end{bmatrix} \in \mathbb{R}^{6k\times4k},$$

$$\mathbf{A} = \mathbf{B}_c^T \begin{bmatrix} \mathrm{Ad}_{\mathbf{T}_{oc_1}^{-1}} & -\mathrm{Ad}_{\mathbf{T}_{hc_1}^{-1}}\mathbf{J} \\ \mathrm{Ad}_{\mathbf{T}_{oc_2}^{-1}} & -\mathrm{Ad}_{\mathbf{T}_{hc_2}^{-1}}\mathbf{J} \\ \vdots & \vdots \\ \mathrm{Ad}_{\mathbf{T}_{oc_k}^{-1}} & -\mathrm{Ad}_{\mathbf{T}_{hc_k}^{-1}}\mathbf{J} \end{bmatrix} \in \mathbb{R}^{4k\times(6+n)},$$

allows us to rewrite the contact constraints (1) in the Pfaffian form $\mathbf{A}\mathbf{v} = \mathbf{0}_{4k}$, where $\mathbf{v} = \left[\mathbf{v}_{so}^T \; \dot{\boldsymbol{\theta}}^T\right]^T \in \mathbb{R}^{6+n}$. We call this equation the differential contact constraints (DCC)

$$\mathbf{B}_{c_i}^T \left(\left[\mathbf{v}_{hc_i},\ \mathrm{Ad}_{\mathbf{T}_{hc_i}}^{-1}\mathbf{v}_{sh}\right] - \left[\mathbf{v}_{oc_i},\ \mathrm{Ad}_{\mathbf{T}_{oc_i}}^{-1}\mathbf{v}_{so}\right] - \mathrm{Ad}_{\mathbf{T}_{hc_i}}^{-1}\dot{\mathbf{v}}_{sh} + \mathrm{Ad}_{\mathbf{T}_{oc_i}}^{-1}\dot{\mathbf{v}}_{so}\right) = \mathbf{0}_4.$$

This can be simplified by considering the following

$$\mathbf{v}_{hc_i} = -\mathrm{Ad}_{\mathbf{T}_{hc_i}^{-1}}\mathbf{v}_{sh} + \mathrm{Ad}_{\mathbf{T}_{oc_i}^{-1}}\mathbf{v}_{so} + \mathbf{v}_{oc_i},$$

so that

$$\left[\mathbf{v}_{hc_i},\ \mathrm{Ad}_{\mathbf{T}_{hc_i}^{-1}}\mathbf{v}_{sh}\right] = \left[\mathrm{Ad}_{\mathbf{T}_{oc_i}^{-1}}\mathbf{v}_{so},\ \mathrm{Ad}_{\mathbf{T}_{hc_i}^{-1}}\mathbf{v}_{st}\right] + \left[\mathbf{v}_{oc_i},\ \mathrm{Ad}_{\mathbf{T}_{hc_i}^{-1}}\mathbf{v}_{st}\right].$$

Plugging this back to the DCC, we get

$$\mathbf{B}_{c_i}^T \left(\left[\mathrm{Ad}_{\mathbf{T}_{oc_i}}^{-1}\mathbf{v}_{so},\ \mathrm{Ad}_{\mathbf{T}_{hc_i}}^{-1}\mathbf{v}_{sh}\right] + \left[\mathbf{v}_{oc_i},\ \mathrm{Ad}_{\mathbf{T}_{hc_i}}^{-1}\mathbf{v}_{sh}\right] - \left[\mathbf{v}_{oc_i},\ \mathrm{Ad}_{\mathbf{T}_{oc_i}}^{-1}\mathbf{v}_{so}\right] - \mathrm{Ad}_{\mathbf{T}_{hc_i}}^{-1}\dot{\mathbf{v}}_{st} + \mathrm{Ad}_{\mathbf{T}_{oc_i}}^{-1}\dot{\mathbf{v}}_{so}\right) = \mathbf{0}_4.$$

Now, since the contact point is fixed in the object frame of reference, we have $\mathbf{v}_{oc_i}^b \equiv \mathbf{0}_6$ for all $i \in \mathfrak{I}$. Therefore, the second and the third brackets in the expression above drop, yielding a simplified differential contact constraint

$$\mathbf{B}_{c_i}^T \left(\left[\mathrm{Ad}_{\mathbf{T}_{oc_i}^{-1}}\mathbf{v}_{so},\ \mathrm{Ad}_{\mathbf{T}_{hc_i}^{-1}}\mathbf{v}_{sh}\right] - \mathrm{Ad}_{\mathbf{T}_{hc_i}}^{-1}\dot{\mathbf{v}}_{sh} + \mathrm{Ad}_{\mathbf{T}_{oc_i}}^{-1}\dot{\mathbf{v}}_{so}\right) = \mathbf{0}_4.$$

Invoking the relationship between the joint velocities and the end-effector velocity, we get

$$\mathbf{B}_{c_i}^T \left(\left[\mathrm{Ad}_{\mathbf{T}_{oc_i}^{-1}}\mathbf{v}_{so},\ \mathrm{Ad}_{\mathbf{T}_{hc_i}^{-1}}\mathbf{v}_{sh}\right] - \mathrm{Ad}_{\mathbf{T}_{hc_i}}^{-1}\dot{\mathbf{J}}\dot{\boldsymbol{\theta}} - \mathrm{Ad}_{\mathbf{T}_{hc_i}}^{-1}\mathbf{J}\ddot{\boldsymbol{\theta}} + \mathrm{Ad}_{\mathbf{T}_{oc_i}}^{-1}\dot{\mathbf{v}}_{so}\right) = \mathbf{0}_4. \quad (2)$$

3 Kinematics

In order to map the tool velocities and forces to the joint velocities and torques, we need to derive the kinematics of the manipulator and the object. Object kinematics are well-known rigid body kinematics. To manipulate the pizza dough, we use a 7 DoF serial robotic manipulator. We want this manipulator to have similar characteristics as the human arm. To that end, we choose the topology of the manipulator to be that of an S-R-S robot. We have developed a humanoid robotic torso mounted on an omnidirectional mobile platform, called RoDyMan, in our laboratory and the S-R-S manipulator corresponds to one of the arms of RoDyMan robot, whose frontal view is shown in Fig. 1. We have chosen to perform the tossing and catching operations with the right arm. This viewpoint lets us identify the first joint with the shoulder, the second joint with the elbow, and the last joint with the wrist.

Again, we follow [5] for the derivation of the kinematics of the S-R-S manipulator. This procedure involves using the exponential coordinates for the end-effector position and therefore maintains the inherent geometric features of the manipulator. The joint axes, expressed in $\mathcal{W}$, are given by $\boldsymbol{\xi}_i = \begin{bmatrix} -\boldsymbol{\omega}_i \times \mathbf{p}_i \\ \boldsymbol{\omega}_i \end{bmatrix}$, for each joint $i \in \{1, \ldots, 7\}$, where, $\boldsymbol{\omega}_i \in \mathbb{R}^3$ is the axis of rotation of each of the ith joint expressed in $\mathcal{W}$ and $\mathbf{p}_i \in \mathbb{R}^3$ is any point on this axis of rotation, written in $\mathcal{W}$.

The position forward kinematics is then given by

$$\mathbf{T}_{sh}(\theta) = e^{\hat{\xi}_1\theta_1} e^{\hat{\xi}_2\theta_2} e^{\hat{\xi}_3\theta_3} e^{\hat{\xi}_4\theta_4} e^{\hat{\xi}_5\theta_5} e^{\hat{\xi}_6\theta_6} e^{\hat{\xi}_7\theta_7} \mathbf{T}_{sh}(0), \tag{3}$$

where $\mathbf{T}_{sh}(0)$ is the initial pose of the end-effector with respect to the base frame and the exponential mapping $exp : \mathfrak{se}(3) \to SE(3)$ is as defined in [5].

The relationship between the joint velocities and the end-effector velocities is given by the body manipulator Jacobian, $\mathbf{J}^b \in \mathbb{R}^{6\times n}$. This is found by the following computation

$$\mathbf{J}^b = \mathbf{J}^b_{sh} = \left[\boldsymbol{\xi}_1^\dagger \; \boldsymbol{\xi}_2^\dagger \; \boldsymbol{\xi}_3^\dagger \; \boldsymbol{\xi}_4^\dagger \; \boldsymbol{\xi}_5^\dagger \; \boldsymbol{\xi}_6^\dagger \; \boldsymbol{\xi}_7^\dagger\right],$$

where

$$\boldsymbol{\xi}_i^\dagger = \mathrm{Ad}^{-1}_{\left(e^{\hat{\xi}_i\theta_i} \cdots e^{\hat{\xi}_7\theta_7}\mathbf{T}_{sh}(0)\right)} \boldsymbol{\xi}_i,$$

with $i = 1, \ldots, 7$. Given the joint velocities, we can then find the end-effector velocities by the relation $\mathbf{v}^b_{sh} = \mathbf{J}^b\dot{\boldsymbol{\theta}}$. Note that away from the singularities, $\mathbf{J}^b$ has rank 6. Given an end-effector velocity, there are infinitely many joint velocities that supply this end-effector velocity. To achieve the desired end-effector velocity, we need a redundancy resolution technique to decide which joint velocity we want to supply.

4 Dynamics

We have to model the natural motion of the pizza dough, which is a deformable object. We assume the dough has a circular shape with a radius that changes due to the various S^1-symmetric forces. The height of the geometric center of the pizza may change due to fluid dynamical forces on the dough.

4.1 Variable Inertia Rigid Body Orientation Dynamics

Although the mass of the dough remains constant, its moment of inertia changes due to the change in its shape. In turn, this variation affects the conventional Euler-Lagrange equations of motion of the rigid body. In this section, we derive the appropriate equations of motion for the orientation dynamics of the dough in the body frame.

The Lagrangian of the pizza dough is given by

$$L(\mathbf{R}_{so}, \dot{\mathbf{R}}_{so}) = -\frac{\det(\mathbf{I}_o)}{2}\mathrm{tr}\Big[\big(\mathbf{I}_o^{-1}\hat{\omega}_o\big)^2\Big] = \frac{1}{2}\omega_o^T\mathbf{I}_o\omega_o, \tag{4}$$

where det () and tr() indicates the determinant and the trace of a square matrix, respectively. The dependence of the Lagrangian (4) on $\mathbf{R}_{so}$ and $\dot{\mathbf{R}}_{so}$ stems from the orientation kinematics, namely, from the equation $\hat{\omega}_{so} = \mathbf{R}_{so}^T\dot{\mathbf{R}}_{so}$.

We would like to calculate the conventional Euler-Lagrange equations for the rigid body with a variable moment of inertia. It is well-known that the rotational dynamics is symmetric with respect to the action of the group $SO(3)$ on itself. Therefore, by the theory of reduction of dynamics [12], the equations of motion drop to the quotient $TSO(3)/SO(3) \cong \mathfrak{so}(3)$. We have

$$\frac{\partial L}{\partial \mathbf{R}_{so}} = -\det(\mathbf{I}_o)\frac{\partial}{\partial \mathbf{R}_{so}}\mathrm{tr}\Big[\big(\mathbf{I}_o^{-1}\mathbf{R}_{so}^T\dot{\mathbf{R}}_{so}\big)^2\Big] = -\det(\mathbf{I}_o)\mathbf{R}_{so}\left[\hat{\omega}_o, \mathbf{I}_o^{-1}\hat{\omega}_o\mathbf{I}_o^{-1}\right]$$
$$\frac{\partial L}{\partial \dot{\mathbf{R}}_{so}} = -\frac{\det(\mathbf{I}_o)}{2}\frac{\partial}{\partial \dot{\mathbf{R}}_{so}}\mathrm{tr}\Big[\big(\mathbf{I}_o^{-1}\mathbf{R}_{so}^T\dot{\mathbf{R}}_{so}\big)^2\Big] = \det(\mathbf{I}_o)\mathbf{R}_{so}\mathbf{I}_o^{-1}\dot{\hat{\omega}}_o\mathbf{I}_o^{-1}.$$

Here, the differentiation with respect to the matrices $\mathbf{R}_{so}$ and $\dot{\mathbf{R}}_{so}$ is performed using the familiar matrix differentiation rules, followed by projecting the result onto the tangent space $T_R SO(3)$.

Pulling the derivatives above back to the Lie algebra via the left translation map yields

$$\frac{\partial L}{\partial \mathbf{R}_{so}} = -\det(\mathbf{I}_o)\left[\hat{\omega}_o, \mathbf{I}_o^{-1}\hat{\omega}_o\mathbf{I}_o^{-1}\right], \quad \frac{\partial L}{\partial \dot{\mathbf{R}}_{so}} = \det(\mathbf{I}_o)\mathbf{I}_o^{-1}\dot{\hat{\omega}}_o\mathbf{I}_o^{-1}.$$

Finally, the Euler-Poincaré equations motion are found to be

$$\mathbf{I}_o\dot{\omega}_o - \mathbf{I}_o\left(\hat{\omega}_o\mathbf{I}_o^{-1}\frac{\mathrm{d}\mathbf{I}_o}{\mathrm{d}t} + \frac{\mathrm{d}\mathbf{I}_o}{\mathrm{d}t}\mathbf{I}_o^{-1}\hat{\omega}_o\right)^{\vee} + \mathrm{tr}\left(\mathbf{I}_o^{-1}\frac{\mathrm{d}\mathbf{I}_o}{\mathrm{d}t}\right)\mathbf{I}_o\omega_o + \omega_o \times \mathbf{I}_o\omega_o = \upsilon. \tag{5}$$

4.2 Rigid Body Translational Dynamics

The translational dynamics of the object are the usual Newton's equations written in the frame $\mathcal{O}$

$$m_o\dot{\mathbf{v}}_o + \omega_o \times m_o\mathbf{v}_o + m_o\gamma\mathbf{R}_{so}^T\mathbf{e}_3 = \mathbf{f}. \tag{6}$$

4.3 RoDyMan Arm Manipulator Dynamics

We follow [5] once again to derive the manipulator dynamics, whose details have been omitted for the sake of simplicity. This procedure yields

$$\mathbf{M}_r(\theta)\ddot{\theta} + \mathbf{C}_r(\theta, \dot{\theta})\dot{\theta} + \mathbf{n}(\theta) = \tau. \tag{7}$$

4.4 Object and Manipulator Combined Dynamics

When the fingers and the object are in contact, there are forces arising from this interaction. In the combined system, the forces on the object may only be imparted via the contact, therefore we have the following object dynamics

$$\mathbf{M}_o\dot{\mathbf{v}}_{so}^b + \phi_o\left(\mathbf{T}_{so}, \mathbf{v}_{so}^b\right) = \mathbf{f}_o, \tag{8}$$

where

$$\mathbf{M}_o = \begin{bmatrix} m_o\mathbf{I}_3 & \mathbf{O}_3 \\ \mathbf{O}_3 & \mathbf{I}_o \end{bmatrix} \in \mathbb{R}^{6\times 6},$$

$$\mathbf{f}_o = \sum_{i=1}^{k} \mathrm{Ad}_{\mathbf{T}_{oc_i}^{-1}}^T \mathbf{B}_{c_i}\mathbf{f}_{c_i} \in \mathbb{R}^6,$$

$$\phi_o = \begin{bmatrix} \omega_o \times m_o\mathbf{v}_o + m_o\gamma\mathbf{R}_{so}^T\mathbf{e}_3 \\ -\mathbf{I}_o\left(\hat{\omega}_o\mathbf{I}_o^{-1}\frac{\mathrm{d}\mathbf{I}_o}{\mathrm{d}t} + \frac{\mathrm{d}\mathbf{I}_o}{\mathrm{d}t}\mathbf{I}_o^{-1}\hat{\omega}_o\right)^{\vee} + \mathrm{tr}\left(\mathbf{I}_o^{-1}\frac{\mathrm{d}\mathbf{I}_o}{\mathrm{d}t}\right)\mathbf{I}_o\omega_o + \hat{\omega}_o\mathbf{I}_o\omega_o \end{bmatrix} \in \mathbb{R}^6,$$

where the dependencies of ϕ_o have been dropped.

Similarly, each finger will experience a reaction force due to Newton's third law. These reaction forces occurring at the contact locations are mapped to the hand frame by the corresponding adjoint transformation. Subsequently, the wrench at the hand frame are mapped to the joint torques by the body manipulator Jacobian. Therefore, the reaction force on the hand frame and on the joints are given by

$$\mathbf{f}_h = -\sum_{i=1}^{k} \mathrm{Ad}^T_{\mathbf{T}^{-1}_{hc_i}} \mathbf{B}_{c_i} \mathbf{f}_{c_i} \in \mathbb{R}^6, \qquad \boldsymbol{\tau}_{\mathrm{reac}} = \mathbf{J}^{b^T} \mathbf{f}_h \in \mathbb{R}^n.$$

Consequently, the manipulator dynamics become

$$\mathbf{M}_r \ddot{\boldsymbol{\theta}} + \underbrace{\mathbf{C}_r(\boldsymbol{\theta}, \dot{\boldsymbol{\theta}})\dot{\boldsymbol{\theta}} + \mathbf{n}(\boldsymbol{\theta})}_{\triangleq \phi_r(\boldsymbol{\theta}, \dot{\boldsymbol{\theta}})} = \boldsymbol{\tau} - \sum_{i=1}^{k} \mathbf{J}^{b^T} \mathrm{Ad}^T_{\mathbf{T}^{-1}_{hc_i}} \mathbf{B}_{c_i} \mathbf{f}_{c_i}.$$

In order to express these equations together, we are going to introduce some more definitions, let

$$\mathbf{M} = \begin{bmatrix} \mathbf{M}_o & \mathbf{O}_{6\times n} \\ \mathbf{O}_{n\times 6} & \mathbf{M}_r \end{bmatrix} \in \mathbb{R}^{(6+n)\times(6+n)},$$

$$\phi = \begin{bmatrix} \phi_o \\ \phi_r \end{bmatrix} \in \mathbb{R}^{6+n},$$

$$\bar{\boldsymbol{\tau}} = \begin{bmatrix} \mathbf{0}_6 \\ \boldsymbol{\tau} \end{bmatrix} \in \mathbb{R}^{6+n}, \quad \mathbf{f}_c = \begin{bmatrix} \mathbf{f}_{c_1} \\ \vdots \\ \mathbf{f}_{c_k} \end{bmatrix} \in \mathbb{R}^{4k}.$$

Then, we have the following system of differential-algebraic equations

$$\mathbf{M}\dot{\bar{\mathbf{v}}} + \phi(\mathbf{T}_{so}, \mathbf{v}_{so}, \boldsymbol{\theta}, \dot{\boldsymbol{\theta}}) = \bar{\boldsymbol{\tau}} + \mathbf{A}^T \mathbf{f}_c \tag{9}$$

$$\mathbf{f}_c = \left(\mathbf{A}\mathbf{M}^{-1}\mathbf{A}^T\right)^{-1} \left(\mathbf{A}\mathbf{M}^{-1}(\phi - \bar{\boldsymbol{\tau}}) - \chi\right), \tag{10}$$

or, equivalently, we have

$$\begin{bmatrix} \mathbf{M} & -\mathbf{A}^T \\ -\mathbf{A} & \mathbf{O}_{4k} \end{bmatrix} \begin{bmatrix} \dot{\bar{\mathbf{v}}} \\ \boldsymbol{\lambda} \end{bmatrix} + \begin{bmatrix} \phi \\ \mathbf{0}_{4k} \end{bmatrix} = \begin{bmatrix} \bar{\boldsymbol{\tau}} \\ \dot{\mathbf{A}}\bar{\mathbf{v}} \end{bmatrix}, \tag{11}$$

with

$$\bar{\mathbf{v}} = \begin{bmatrix} \mathbf{v}_{so}^{b^T} & \dot{\theta}^T \end{bmatrix}^T \in \mathbb{R}^{6+n},$$

$$\mathbf{A}^T = \begin{bmatrix} \mathrm{Ad}_{\mathbf{T}_{oc_1}^{-1}}^T \mathbf{B}_{c_1} & \cdots & \mathrm{Ad}_{\mathbf{T}_{oc_k}^{-1}}^T \mathbf{B}_{c_k} \\ -\mathbf{J}^{b^T} \mathrm{Ad}_{\mathbf{T}_{hc_1}^{-1}}^T \mathbf{B}_{c_1} & \cdots & -\mathbf{J}^{b^T} \mathrm{Ad}_{\mathbf{T}_{hc_k}^{-1}}^T \mathbf{B}_{c_k} \end{bmatrix} \in \mathbb{R}^{(6+n)\times 4k}.$$

5 Trajectory Generation

In order to render the catching problem as easy as possible, it may be desirable to toss the dough in the air in such a way that it neither has any lateral linear velocity nor any angular velocity along the roll and pitch axes. As desirable as it might be, unmodeled dynamics due to aerodynamic forces and the compliance of the dough inevitably imparts such velocities while the dough is in flight. We assume the presence of an estimator that predicts the dough's motion while it is in the air and updates its prediction at regular intervals. To begin with, we intend to consider the visual tracking problem of a dough of a particular rigid shape, concentrating on obtaining the motion of its center of mass using our prior experience in the estimation of such objects [3, 4]. Afterwards, we intend to develop visual tracking algorithms to track the freely falling object's deformation and orientation. Preliminary results on tracking deformation and orientation can be found in the previous chapters. In this section, we tackle the problem of generating an optimal trajectory for the end-effector of the robotic hand, given the prediction of the motion of the dough.

5.1 Theory

We are interested in trajectories for which we can specify the initial position, velocity, acceleration, and final velocity and acceleration of the motion. The final position is going to be dictated by the final position of the dough. This final position of the dough is a function of the final time, which will be a parameter to be determined by the optimization process. Note that the motion may be specified in either the joint space, a torus, or the task space, the special Euclidean group of three dimensions over the reals. At this stage, we are going to evade this distinction by assuming that the path we would like to generate is on an arbitrary Riemannian manifold [13] Q. Let us call this path $c : (a, b) \to Q$, and the metric on Q is denoted by $\langle \cdot, \cdot \rangle$. We let $\mathbf{f} : (-\epsilon, \epsilon) \times (a, b) \to Q$ be a variation of c, with $\epsilon > 0$,

$$\mathbf{f}(0, t) = \mathbf{c}(t), \quad \forall t \in [a, b] \quad \text{and} \quad f(s, a) = c(a).$$

We have two vector fields of importance along the path $\mathbf{c}$. The first one is called the *variation field*, defined by

$$\mathbf{s}_{c(s)} = \frac{\partial \mathbf{f}(s,t)}{\partial s} = \frac{\mathrm{d}\mathbf{f}_t(s)}{\mathrm{d}s},$$

and second one is the velocity vector field of $\mathbf{c}$, given by

$$\mathbf{v}_{c(s)} = \frac{\mathrm{d}\mathbf{c}(t)}{\mathrm{d}t} = \frac{\partial \mathbf{f}(s,t)}{\partial t} = \frac{\mathrm{d}\mathbf{f}_s(t)}{\mathrm{d}t},$$

In order to perform calculus on the curves of this Riemannian manifold, we introduce the Levi-Civita connection, ∇. Given a curve $\mathbf{c}(t)$ and a connection, there exists a covariant derivative, which we denote by $\frac{\mathrm{D}}{\mathrm{d}t}$.

Let $\mathfrak{X}(Q)$ denote the set of vector fields on Q. The curvature R of a Riemannian manifold Q is a correspondence that associates to every pair $X, Y \in \mathfrak{X}(Q)$ a mapping $R(X, Y) : \mathfrak{X}(Q) \to \mathfrak{X}(Q)$ given by

$$R(X, Y)Z = \nabla_Y \nabla_X Z - \nabla_X \nabla_Y Z + \nabla_{[X,Y]} Z, \quad Z \in \mathfrak{X}(M).$$

Many paths satisfy the conditions on initial and final positions, velocities, and accelerations. Since we would like to specify the initial position, velocity, acceleration, and final velocity and acceleration, we need a sixth-order differential equation to plug these constraints as boundary conditions. Such a differential equation is what we end up with when we minimize the jerk functional, for example. On the other hand, the jerk functional is not necessarily a measure of how much effort is expended as it is a measure of vibrations within the system. If we would like to minimize the end-effector wrenches needed to catch the dough, we would have to minimize the acceleration functional. This approach, however, yields a fourth-order differential equation and so does not lend itself to imposing the desired boundary conditions.

In order to overcome this quandary, we propose to minimize not just the acceleration nor just the jerk, but a convex combination of the two. As long as we keep the weight of the jerk functional away from zero, we shall still end up with a sixth-order differential equation and will be able to impose the desired boundary conditions. Furthermore, we can tune the weight of the acceleration functional so that it is arbitrarily close to unity, which would practically ignore the effect of the jerk functional and yield an almost optimal minimum acceleration path that successfully catches the dough. With this motivation, we define the cost functional to be minimized to be

$$L_{\mathbf{c}}(s) := \int_{t_0}^{t_f + s\delta t_f} \alpha \left\langle \frac{\mathrm{D}^2 \mathbf{v}_{c(s)}}{\partial t^2}, \frac{\mathrm{D}^2 \mathbf{v}_{c(s)}}{\partial t^2} \right\rangle + \beta \left\langle \frac{\mathrm{D}\mathbf{v}_{c(s)}}{\partial t}, \frac{\mathrm{D}\mathbf{v}_{c(s)}}{\partial t} \right\rangle \mathrm{d}t \qquad (12)$$

where the weights α and β satisfy $\alpha + \beta = 1$. Here, we consider only the case where the final position is left free and is part of the minimization problem. We calculate the first variation of this functional using analogous calculations as in [10] and, in addition, taking special care of the free endpoint conditions. This yields

$$
\begin{aligned}
\frac{1}{2}\frac{\mathrm{d}}{\mathrm{d}s}L_{\mathbf{c}}(s) = \int_{t_0}^{t_f} \Bigg\langle \alpha \Bigg(-\frac{\mathrm{D}^5\mathbf{v}_{c(s)}}{\partial t^5} - R\left(\mathbf{v}_{c(s)}, \frac{\mathrm{D}^3\mathbf{v}_{c(s)}}{\partial t^3}\right)\mathbf{v}_{c(s)} \\
+ R\left(\frac{\mathrm{D}\mathbf{v}_{c(s)}}{\partial t}, \frac{\mathrm{D}^2\mathbf{v}_{c(s)}}{\partial t^2}\right)\mathbf{v}_{c(s)}\Bigg) \\
+ \beta\left(\frac{\mathrm{D}^3\mathbf{v}_{c(s)}}{\partial t^3} + R\left(\mathbf{v}_{c(s)}, \frac{\mathrm{D}\mathbf{v}_{c(s)}}{\partial t}\right)\mathbf{v}_{c(s)}\right), \ \mathbf{s}_{c(s)}\Bigg\rangle \mathrm{d}t \\
+ \left(\frac{1}{2}\alpha\left\langle\frac{\mathrm{D}^2\mathbf{v}_{c(s)}}{\partial t^2}, \frac{\mathrm{D}^2\mathbf{v}_{c(s)}}{\partial t^2}\right\rangle + \left\langle\alpha\frac{\mathrm{D}^4\mathbf{v}_{c(s)}}{\partial t^4} - \beta\frac{\mathrm{D}^2\mathbf{v}_{c(s)}}{\partial t^2}, \frac{\mathrm{d}\zeta}{\mathrm{d}t}\right\rangle\right)\Bigg|_{t_f} \delta t_f.
\end{aligned}
$$

Consequently, the necessary conditions for the minimization are

$$
\begin{aligned}
\alpha\left(-\frac{\mathrm{D}^5\mathbf{v}_{c(s)}}{\partial t^5} - R\left(\mathbf{v}_{c(s)}, \frac{\mathrm{D}^3V}{\partial t^3}\right)\mathbf{v}_{c(s)} + R\left(\frac{\mathrm{D}\mathbf{v}_{c(s)}}{\partial t}, \frac{\mathrm{D}^2\mathbf{v}_{c(s)}}{\partial t^2}\right)\mathbf{v}_{c(s)}\right) \\
+ \beta\left(\frac{\mathrm{D}^3\mathbf{v}_{c(s)}}{\partial t^3} + R\left(\mathbf{v}_{c(s)}, \frac{\mathrm{D}\mathbf{v}_{c(s)}}{\partial t}\right)\mathbf{v}_{c(s)}\right) = \mathbf{0}_3, \quad \forall t_0 \le t \le t_f \qquad (13a)
\end{aligned}
$$

$$
\left(\frac{1}{2}\alpha\left\langle\frac{\mathrm{D}^2\mathbf{v}_{c(s)}}{\partial t^2}, \frac{\mathrm{D}^2\mathbf{v}_{c(s)}}{\partial t^2}\right\rangle + \left\langle\alpha\frac{\mathrm{D}^4\mathbf{v}_{c(s)}}{\partial t^4} - \beta\frac{\mathrm{D}^2\mathbf{v}_{c(s)}}{\partial t^2}, \frac{\mathrm{d}\zeta}{\mathrm{d}t}\right\rangle\right)\Bigg|_{t_f} = 0 \qquad (13b)
$$

$$
\mathbf{c}(t_0) = \gamma_0, \quad \mathbf{v}_{c(s)}(t_0) = \mathbf{v}_0, \quad \frac{\mathrm{D}\mathbf{v}_{c(s)}}{\partial t}(t_0) = \mathbf{a}_0, \qquad (13c)
$$

$$
\mathbf{c}(t_f) = \zeta(t_f), \quad \mathbf{v}_{c(s)}(t_f) = \frac{\mathrm{D}\mathbf{v}_{c(s)}}{\partial t}(t_f) = \mathbf{0}_3, \qquad (13d)
$$

where suitable parametrized initial and final conditions terms have been introduced.

5.2 *Generating Hand Frame Trajectories*

In this section, we discuss how we generate optimal trajectories for the motion of the end-effector, given the initial position $\gamma_o = \left[x_0 \ y_0 \ z_0\right]^T \in \mathbb{R}^3$ and velocity $\mathbf{v}_{c(s)} = \left[v_{d0,x} \ v_{d0,y} \ v_{d0,z}\right]^T \in \mathbb{R}^3$ of the dough at the moment it is tossed by the end-effector. With this data, we can integrate the dynamics of the center of mass of the dough to find the path it takes

$$
\mathbf{p}_d(t) = \begin{bmatrix} x_0 + v_{d0,x}t \\ y_0 + v_{d0,y}t \\ z_0 + v_{d0,z}t - \frac{1}{2}\gamma t^2 \end{bmatrix} \in \mathbb{R}^3.
$$

Moreover, we can also numerically integrate the rotational dynamics of the dough, derived in Sect. 4.1, to find the rotational path the dough takes. We then use this information to find the extremal of the cost functional defined in the previous Sect. 5.1. In particular, we break this problem into two. We compute the optimal translational path using the necessary conditions (13). The final time generated from this procedure is then used to determine the orientation the tool needs to have by evaluating the orientational trajectory of the dough at this instant. Without breaking the problem into two, one could also find the optimal trajectory directly in $SE(3)$. However, in this case, the rotational motion of the dough must be interpolated using one of the available techniques [10]. Moreover, the free endpoint condition presented in the necessary conditions (13) has to be solved iteratively, injecting a fair deal of complexity into the problem.

The necessary conditions presented in Eq. (13) yields the differential equation $-\alpha \mathbf{p}^{(6)} + \beta \mathbf{p}^{(4)} = \mathbf{0}_3$, with $\mathbf{p}(t) \in \mathbb{R}^3$ the position of the end effector at time t, in addition to the following concrete boundary condition equations for the translational motion of the end-effector

$$\begin{bmatrix} \mathbf{p}(t_0) - \mathbf{p}_0 \\ \dot{\mathbf{p}}(t_0) - \mathbf{v}_0 \\ \ddot{\mathbf{p}}(t_0) - \mathbf{a}_0 \\ \mathbf{p}(t_f) - \mathbf{p}_d(t_f) - l_h \mathbf{R}_d(t_f)\mathbf{e}_3 \\ \dot{\mathbf{p}}(t_f) \\ \ddot{\mathbf{p}}(t_f) \\ \mathbf{p}^{(5)}(t_f)^T \dot{\mathbf{p}}_d(t_f) + \frac{1}{2}\mathbf{p}^{(3)}(t_f)^T \mathbf{p}^{(3)}(t_f) \end{bmatrix} = \mathbf{0}_{19},$$

where the first three equations are self-explanatory. They are boundary conditions to match the hand frame's initial position, velocity, and acceleration with that of the dough. Assuming the body z-axis of the dough points orthogonal to the plane of the dough and $l_h > 0$ is a constant, the fourth equation states that the end-effector should be placed directly underneath the dough at the final time. The fifth and sixth conditions impose the design choice that the final translational velocity of the tool frame vanishes. Upon experimentation, it was empirically observed that no slippage occurs if the hand is stopped at the time of contact (catching). Lastly, the final equation is the free endpoint condition from which we solve for the final time at which the catching should occur.

The general solution to the differential equation is given by

$$\mathbf{p}(t) = \begin{bmatrix} x(t) \\ y(t) \\ z(t) \end{bmatrix}$$

$$= \sum_{n=0}^{3} \begin{bmatrix} a_n t^n \\ b_n t^n \\ c_n t^n \end{bmatrix} + \left(\frac{\alpha}{\beta}\right)^2 \begin{bmatrix} a_4 \cosh\left(\sqrt{\frac{\beta}{\alpha}} t\right) + a_5 \sinh\left(\sqrt{\frac{\beta}{\alpha}} t\right) \\ b_4 \cosh\left(\sqrt{\frac{\beta}{\alpha}} t\right) + b_5 \sinh\left(\sqrt{\frac{\beta}{\alpha}} t\right) \\ c_4 \cosh\left(\sqrt{\frac{\beta}{\alpha}} t\right) + c_5 \sinh\left(\sqrt{\frac{\beta}{\alpha}} t\right) \end{bmatrix}.$$

We use this analytic expression in the boundary conditions presented above along with a nonlinear root finder to solve for the 19 unknowns, namely, the coefficients a_0 through c_5 and the final time t_f.

The next task is to find the optimal rotational path the end-effector is desired to follow. The necessary conditions in Eq. (13) yields the differential equation

$$-\alpha\left(\omega^{(5)} + 2\omega \times \omega^{(4)} + \frac{5}{4}\omega \times \left(\omega \times \omega^{(3)}\right) + \frac{5}{2}\dot{\omega} \times \omega^{(3)} + \frac{1}{4}\omega \times (\omega \times (\omega \times \ddot{\omega})) + \frac{3}{2}\omega \times (\dot{\omega} \times \ddot{\omega}) - (\omega \times \ddot{\omega}) \right.$$
$$\left. \times \dot{\omega} - \frac{1}{4}(\omega \times \dot{\omega}) \times \ddot{\omega} - \frac{3}{8}\omega \times ((\omega \times \dot{\omega}) \times \dot{\omega}) - \frac{1}{8}(\omega \times (\omega \times \dot{\omega})) \times \dot{\omega}\right) + \beta\left(\omega^{(3)} + \omega \times \ddot{\omega}\right) = \mathbf{0}_3,$$

along with the boundary conditions

$$\begin{bmatrix} \mathbf{I}_3 - \mathbf{R}_{sh}(t_0)^T \mathbf{R}_{so}(t_0) \\ \hat{\boldsymbol{\omega}}_{sh}(t_0) - \mathbf{R}_{oh}^{-1}(t_0)\hat{\boldsymbol{\omega}}_{so}(t_0)\mathbf{R}_{oh}(t_0) \\ \hat{\boldsymbol{\alpha}}_{sh}(t_0) - \mathbf{R}_{oh}^{-1}(t_0)\hat{\boldsymbol{\alpha}}_{so}(t_0)\mathbf{R}_{oh}(t_0) \\ \mathbf{I}_3 - \mathbf{R}_{sh}(t_f)^T \mathbf{R}_{so}(t_f) \\ \hat{\boldsymbol{\omega}}_{sh}(t_f) \\ \hat{\boldsymbol{\alpha}}_{sh}(t_f) \end{bmatrix} = \mathbf{O}_{18\times 3},$$

where $\boldsymbol{\alpha} \in \mathbb{R}^3$ is the angular acceleration. The first three equations match the initial orientation, angular velocity, and angular acceleration of the hand frame and the dough, taking into account the constant rotational offset between them due to the choice of initial conditions of the reference frames. The fourth equation is a condition that the final orientations of the end effector and the dough should match, where the yaw rotation of the dough has been eliminated. The final two equations dictate that the final angular velocity and acceleration of the hand vanish.

This boundary value problem has no analytical solution and needs to be found using a numerical boundary-value problem solver. We have used Matlab's "*bvp4c*" function to accomplish this task.

Table 2 Projectile motion

Time of flight	t_f	$\frac{2\mathbf{v}_o^T \mathbf{e}_3}{\gamma}$
Max projectile height	h_d	$\frac{(\mathbf{v}_o^T \mathbf{e}_3)^2}{2\gamma}$
No. of revolutions	n_d	$\frac{\omega_o^T \mathbf{e}_3 t_f}{2\pi}$

6 Control Law

There are two distinct control objectives that we are pursuing in this work. The first one is to toss the pizza dough such that it reaches a certain height, $h_d > 0$, and by the time it comes back to the level of the fingers it has spun a certain number of times, $n_d > 0$. The second objective, which begins after the tossing phase, is to catch the pizza dough. We assume that we have generated an optimal trajectory for the S-R-S manipulator's end-effector to follow to intercept the dough. The second control objective is to have the end-effector follow this generated trajectory.

6.1 *Tossing*

The simplest plan to achieve the tossing goal may be excogitated by assuming that the center of mass of the dough behaves like a point mass during the free-fall phase. We can analytically determine how long the flight takes and the maximum height that the center of mass reaches with this assumption. These are the familiar formulas from elementary mechanics, cf. Table 2.

From the maximum height formula, we can derive an expression for the desired velocity of the dough at the time of release: $\mathbf{v}_{o,d} = \left[0\ 0\ \sqrt{2\gamma h_d}\right]^T$. Substituting the formula for the time of flight into the formula for n_d, we find a formula for the desired angular velocity of the dough at the time of release

$$n_d = \frac{\omega_o^T \mathbf{e}_3 t_f}{2\pi} = \frac{\omega_o^T \mathbf{e}_3 \mathbf{v}_o^T \mathbf{e}_3}{\gamma\pi} = \frac{\omega_o^T \mathbf{e}_3}{\pi}\sqrt{\frac{2h_d}{\gamma}} \implies \omega_{o,d} = \left[0\ 0\ \pi n_d \sqrt{\frac{\gamma}{2h_d}}\right]^T.$$

Combining these last two expressions for the desired linear and angular velocities, we get the desired velocity for the dough as

$$\mathbf{v}_{so,d} = \left[0\ 0\ \sqrt{2\gamma h_d}\ 0\ 0\ \pi n_d \sqrt{\frac{\gamma}{2h_d}}\right]^T.$$

In order to achieve this velocity, we need to act at the acceleration level and control the velocity accordingly. We go back to Eq. (8) and compute the contact forces, $\mathbf{f}_c$, that will impose an exponentially stable velocity dynamics to the desired velocity. To that end, let us prescribe

$$\sum_{i=1}^{k} \mathrm{Ad}^T_{\mathbf{T}^{-1}_{oc_i}} \mathbf{B}_{c_i} \mathbf{f}_{c_i} = -\mathbf{M}_o \mathbf{K}_d \left(\mathbf{b}_{so} - \mathbf{v}_{so,d}\right) + \phi_o, \tag{14}$$

where $\mathbf{K}_d \in \mathbb{R}^{6\times 6}$ is a symmetric, positive-definite matrix. If we can find forces to apply to our fingers such that the contact forces are as given above, then the object velocity will converge to the desired velocity exponentially fast with the rate of convergence given by the eigenvalues of $\mathbf{K}_d$. To find suitable forces that we can apply to the fingers in order to generate the desired contact forces (14), we go back to the defining equation of the contact forces given by the second equation in (10). Rearranging this equation gives

$$\mathbf{AM}^{-1}\bar{\tau} = \mathbf{AM}^{-1}\phi - \mathbf{AM}^{-1}\mathbf{A}^T\mathbf{f}_c - \dot{\mathbf{A}}\bar{\mathbf{v}}.$$

However, we need to have the first six components of the vector $\bar{\tau}$ to vanish because those are the forces that are imposed on the object directly and there is no such physical for at our disposal. Imposing this constraint yields the following equation

$$\underbrace{-\mathbf{B}_c^T \left[\mathrm{Ad}^T_{\mathbf{T}^{-1}_{tc_1}} \; \mathrm{Ad}^T_{\mathbf{T}^{-1}_{tc_2}} \cdots \mathrm{Ad}^T_{\mathbf{T}^{-1}_{tc_k}}\right]^T \mathbf{J}^b \mathbf{M}_r^{-1}}_{\triangleq \mathbf{G}} \tau = \mathbf{AM}^{-1}\phi - \mathbf{AM}^{-1}\mathbf{A}^T\mathbf{f}_c - \dot{\mathbf{A}}\bar{\mathbf{v}}.$$

Therefore, one possible set of joint torques that would yield the desired contact forces is given by

$$\tau = \mathbf{G}^{\dagger}\left(\mathbf{AM}^{-1}\phi - \mathbf{AM}^{-1}\mathbf{A}^T\mathbf{f}_c - \dot{\mathbf{A}}\bar{\mathbf{v}}\right), \tag{15}$$

where $\mathbf{G}^{\dagger}$ is the pseudo-inverse of $\mathbf{G}$ and $\mathbf{f}_c$ is any solution of the underdetermined Eq. (14).

6.2 Catching

The strategy we employ to catch the dough is to move the end-effector directly underneath the dough. This direction is determined by the body negative z-axis of the object frame. Moreover, we orient the end-effector such that the z-axes of the dough and the end-effector are parallel. The latter behavior is imposed so that all the fingers come in contact with the dough almost simultaneously, which reduces the likelihood that slippage occurs during the catching phase.

We refer to the optimal trajectory, generated in Sect. 5 by $\boldsymbol{\zeta}^*(t) = (\mathbf{R}^*(t), \mathbf{p}^*(t)) \in SE(3)$ and to the actual end-effector trajectory by $\boldsymbol{\zeta}(t) = (\mathbf{R}(t), \mathbf{p}(t)) \in SE(3)$. In order to follow the desired optimal trajectory, we find the desired velocity field $\mathbf{v}_{\text{des}} \in \mathfrak{X}(SE(3))$ whose integral curves converge to the desired optimal trajectory. In the next step, we control the joints of the manipulator so that the velocity of the end-effector exponentially converges to this desired velocity. We utilize the redundancy of the manipulator by adding an auxillary control term, which is the orthogonal projection onto the null space of the manipulator Jacobian of the gradient vector field of the dynamic manipulability index.

The desired vector field $\mathbf{v}_{\text{des}}$ is found as the gradient vector field of the potential function $\Psi : SE(3) \to \mathbb{R}$, given by

$$\Psi(\mathbf{R}, \mathbf{p}) = k_R \| \left(\mathbf{I}_3 - \mathbf{R}^{*T}\mathbf{R}\right) \mathbf{e}_3 \|^2 + \frac{k_p}{2} \| \mathbf{R}^T \left(\mathbf{p} - \mathbf{p}^*\right) \|^2, \tag{16}$$

where $k_R, k_p > 0$ are rotational and translational proportional gains, respectively. The gradient of this potential function is then computed to be

$$\mathbf{v}_{\text{des}}(\mathbf{R}, \mathbf{p}) = \left(-k_R \mathbf{R} \left(\mathbf{R}^T \mathbf{R}^* \mathbf{e}_3 \mathbf{e}_3^T - \mathbf{e}_3 \mathbf{e}_3^T \mathbf{R}^{*T} \mathbf{R}\right), \quad -k_p \mathbf{R}^T (\mathbf{p} - \mathbf{p}^*)\right).$$

Note that, the integral curves of $\mathbf{v}_{\text{des}} \in \mathfrak{X}(SE(3))$ are such that $(\mathbf{R}\mathbf{e}_3, \mathbf{p}) \xrightarrow[t \to \infty]{} (\mathbf{R}^*\mathbf{e}_3, \mathbf{p}^*)$, exponentially fast. The final step in the control design procedure is then to control the velocity $\mathbf{v}$ of the end-effector such that $\mathbf{v} \xrightarrow[t \to \infty]{} \mathbf{v}_{\text{des}}$, exponentially fast. If this is accomplished then by the vanishing perturbation theory [14], we can claim that $(\mathbf{R}\mathbf{e}_3, \mathbf{p}) \xrightarrow[t \to \infty]{} (\mathbf{R}^*\mathbf{e}_3, \mathbf{p}^*)$, exponentially fast.

We solve this problem by differentiating the relation $\mathbf{J}^b \dot{\boldsymbol{\theta}} = \mathbf{v}$ to get $\dot{\mathbf{J}}^b \dot{\boldsymbol{\theta}} + \mathbf{J}^b \ddot{\boldsymbol{\theta}} = \dot{\mathbf{v}}$. Substituting for $\ddot{\boldsymbol{\theta}}$ from Eq. (7), we arrive at the relation

$$\dot{\mathbf{v}} = \mathbf{J}^b \mathbf{M}_r^{-1} \left(\boldsymbol{\tau} - \mathbf{C}_r(\boldsymbol{\theta}, \dot{\boldsymbol{\theta}})\dot{\boldsymbol{\theta}} - \mathbf{n}(\boldsymbol{\theta})\right) + \dot{\mathbf{J}}^b \dot{\boldsymbol{\theta}}.$$

If we set the right hand side equal to $-k_d \left(\mathbf{v} - \mathbf{v}_{des}\right)$, then from the linear control theory, we know that $\mathbf{v} \xrightarrow[t \to \infty]{} \mathbf{v}_{\text{des}}$. For that purpose, we first set $\boldsymbol{\tau} = \mathbf{C}_r(\boldsymbol{\theta}, \dot{\boldsymbol{\theta}})\dot{\boldsymbol{\theta}} + \mathbf{n}(\boldsymbol{\theta}) + \boldsymbol{\tau}'$ and we then set

$$\boldsymbol{\tau}' = \tilde{\mathbf{J}}^T \left(\tilde{\mathbf{J}}\tilde{\mathbf{J}}^T\right)^{-1} \left(-k_d \left(\mathbf{v} - \mathbf{v}_{\text{des}}\right) - \dot{\mathbf{J}}^b \dot{\boldsymbol{\theta}}\right) + \left(\mathbf{I}_n - \tilde{\mathbf{J}}^T \left(\tilde{\mathbf{J}}\tilde{\mathbf{J}}^T\right)^{-1} \tilde{\mathbf{J}}\right) \boldsymbol{\eta},$$

where $\tilde{\mathbf{J}} = \mathbf{J}^b \mathbf{M}_r^{-1} \in \mathbb{R}^{6 \times n}$ is an auxiliary control term that is designed to maximize the dynamic manipulability index, $\boldsymbol{\eta} = \dfrac{\partial \sqrt{\det\left(\tilde{\mathbf{J}}\tilde{\mathbf{J}}^T\right)}}{\partial \boldsymbol{\theta}} \in \mathbb{R}^n$.

7 Simulation

In the RoDyMan prototype that we have developed (see Fig. 1), the $n = 7$ axes of rotation ω_i and the points $\mathbf{q}_i$ on these axes of rotation are given by

$$\omega_1 = \left[0 - \sin\left(\frac{\pi}{3}\right)\cos\left(\frac{\pi}{3}\right)\right]^T, \quad \omega_2 = \left[-1\ 0\ 0\right]^T,$$
$$\omega_3 = \omega_4 = \omega_6 = \left[0\ -1\ 0\right]^T, \quad \omega_5 = \omega_7 = \left[0\ 0\ -1\right]^T$$
$$\mathbf{q}_1 = \mathbf{q}_2 = \mathbf{q}_3 = \left[0\ -l_{0y}\ l_{0z}\right]^T, \ \mathbf{q}_4 = \left[0\ -l_{0y}\ l_{0z} - l_1\right]^T,$$
$$\mathbf{q}_5 = \mathbf{q}_6 = \mathbf{q}_7 \left[0\ -l_{0y}\ l_{0z} - l_1 - l_2\right]^T,$$

with $l_{0y}, l_{0z}, l_1, l_2, l_3 > 0$ proper lengths. To complete the derivation of position level forward kinematics, we have selected the initial pose of the end-effector with respect to the base frame to be the following homogeneous matrix

$$\mathbf{T}_{st}(0) = \begin{bmatrix} 1 & 0 & 0 & 0 \\ 0 & -1 & 0 & -l_{0y} \\ 0 & 0 & -1 & l_{0z} - l_1 - l_2 - l_3 \\ 0 & 0 & 0 & 1 \end{bmatrix}.$$

Using the dynamics of the combined system introduced in Sect. 4, we have tested the controllers developed in Sect. 6. When integrating the equations of motion, we use the constraints due to the first finger and the z-components of the second and third fingers since these constraints form a maximally independent set of constraints. For the tossing objective, we have set $h_d = 0.5$ m and $n_d = 5$. With these numbers, the desired object velocity at the instant of release reads $\mathbf{v}_d = \left[0\ 0\ 3.1321\ 0\ 0\ 49.199\right]^T$, in which the first three components have m/s as unit measure, and rad/s for the last three components. However, in order to account for unmodelled dynamics and disturbances, we intentionally impart nonzero lateral linear velocity and roll-pitch angular velocities to the dough by setting $\mathbf{v}_d = \left[0.1\ {-0.1}\ 3.1321\ 0.75\ {-0.5}\ 49.199\right]^T$. We consider the situation where there are $k = 3$ fingers in contact with the dough.

In this simulation the tossing phase lasts about 0.08 s. The plots in Fig. 2a show the yaw velocity and angle of the object along with its velocity and position along the inertial z-axis. We observe in the top-left subfigure that the vertical velocity of the object has reached 3.132 m/s and in the bottom-left subfigure its yaw velocity has reached 49.198 rad/s at the time of release. In the top-right figure we observe the parabolic profile of the z-axis position of the object, and in the bottom-right figure, we see that after the time of release (around 0.08 s), the object has rotated 5 times by the time it is caught.

After the release has taken place, the optimal trajectory generation is performed and the manipulator is commanded with the relevant control law. It turns out that the best time to intercept the pizza is $t_f = 0.729$ s at which instant the translation of the center of mass of the dough and the orientation of the dough read

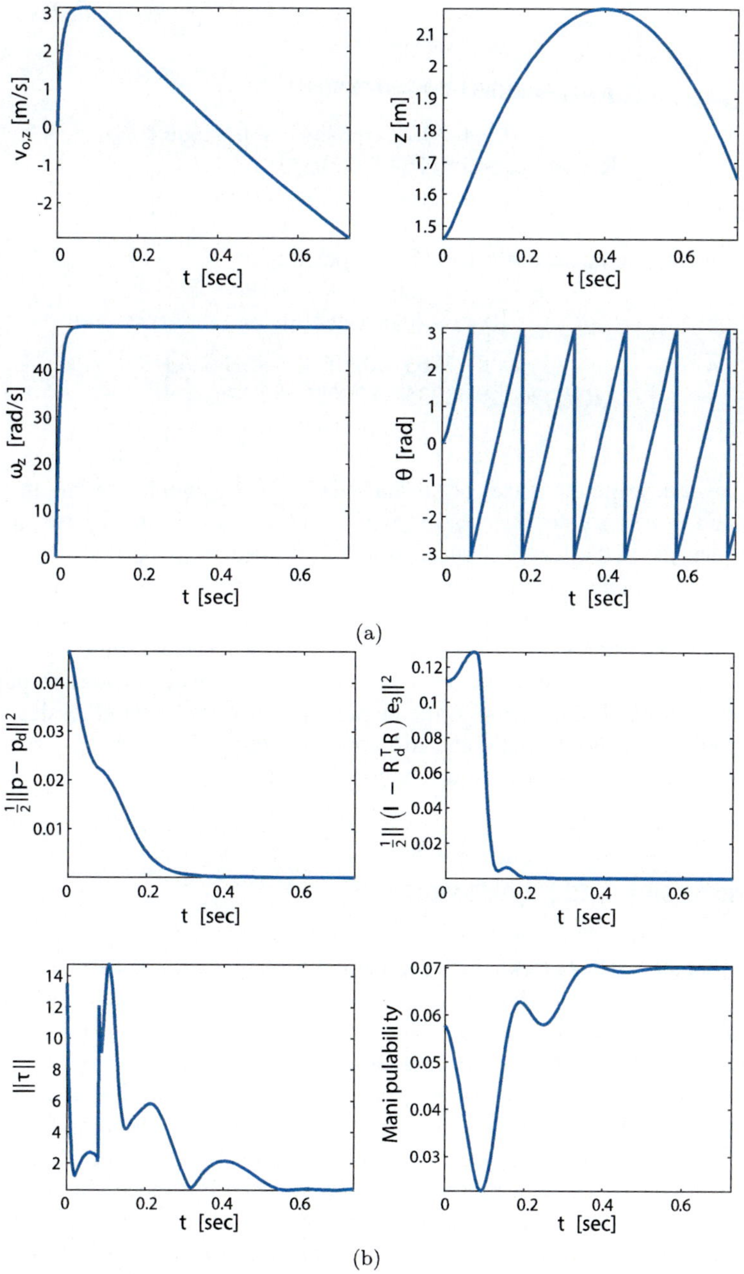

Fig. 2 **a** Tossing: Object motion; **b** Catching: End-effector motion

$$\mathbf{p}_d(t_f) = \begin{bmatrix} 0.31331 & -0.26305 & 1.6793 \end{bmatrix} \text{ m},$$

$$\mathbf{R}_d(t_f) = \begin{bmatrix} -0.3925 & 0.84341 & -0.36686 \\ -0.91162 & -0.30383 & 0.27684 \\ 0.12203 & 0.4431 & 0.88813 \end{bmatrix},$$

and the interception pose of the hand is computed to be

$$\mathbf{p}(t_f) = \begin{bmatrix} 0.48 & -0.38 & 1.46 \end{bmatrix} \text{ m},$$

$$\mathbf{R}(t_f) = \begin{bmatrix} 0.77963 & -0.50743 & -0.36686 \\ 0.61235 & 0.74053 & 0.27684 \\ 0.13116 & -0.44048 & 0.88813 \end{bmatrix}.$$

As desired, while the z-axes of the end-effector and dough frames are matched, the x- and y-axes are not necessarily parallel. We observe in the top two plots of Fig. 2b that the end-effector is successfully controlled to its desired pose, as the positive semidefinite potential function Ψ presented in Eq. (16) is bounded below and above by a positive constant. In the usual Euclidean norm, the bottom left plot depicts the total amount of torque expended at each instant in time. Finally, the bottom right plot shows how the dynamic manipulability index evolves throughout the manipulation. The decrease at the beginning is due to the rapid tossing of the pizza. As the redundancy is exploited to favour the manipulability during the catching phase, we see that it climbs back up slightly above its starting value.

8 Discussion and Conclusion

In this chapter, we have tackled the problem of tossing and catching a pizza dough with a humanoid robot. We have developed the models of grasping the dough with robotic fingers, the kinematics and dynamics of the robotic manipulator, the deformable pizza dough, and the combined system. We have developed a control law that achieves the desired tossing motion using these kinematic and dynamic models. In order to plan the catching of the dough as efficiently as possible, we have found an optimal path that minimizes a convex combination of the acceleration and jerk functionals. Once the trajectory to be taken has been decided, the control law we developed for the catching phase makes sure the robotic manipulator follows this desired path. We have then presented simulation results in plots of interesting quantities.

References

1. A.C. Satici, F. Ruggiero, V. Lippiello, B. Siciliano, Coordinate-free framework for robotic pizza tossing and catching. In: 2016 IEEE International Conference on Robotics and Automation, pp. 3932–3939, Stockholm, S (2016)
2. M.M. Schill, F. Gruber, M. Buss, Quasi-direct nonprehensile catching with uncertain object states, in *2015 IEEE International Conference on Robotics and Automation*, pp. 2468–2474 (2015)
3. V. Lippiello, F. Ruggiero, B. Siciliano, 3D monocular robotic ball catching. Robot. Auton. Syst. **61**(12), 1615–1625 (2013)
4. P. Cigliano, V. Lippiello, F. Ruggiero, B. Siciliano, Robotic ball catching with an eye-in-hand single-camera system. IEEE Trans. Control Syst. Technol. **23**(5), 1657–1671 (2015)
5. R.M. Murray, Z. Li, S.S. Sastry, *A Mathematical Introduction to Robotic Manipulation* (CRC press, 1994)
6. K.M. Lynch, M.T. Mason, Dynamic nonprehensile manipulation: Controllability, planning, and experiments. Int. J. Robot. Res. **18**(1), 64–92 (1999)
7. J.-C. Ryu, F. Ruggiero, K.M. Lynch, Control of nonprehensile rolling manipulation: balancing a disk on a disk. IEEE Trans. Robot. **29**(5), 1152–1161 (2013)
8. K.-C. Liu, J. Friend, L. Yeo, Rotating bouncing disks, tossing pizza dough, and the behavior of ultrasonic motors. Phys. Rev. E **80**(4), 046201–1–046201–11 (2009)
9. M. Zefran, V. Kumar, C.B. Croke, On the generation of smooth three-dimensional rigid body motions. IEEE Trans. Robot. Autom. **14**(4), 576–589 (1998)
10. P. Crouch, F.S. Leite, The dynamic interpolation problem: on Riemannian manifolds, Lie groups, and symmetric spaces. J. Dyn. control Syst. **1**(2), 177–202 (1995)
11. A.M. Bloch, P.E. Crouch, Nonholonomic control systems on Riemannian manifolds. SIAM J. Control Optim. **33**(1), 126–148 (1995)
12. J.E. Marsden, T.S. Ratiu, *Introduction to Mechanics and Symmetry: A Basic Exposition of Classical Mechanical Systems*. Texts in Applied Mathematics (Springer, 1999)
13. M.P. do Carmo, *Riemannian Geometry*. Mathematics (Boston, Mass.) (Birkhäuser, 1992)
14. H.K. Khalil, *Nonlinear Systems* (Prentice Hall, 2002)

Planning Framework for Robotic Pizza Dough Stretching with a Rolling Pin

Jung-Tae Kim, Fabio Ruggiero, Vincenzo Lippiello, and Bruno Siciliano

Abstract Stretching a pizza dough with a rolling pin is a nonprehensile manipulation. Since the object is deformable, force closure cannot be established, and the manipulation is carried out in a nonprehensile way. The framework of this pizza dough stretching application that is explained in this chapter consists of four sub-procedures: (i) recognition of the pizza dough on a plate, (ii) planning the necessary steps to shape the pizza dough to the desired form, (iii) path generation for a rolling pin to execute the output of the pizza dough planner, and (iv) inverse kinematics for the bi-manual robot to grasp and control the rolling pin properly. Using the deformable object model described in Chap. 3, each sub-procedure of the proposed framework is explained sequentially.

1 Brief Introduction

Making a pizza is a wonderful art, and it requires delicate skills like preparing a pizza dough mixed with wheat powder, water, salt, and other ingredients. An accurate ratio is put in the preparation, stretching it dynamically and quickly into a disk-shaped, saucing or dressing it with proper ingredients and quantity, and finally burning it evenly and sufficiently in a wood-burning oven.

J.-T. Kim (✉)
MIKNMEK Inc., CheongAm-ro 87, Nam-gu, Pohang-si, 37673 Gyeongsangbuk-do, Republic of Korea

F. Ruggiero · V. Lippiello · B. Siciliano
Department of Electrical Engineering and Information Technology, PRISMA Lab, CREATE Consortium & University of Naples Federico II, Via Claudio 21, 80125 Naples, Italy
e-mail: fabio.ruggiero@unina.it

V. Lippiello
e-mail: vincenzo.lippiello@unina.it

B. Siciliano
e-mail: bruno.siciliano@unina.it

B. Siciliano and F. Ruggiero (eds.), *Robot Dynamic Manipulation*,
Springer Tracts in Advanced Robotics 144,
https://doi.org/10.1007/978-3-030-93290-9_9

Fig. 1 Application: stretching a pizza dough with a rolling pin

Table 1 Main symbols used in this chapter

Definition	Symbol
Center of the pizza dough	$\mathbf{x}_0 = \begin{bmatrix} c_x & c_y \end{bmatrix}^T \in \mathbb{R}^2$
Generic point of the dough	$\mathbf{x} \in \mathbb{R}^2$
Boundary of the pizza dough	∂S
Thickness of the dough	$h > 0$
Frame associated to the plate	$\mathcal{D}$
Angle between the longest-axis of the dough shape and the x-axis of $\mathcal{D}$	$\theta \in \mathbb{R}$

Among these technical processes, this chapter focuses on stretching a pizza dough with a rolling pin (Fig. 1). This process includes two critical techniques: (i) the manipulation of a deformable object, and (ii) the control of the rolling pin to execute the proper actions. Regarding the former, difficulties arise during the manipulation planning of a deformable object because of the absence of a precise model. Indeed, typical deformable objects can not be described with just one property but have multiple properties like viscosity, elasticity, plasticity, or others, and mixed of them. Even though several proposed models represent these properties of a deformable object, it is not easy to find a suitable one for the pizza dough. In this chapter, the SPH formulation explained in Chap. 3 is used to model a highly viscous deformable object like the pizza dough, which is eligible to describe highly deformable objects, even for liquids. Regarding the latter, researches about controlling a tool with a robotic system are well established. However, nonprehensile manipulation is still a relevant and challenging topic. A general inverse-kinematics manipulation planning for the RoDyMan robot equipped with a rolling pin has been used in this pizza dough stretching application (Table 1).

Both techniques are explained in this chapter. The outline of the chapter consists of this introduction and a survey about the related state of the art (Sect. 2). The sketch of the proposed framework for the pizza dough stretching is depicted in Sect. 3, while the explanation of each sub-procedure is given from Sect. 4 to Sect. 7. Simulation results are explained in Sect. 8. Finally, Sect. 9 concludes the chapter.

2 Related Research

Manipulation of a deformable or a rheological object (i.e., the pizza dough) requires an understanding of the object's properties like viscosity, elasticity, and plasticity. For example, a bread dough consists of gluten proteins and various minor ingredients, including minerals. The gluten proteins play a crucial role in determining the unique baking quality of wheat by conferring water absorption capacity, cohesivity, viscosity, and elasticity. Gluten proteins can be divided into two main fractions according to their solubility in aqueous alcohols: the soluble gliadins and the insoluble glutenins [1]. It is widely accepted that gliadin accounts for the viscous properties and glutenin imparts the strength and elasticity that are necessary to hold the gases that are produced during fermentation and baking [2].

Many researchers have tried to characterize the bread dough's fundamental properties and analyze the influences of the substances. The densities of various doughs were measured in [3], while the viscoelasticity of bread dough was examined in [4–6]. The main substances for the bread dough that are H_2O (water), D_2O (heavy water), esterifying agents for glutamine residues, urea, salts, agents affecting disulfide bonding, and the protein subunits, were listed up in [7], also characterizing the influence of the substances for the rheology of the bread dough. Seventeen commercially available European wheat cultivars were sampled in [2]. Through these samples, the authors had the creep-recovery experiments and analyzed for a set of chemical and rheological parameters and baking quality using the PCA method. The dynamic rheological properties of glutens fractions with two English-grown wheat cultivars, Hereward and Riband, were studied in [8]. The authors confirmed that the viscoelasticity of the glutenin sub-fraction of gluten and differences in the ratio of gliadin to glutenin are the main factors governing inter-cultivar differences in the viscoelasticity of wheat gluten. Similarly, in [9], the authors experimented with the uniaxial elongational and shear rheology properties of doughs affected by the protein contents or glutenin-to-gliadin ratio. The conclusion was that increasing protein content lowered the maximum shear viscosity while increasing the glutenin-to-gliadin ratio increased the maximum shear viscosity. Stress and strain of the dough related to the ratio of the feed sheet thickness of the roller gap and the roller's speed in the sheeting system were studied in [10]. The authors applied the lubrication approximation for the equation of motion and used an inelastic power-law model for the dough rheology. The relationship between the rheological properties for static dough and dynamic rheological properties for dough crumbs was also investigated in [11]. The former was evaluated by texture profile analysis, like uni-axis (stretching) and bi-axial extension (inflation), and gluten index in static compression. The latter was evaluated through dynamic mechanical analysis and thermal mechanical analysis in dynamic compression. In [12], the authors described the influence of various substances in a dough. They introduced widely used modeling methods for dough rheology like power-law, Maxwell model, Lethersich model, Peleg model, and listing the various measurement methods for dough properties like farinograph, mixograph, rheomixer, extensigraph, alveograph, amylograph, maturograph, and so on. Base on these known properties, several visco-elastoplastic models were proposed, as the Herschel-Bulkley model and the K-BKZ model [13]. Bingham model is also one of the well-known models for representing plastic properties [14].

The baking industry uses rolling (or sheeting) process using counter-rotating rolls as a dough forming process for various products, such as cookies, crackers, pizza, bread, and pastry. The rolling process is akin to calendering, which is used in many industries, such as the paper, plastics, rubber, and steel industries [15, 16]. An overall process for stretching of bread doughs was designed and implemented in [17].

There are more general researches for acquiring the properties of deformable objects. A four-element model for characterizing the viscoelasticity was proposed in [18]. In [19], a neural network model was employed to estimate an object's elastic properties. A FEM model was instead employed in [20] to estimate such properties. Interactive approaches to get object's elastic properties, even without the use of any specific model, were used in [21–25].

3 Framework for a Pizza Dough Stretching Behaviour

The application handled by this chapter roughly consists of three main components: (i) a robotic system grasping a tool, (ii) a deformable object, and (iii) the tool itself. In this application, the employed robotic system is the RoDyMan robot, the deformable object is a pizza dough, and the tool is a rolling pin (see Fig. 1).

The proposed framework is described in this section. Input data from a sensor device are acquired. This generates a proper action sequence for the robot. The employed 9 sensor device is a RGB-D Kinect camera mounted on the head of the RoDyMan robot.

The devised framework can be split into four components: (i) (deformable) object recognition, (ii) planning actions on the deformable object; (iii) planning actions for the tool, and (iv) robot manipulaiton planning. As evident from Fig. 2, these components are concatenated each other.

The first component, which is the object recognition module, gets sensor data as input. Its output is the status of the recognized object. In this application, an RGB-D Kinect camera takes pictures of the pizza dough on a plate. Afterwards, this component separates the area of the pizza dough and the background. It reconstructs the 3D shape of the pizza dough based on the 2D image data and some additional information about the pizza dough. Finally, this object recognition module describes the shape of the pizza dough through numerical data. This description indicates the current status of the pizza dough.

The second component, which is the one in charge of planning the actions on the deformable object, gets the output of the previous component, the desired final shape of the dough, and some additional information (i.e., the transform look-up-table, which will be illustrated in the following sections). Its output is the planned sequence of actions on the deformable object. In this application, the current shape of the pizza dough, the desired shape of the pizza dough, and information about the dough deformation (i.e., how a particular action of the rolling pin deforms the object into another shape) are given. Then, this component finds out the best sequence of actions of the rolling pin to get the desired shape.

Similarly, the third component gets the output of the previous one and generates a continuous motion of the tool to realize the desired actions on the deformable object. Each action from the previous component is disconnected from the other. Therefore, it is necessary to generate a smooth continuous motion, which is the output of this module.

Finally, the last component is in charge of the robot planning to realize the output of the previous component. Considering the kinematic information of the employed robot and all the constraints, this component generates a smooth motion sequence for the robot to suitably moving the tool as specified by the previous component.

Each component is explained in detail wtihin the sequel sections.

4 Pizza Dough Recognition

The pizza dough recognition component consists of two procedures: (i) digital image processing for raw data from the sensor device, and (ii) how to describe the status of the flatted deformable object.

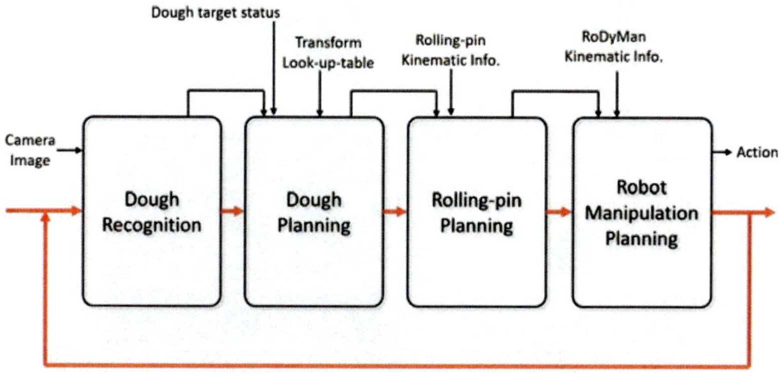

Fig. 2 Sketch of the overall process that consists of four concatenated components

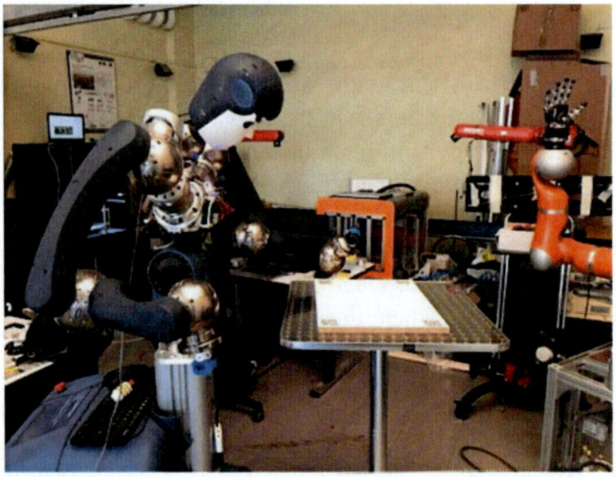

(a)

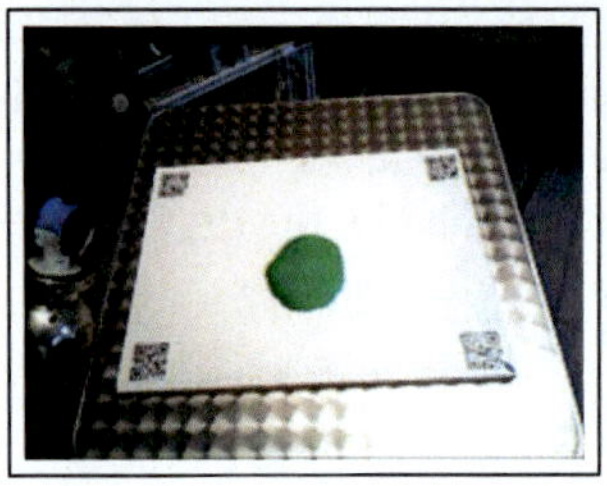

(b)

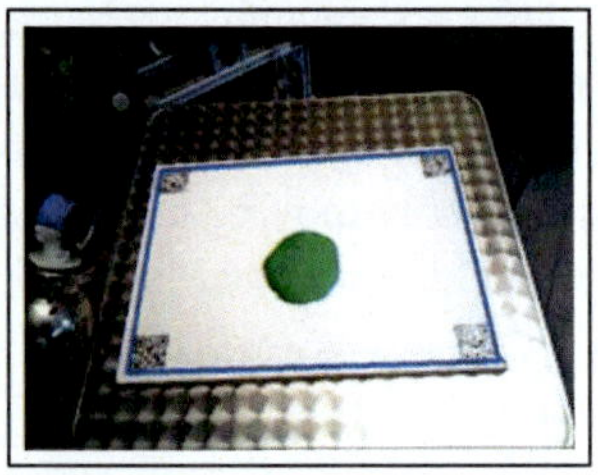

(c)

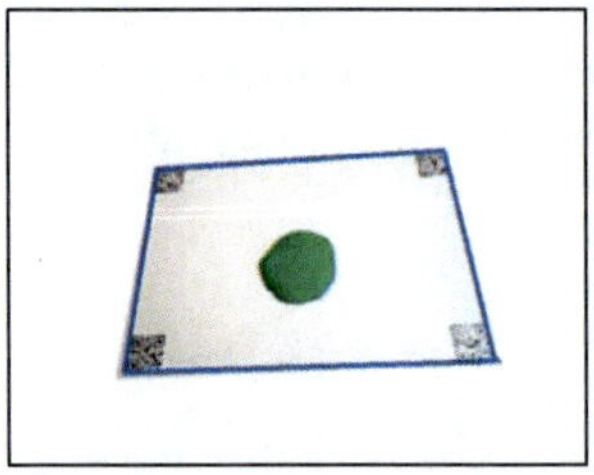

(d)

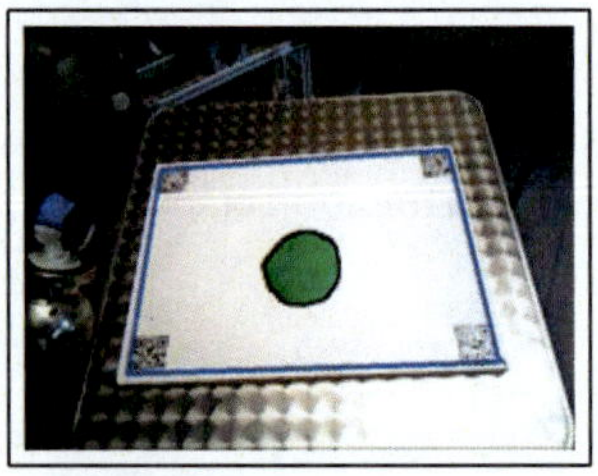

(e)

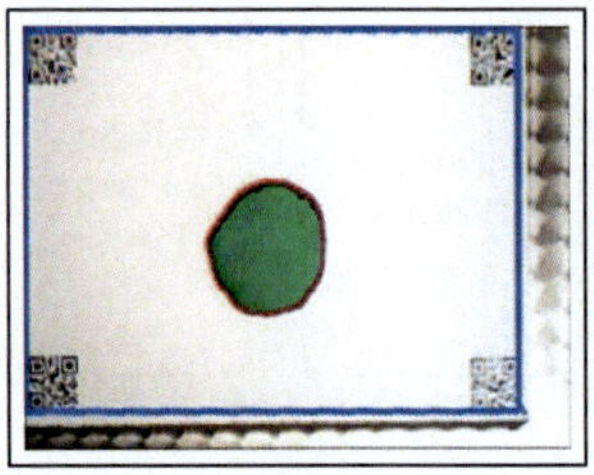

(f)

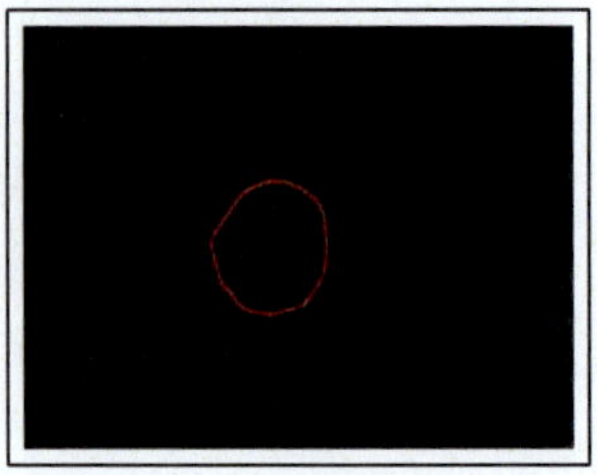

(g)

Fig. 3 Detection of the pizza dough on a white rectangular plate

4.1 Image Processing for Sensor Data

The perception procedure for a deformable object roughly consists of two parts in this application; one is the image processing for the camera sensor data, and the other is the representation of the status of the pizza dough. The perception method depends highly on the application; therefore, in this case, we analyse the stretching action of a pizza dough using the RoDyMan robot (Fig. 3a), and the recognition procedure naturally depends on a specific detector for the pizza dough. To reduce the difficulties of the perception based on image sensor data, some artificial restrictions are applied: restriction to the workspace by defining the boundary area (i.e., a plain rectangular plate) and the usage of a particular coloured deformable object (i.e., the blue colour) contrasted to the background colour (i.e., white colour). However, these restrictions are not critical but can be addressed with additional image processing.

With the help of the absolute position information of at least three corner markers in a rectangular plate and the kinematic information provided by the RoDyMan robot, it is not difficult to induce the frame transformation between a point in a given 2D camera image and the corresponding absolute position in the world frame [26]. In the carried out experiments, QR codes were placed at each corner of the rectangular plate, and then using an image matching with SIFT [27] the corner points were detected. Other feature detectors like the SURF [28] or the FAST [29] are also available. In order to easier detection of the corners, AR codes and the corresponding code detectors (i.e., ARToolkit [30]) can be used. The primary purpose of corner detection is to remove the dough plate's outlier and induce a transformation $\mathcal{T}$ from a 2D view image to a top-down viewed 2D space.

For the initial pizza dough, assuming that the ball- or bell-shaped pizza dough is symmetric to the vertical rotation axis, the reconstruction of the 3D shape is obtainable with partial camera views or depth sensors [31, 32]. Suppose the pizza dough is shallow after being pushed by a rolling pin so that the height difference is ignorable. In that case, the transformation $\mathcal{T}$ can be directly applied to the detected pizza dough as well as the plate. However, the initial pizza dough is more like a ball- or bell-shaped than a shallow disk. We do not commit much error in seeing the pizza dough as a 2D figure on a plate with uniform thickness: indeed, the more shallow the thickness of the dough, the minor error it occurs regarding a cylinder shape.

An example of the image processing steps to identify the pizza dough on the plate is given in Fig. 3b–g. With reference to the labels in the figure, the steps now briefly described. *b*) The raw image data from the camera mounted on the head of the RoDyMan robot is shown. With default information, we assume that the camera's view covers the whole area of the dough plate. *c*) There are four corner marks on the white dough plate. After detecting the marks, the area of the dough plate is outlined with blue lines. *d*) The outer area of the dough plate is removed from the image. *e*) Within the dough plate, the dough clay (the green object) is detected, and the outline is extracted through an edge detector [33], like the Canny edge detector [34]. *f*) The raw 2D camera image is deformed into the orthogonal top-down view with

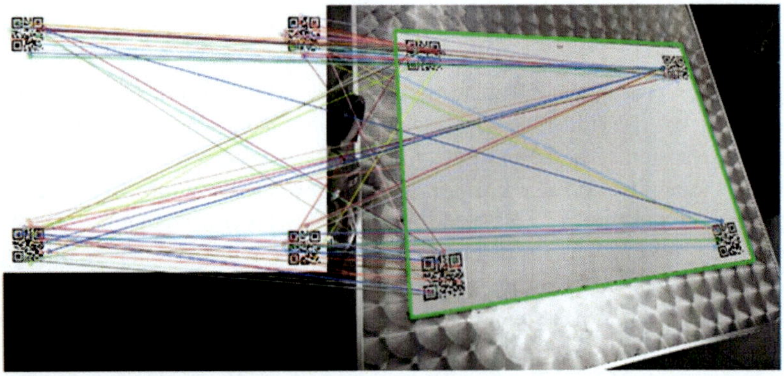

Fig. 4 Feature matching on the corners of the plate through the SURF method. On the left, the template image. On the right, the camera image. The green lines indicate the boundary of the detected plate

predefined width and height. *g*) The deformed outline of the detected pizza dough is finally obtained.

4.2 Description for a Status of a Pizza Dough

Pizza dough has various shapes. However, according to general experiences and our experiments in making a pizza, there is no problem assuming that the pizza dough's shape is convex. Hence, we can define the state of the pizza dough as the set of distances between the center of the dough and its boundary and the related tickness

$$\mathcal{C}_X : \left\{ \begin{bmatrix} \|\mathbf{x} - \mathbf{x}_0\| \\ h \end{bmatrix} \in \mathbb{R}^2 : \mathbf{x} \in \partial S \right\}. \tag{1}$$

The dough center is intended as the geometric center of its shape. During the implementation, we discretized the configuration as a set of angle-equally sampled distances (see Fig. 5a) and vectorized them clock-wisely. The histogram of the vectorized configuration is shown in Fig. 5b.

When the dough shape is deformed through the action of the rolling pin, the centre position and the rotated angle of the dough are less important. Indeed, the relative angle between the rolling pin and the pizza dough would affect the deformation. To get a rotation-invariant configuration, the order of the histogram elements is rearranged so that the longest distance vector is the first one in the set (see Fig. 5c). This method has been frequently used in object recognition algorithms of image processing, *e.g.*, SIFT [35], SURF [28], MSER [36], FAST [29], and BRISK [37].

An extended configuration space $\mathcal{C}_Y$ is introduced to include the position of the centre and the angle between the longest distance vector (the first one in the rearranged

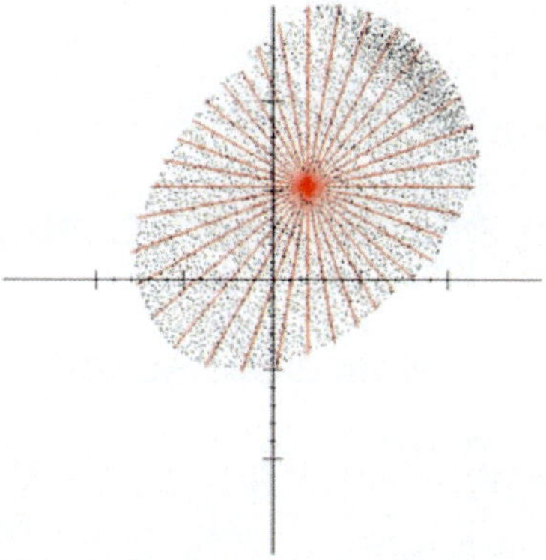

(a) Deformable object model projected onto the 2D space.

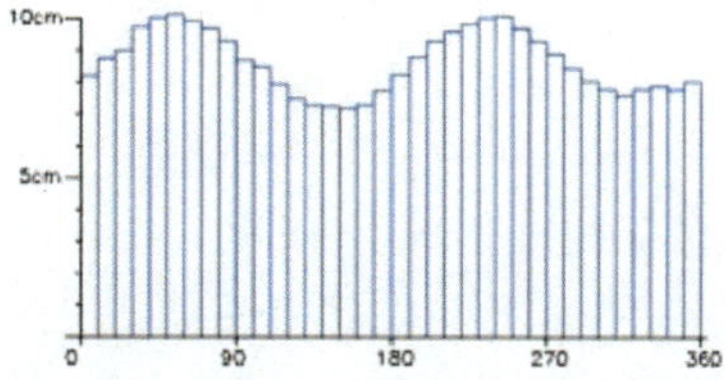

(b) Histogram of the distance from the deformable object's centre.

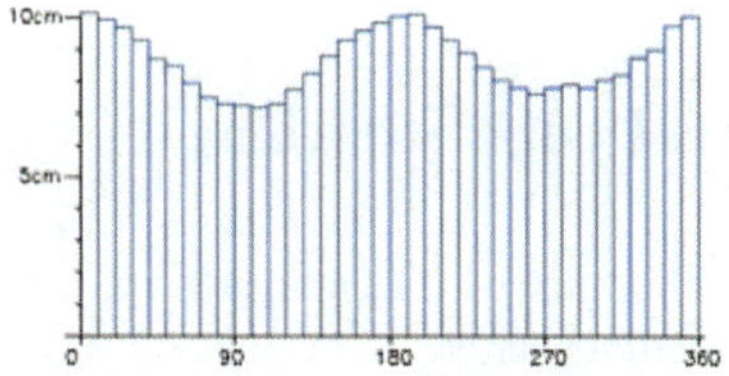

(c) Rearranged histogram for the longest one to be the first bin following by sequel bins.

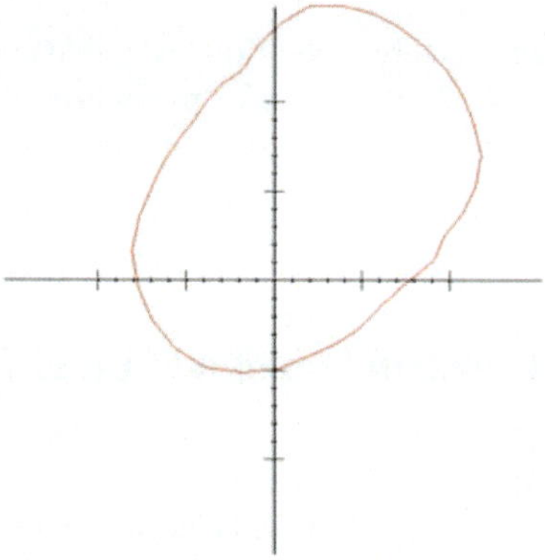

(d) Reconstructed shape from the configuration $\mathbf{y}$.

Fig. 5 Configuration for a dough state. The red radial lines in **a** indicate equal-angle sampled distance from the centre of the pizza dough shape to the boundary. Blue line is the longest distance from the centre. The sampled distances are ordered clock-wisely

histogram) and the x-axis of the frame $\mathcal{D}$

$$\mathcal{C}_Y : \left\{ \begin{bmatrix} \chi \\ c_x \\ c_y \\ \theta \end{bmatrix} \in \mathbb{R}^5 : \chi \in \mathcal{C}_x \right\}. \tag{2}$$

It can be seen in Fig. 5d how the reconstructed shape by the extended configuration $\mathcal{C}_Y$ is similar to the original shape. The original configuration space $\mathcal{C}_X$ of the pizza

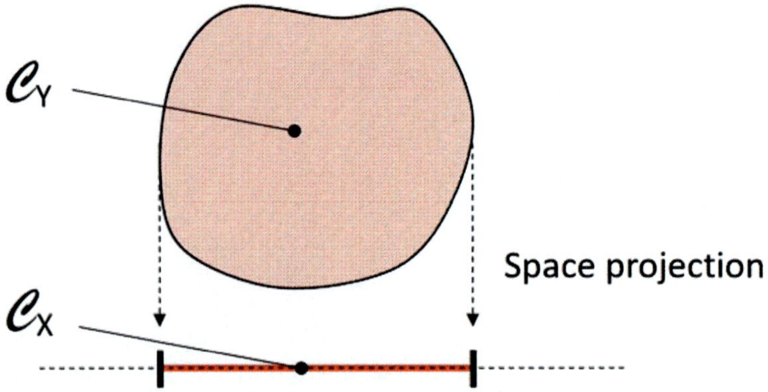

Fig. 6 The configuration space $\mathcal{C}_X$ and its extended configuration space $\mathcal{C}_Y$

dough is a subspace of the extended configuration space $\mathcal{C}_Y$: their relation is drawn in Fig. 6. Through space projection, the extended configuration space $\mathcal{C}_Y$ can be submerged into the original configuration space $\mathcal{C}_X$, that is, $\mathcal{C}_Y/\mathcal{I} = \mathcal{C}_X$ where $\mathcal{I} = \{c_x = 0, c_y = 0, \theta = 0\}$.

5 Construction of a Planner for Pizza Dough Stretching

Given a generic configuration space $\mathcal{C}$, a initial status $q_I \in \mathcal{C}$, and a final status $q_G \in \mathcal{C}$, the planner finds the sequence of intermediate status

$$q_I \to q_1 \to \cdots \to q_{n-1} \to q_G, \tag{3}$$

to reach q_G from q_I, and where $q_1, \ldots, q_{n-1} \in \mathcal{C}$. Each status is referred to as a configuration q in a given configuration space $\mathcal{C}$.

A way to find such a sequence of intermediate configurations is to associate to each of them a cost value relative to q_G. Let $V(q_i) > 0$ be a cost value related to the distance from $q_i \in \mathcal{C}$ to q_G, then a gradient descent method might be used to find the sequence of interemdiate configurations such that

$$V(q_I) \geq V(q_1) \geq \cdots \geq V(q_{n-1}) \geq V(q_G) = 0. \tag{4}$$

A movement from one configuration to another is called a transition $\mathcal{T}$. The transition occurs directly or indirectly through an action $\alpha \in \mathcal{A}$, where $\mathcal{A}$ is the set of admissible actions. Therefore, we can move from a configuration $q \in \mathcal{C}$ to a configuration $q' \in \mathcal{C}$ as

$$q \to q' = \mathcal{T}(q, \alpha). \tag{5}$$

The robotic system must execute the action α through a proper control design, taking into account possible errors during the execution.

Concerning the pizza stretching application, the configuration of the pizza dough is given by the sets $\mathcal{C}_X$ and $\mathcal{C}_Y$. The cost function $V(\cdot)$, the actions $\alpha \in \mathcal{A}$ and the transitions $\mathcal{T}$ to achieve the sough task are explained in the following.

5.1 Cost Value Function

In this application, the cost value function $V(\cdot)$ is defined as how the shape of the pizza in the configuration q is different from the shape of the pizza in the desired configuration q_G. Hence, the more similar the shape of q is to q_G, the smaller the cost value V is.

A comparison between two 3D objects is typically made by checking the respective volumes. However, the comparison can be simplified in this application by projecting the two 3D volumes into a 2D plane. In fact, the target object has a shallow disk, and the error between the approximated cylinder shape and the original dough shape decreases as deformations of rolling pin actions are made.

The comparison is made by calculating the ratio of the occupied 2D area of the current dough shape over the 2D area of the target one while ignoring the space out of the target 2D area (see Fig. 7). The areas can be measured by counting the occupied grid cells after discretizing the 2D plate.

The designed cost value function for this application is

$$V(q) = 1 - \frac{\text{area}(q)}{\text{area}(q_G)}, \tag{6}$$

where $q, q_G \in \mathcal{C}_Y$ are the current and the target configurations of the pizza dough shape, respectively. The operator to compute the area is defined as

$$\text{area}(q) = \sum_{ij} \text{occ}(q, i, j),$$

where i and j are the indexes of discretized 2D space, and the occupancy function $\text{occ}(q, i, j)$ is defined as

$$\text{occ}(q, i, j) = \begin{cases} 1 & \text{if } (i, j) \in A(q) \cap A(q_G) \\ -\kappa & \text{if } (i, j) \in A(q) \cap \neg A(q_G) \\ 0 & \text{otherwise} \end{cases}, \tag{7}$$

where $A(q)$ is the part of the plate occupied by the dough in the configuration q, $\neg A(q)$ is the complement of $A(q)$, and $\kappa > 0$ is a penalty weight. In particular, $\kappa = 0$ means that we do not care if the current shape of the pizza dough is outside the target

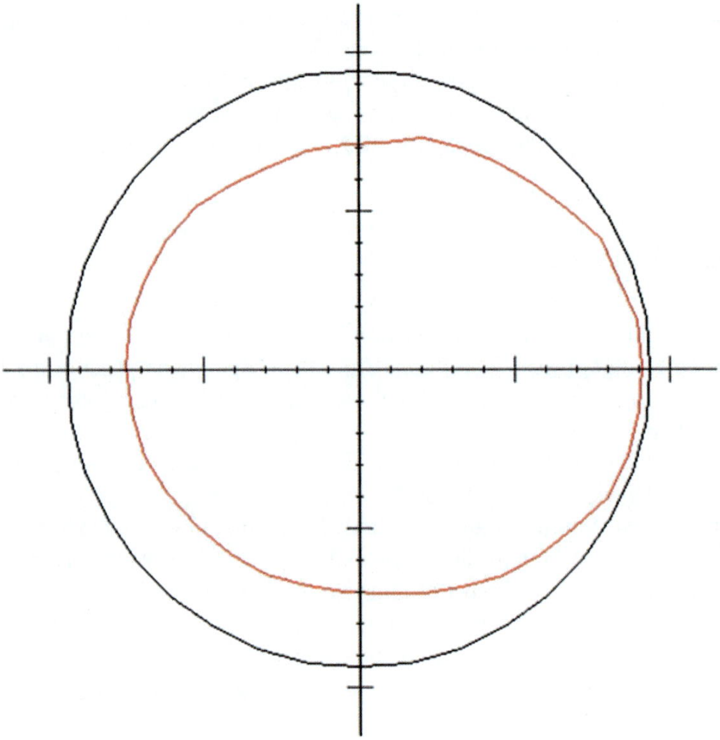

Fig. 7 An comparison between the current shape (red line) and the target shape (black line)

one, while $\kappa \gg 0$ indicates that we strongly penalize the case in which the current dough is outside the target one even partially. As it is built, the cost function assumes values in the range [0, 1]. When the cost function is zero, the current dough shape covers all the target one. In practice, we verified that it is difficult and inefficient to cover the target shape area completely: we thus recommend saturating to zero the cost function when its value is under 0.1, corresponding to the fact that about the 90% of the target shape area is covered.

5.2 *Actions for a Deformable Object*

There are a lot of possible actions to modify a pizza dough employing a rolling pin. For simplicity, the available actions in this application are limited. First, slanted or downing movements are prohibited. Besides, an action cannot change its angle during the movement. Then, the distance between the rolling pin and the plate is constant during all the movement. Finally, the rolling pin's movement is always done in contact with the dough.

Following the above constraints, the action set is defined as follows

$$\mathcal{A}: \left\{ \begin{bmatrix} \delta \\ \phi \end{bmatrix} \in \mathbb{R}^2 \right\}, \tag{8}$$

where $\delta > 0$ is the height between the rolling pin and the table and $\phi \in [0°, \ldots, 180°]$ is the angle between the longest axis of the pizza dough and the x-axis in $\mathcal{D}$. Because the rolling pin's forward and backward movements are not discriminated in this application, the provided interval for ϕ is enough. During simulations, the action set $\mathcal{A}$ consisting of a height δ and an angle ϕ was discretized. The number of heights and angles used in our simulations are provided in Sect. 8.

Notably, the deformations caused by the actions defined above are relative to the local configuration space $\mathcal{C}_X$ and not the global configuration space $\mathcal{C}_Y$. This means that the deformation of the pizza after an action $\boldsymbol{\alpha} \in \mathcal{A}$ is not affected by the global status $\mathbf{y} \in \mathcal{C}_Y$. However, it depends on $\boldsymbol{\chi} \in \mathcal{C}_X$ and the relative angle between the pizza dough and the rolling pin.

5.3 *Transition Originated from an Action*

A transition changes the status of the dough to another thanks to an action. The transition function $\mathbf{T} : \mathcal{C}_Y \times \mathcal{A} \rightarrow \mathcal{C}_Y$ can be defined as

$$\mathbf{y}' = T(\mathbf{y}, \boldsymbol{\alpha}), \tag{9}$$

where $\mathbf{y} \in \mathcal{C}_Y$ is the current state of the pizza dough and $\mathbf{y}' \in \mathcal{C}_Y$ is the one obtained after the execution of the action $\boldsymbol{\alpha} \in \mathcal{A}$.

As previously mentioned, the deformation of the pizza dough through an action is more related to the relative angle between the rolling pin and the longest axis rather than the absolute pose of the pizza dough. Therefore, the Eq. (9) can be rewritten as

$$\begin{aligned} \mathbf{y}' &= T(\mathbf{y}, \boldsymbol{\alpha}) \\ &= T(\left[\mathbf{x}^T\ \mathbf{0}_3^T\right]^T, \left[\delta\ \phi - \theta\right]^T) + \left[\mathbf{0}_2^T\ \mathbf{x}_0^T\ \theta\right]^T. \end{aligned} \tag{10}$$

The new action $\boldsymbol{\alpha}' = \left[\delta\ \phi - \theta\right]^T$ is already included within $\mathcal{A}$; hence, there is no change in the size of the action set $\mathcal{A}$. On the other hand, the state $\left[\mathbf{x}^T\ \mathbf{0}_3^T\right]^T$ might be interpreted as adimension reduction by projection (Fig. 6). In this way, the domain of the transition function T can be drastically reduced.

In the next section, how to generate such transitions will be explained.

5.4 LUT Method

Even though the deformation of an object due to an action is very well simulated through the SPH models (see Chap. 3.1), the problem is the computation time for each transition from one state to another state. The time depends on the resolution of the deformable object model and other conditions. In the chapter, the carried-out simulations took several minutes to hours for one transition. Considering that a planner requires at least several dozens, thousands, or much more transitions, this object model is not suitable for real-time or short-time planning.

LUT method is well-known for overcoming this computing-time problem. Possible transitions are calculated previously off-line, and then only the results are recorded into a transition database. Working online, the planner looks for a suitable transition in the database and uses it.

As explained in the previous Sect. 5.3, a transition T is independent from the position and rotation angle of the current state $\mathbf{y} \in \mathcal{C}_Y$. Therefore, LUT contains only transitions from a state $\boldsymbol{\chi} \in \mathcal{C}_X$ at centre $\mathbf{x}_0 = \mathbf{0}_2$ and zero rotation angle, $\theta = 0$, with actions $\boldsymbol{\alpha} \in \mathcal{A}$.

A transition space is continuous, but it is infeasible to generate and store all possible transitions. Therefore, discretizing the transition space is necessary, which requires storing selected transitions into a database and a method to find a proper transition within it. The conventional method to find a proper transition is to look for the most similar one and use it [38], or to use an interpolating method to estimate unknown transitions from a given state $\boldsymbol{\chi} \in \mathcal{C}_X$ with the neighbour transitions. In the following subsections, we investigate how to find similar transitions into the database and interpolating them.

5.4.1 Similarity

In order to find similar transitions within the database, similarity measure functions between two states $\boldsymbol{\chi}_1, \boldsymbol{\chi}_2 \in \mathcal{C}_X$ and between two actions $\boldsymbol{\alpha}_1, \boldsymbol{\alpha}_2 \in \mathcal{A}$ are needed. We use the diagonal Mahalanobis distance $\text{sim}_X : \mathcal{C}_X \times \mathcal{C}_X \to \mathbb{R}^{\geq 0}$ as a function for the pizza dough configurations, whose definition is given below

$$\text{sim}_X(\boldsymbol{\chi}_1, \boldsymbol{\chi}_2) = \sqrt{\|\boldsymbol{\chi}_1 - \boldsymbol{\chi}_2\|_{\mathbf{S}_X^{-1}}}, \tag{11}$$

where $\mathbf{S}_X = \text{diag}\left(\begin{bmatrix} 1 & \frac{1}{\beta} \end{bmatrix}\right) \in \mathbb{R}^{2\times 2}$, with $\beta > 1$. We use the function $\text{sim}_A : \mathcal{A} \times \mathcal{A} \to \mathbb{R}^{\geq 0}$ for the actions as

$$\text{sim}_A(\alpha_1, \alpha_2) = \sqrt{\|\boldsymbol{\alpha}_1 - \boldsymbol{\alpha}_2\|_{\mathbf{S}_A^{-1}}}, \tag{12}$$

where $\mathbf{S}_A = \text{diag}\left(\begin{bmatrix}\frac{1}{\gamma} & 1\end{bmatrix}\right) \in \mathbb{R}^{2\times 2}$, with $\gamma > 1$.

5.4.2 Interpolation

Like the basic idea of SPH formation in Sect. 3.2.2, we find the neighbour transitions within a specific distance, and then we interpolate them by substituting the transition function T in (9) with a new transition function $\overline{T} : \mathcal{C}_Y \times \mathcal{A} \rightarrow \mathcal{C}_Y$ as

$$\mathbf{y}' = \overline{T}(\mathbf{y}, \boldsymbol{\alpha}) = \frac{\sum_i \overline{T}_a(\mathbf{y}_i, \boldsymbol{\alpha})\, W(\text{sim}_X(\bar{\boldsymbol{\chi}}, \bar{\boldsymbol{\chi}}_i), h_y)}{\sum_i W(\text{sim}_X(\bar{\boldsymbol{\chi}}, \bar{\boldsymbol{\chi}}_i), h_y)} \tag{13}$$

and

$$\overline{T}_a(\mathbf{y}, \boldsymbol{\alpha}) = \frac{\sum_j T(\mathbf{y}, \boldsymbol{\alpha}_j) W(\text{sim}_A(\boldsymbol{\alpha}, \boldsymbol{\alpha}_j), h_a)}{\sum_j W(\text{sim}_A(\boldsymbol{\alpha}, \boldsymbol{\alpha}_j), h_a)}, \tag{14}$$

where the kernel function $W : \mathbb{R}^{\geq 0} \times \mathbb{R}^{+} \rightarrow \mathbb{R}^{\geq 0}$ and the kernel ranges $h_y, h_a \in \mathbb{R}$ have been used. The denominator of the previous expressions is used for normalization purposes, similarly to the SPH formation in Sect. 3.2.2. Notice that $\mathbf{y} = \left[\bar{\boldsymbol{\chi}}^T \; \mathbf{0}_3^T\right]^T$.

After execution of the action by the robot, if the deformed status is the same or similar to the expected one, then the following action will be executed sequentially. Otherwise, a new step by the planner is required.

6 Path Generation for a Rolling Pin

One of the particular features in the proposed planning framework is the independent planning for the tool itself. In contrast, most of the other planning frameworks integrate the planning of the tool and the planning of the robot manipulator. There are benefits and drawbacks to this planning separation. The separation makes the robot manipulation planning manageable. At the same time, there is a need to treat the infeasible action sequences that are a problem when the robot manipulator generates its action sequences to follow the trajectory of the tool. In this application, the preference of simplicity of the manipulation planning makes us separate the planning for a tool from the planning for the robot manipulator.

From the previous sections, the actions to stretch the dough towards the desired shape are generated. A possible sequence of actions is something like: stretch forward first, then stretch 30°, and so on, all with specified heights from the plate. There are two issues for generating a proper action sequence for the rolling pin: the generation of the action itself (called primitive action) and the connection to the following action (called connecting action).

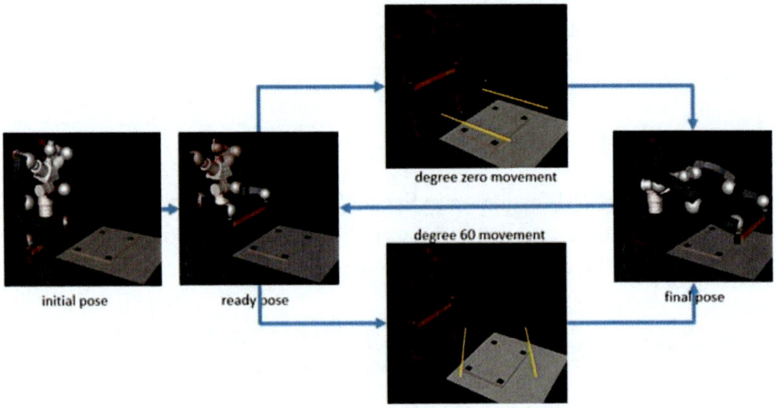

Fig. 8 Example of rolling-pin planning for two given actions on the dough

The primitive action is direct and intuitive. We must keep the desired height of the rolling pin from the plate and try to act at the specified angle by passing through the dough's center. The connecting action is instead artificial and specific to this application. First, two proper poses for the rolling pin are defined, namely the ready pose and the final pose. Then, a routine is repeated, starting with a connecting action from the initial pose to the ready pose until the generated action is completed. The routine steps are as follows: (*a*) doing a connecting action from the ready pose to the initial pose of a primitive action; (*b*) doing the primitive action; (*c*) doing a connecting action from the final pose of the primitive action to the defined final pose; (*d*) doing the connecting action from the defined final pose to the ready pose. This routine is capable of covering all generated actions from the previous dough planning.

There is an example of this process in Fig. 8 with two given actions from the previous dough planning, that is, a zero degree movement and a 60° movement. The generated action sequence consists of (1) a connecting action from the initial pose to the ready pose; (2) a connecting action from the ready pose to the initial pose of the zero-degree movement; (3) the primitive action related to the zero-degree movement; (4) a connecting action from the final pose of the zero-degree action to the final pose; (5) a connecting action from the final pose to the ready-pose; (6) a connecting action from the ready pose to the initial pose of the 60° movement; (7) the primitive action related to the 60° movement; (8) a connecting action from the final pose of the 60° movement to the final pose; (9) a connecting action from the final pose to the ready pose. The yellow lines indicate the trajectories of two holding points of the rolling pin while doing a primitive action.

The left figure in Fig. 9 shows the trajectory of the steps above from 1 to 5. The right figure in Fig. 9 shows the trajectory of the steps above from 6 to 9, in which the yellow lines and the red lines are those of the primitive and connecting actions, respectively.

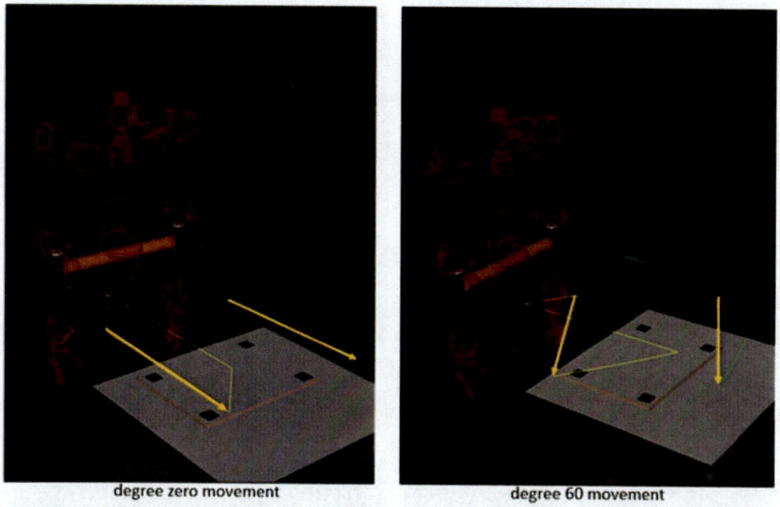

Fig. 9 Example trajectories of two holding points of a rolling pin for a given action on the dough. On the left, a zero-degree action. On the right, a 60° action.The yellow and red lines indicate the trajectories of the primitive and the connecting actions, respectively

7 Inverse Kinematics for the RoDyMan Robot

RoDyMan is the employed robot platform that consists of a mobile base, a 2-DoFs torso, a 2-DoFs neck, and two 7-DoFs arms. This process step aims to make the arms grasp the rolling pin properly and solve the inverse kinematics problem to plan the joint movements.

Many inverse kinematics algorithms are well established in the literature, as those using the LM algorithm or the damping least-squares method [39]. The employed method follows the closed-loop inverse kinematics algorithm with redundancy management explained in [39]. Redundancy is exploited to avoid unnatural postures of the robot.

Figure 10 shows the ready pose of RoDyMan before and after an action. Figure 10b shows the unnatural behaviour that can be avoided through a proper redundancy management, as shows in Fig. 11. Details are omitted here for brevity since this part is well established in the literature and does not bring any new insight into the problem faced by this chapter.

8 Simulations

For simulation purposes, the employed system consists of an Intel®Core™ i7-6500U CPU@2.50 GHz, Memory 8.0 GB with Windows®10 x64 operating system. We used

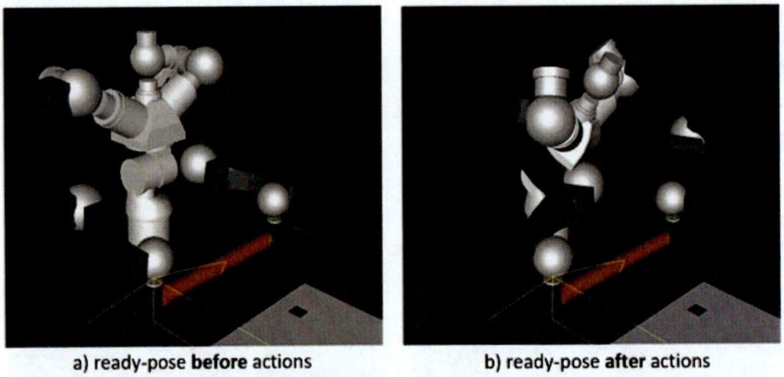

Fig. 10 Ready pose of RoDyMan before and after an action

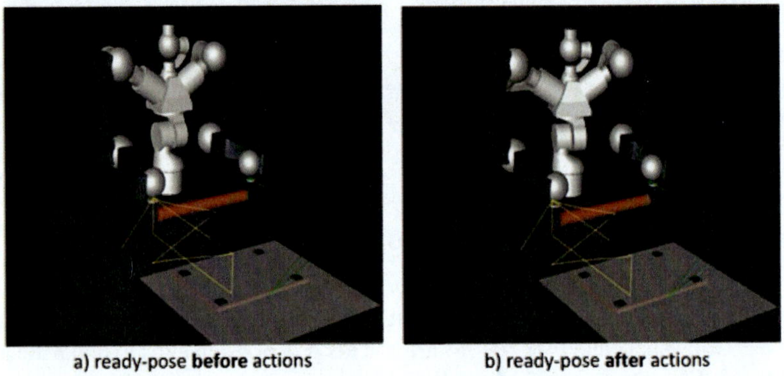

Fig. 11 Improved ready pose of RoDyMan before and after an action thanks to the redundancy management

the Microsoft Foundation Class (MFC) library of Microsoft®, the Qt library,[1] Boost,[2] Eigen,[3] and OpenCV[4] for 2D graphic or OpenSceneGraph[5] for 3D graphic libraries based on C++11 programming language. Additionally, Houdini™ of SideFX®and Blender™ were used to reconstruct the mesh from particles and for graphical rendering, respectively. 3D reconstruction is done with VisualSFM[6] and MeshLab.[7]

---

[1] https://www.qt.io.

[2] https://www.boost.org.

[3] http://eigen.tuxfamily.org.

[4] https://opencv.org.

[5] http://www.openscenegraph.org.

[6] http://ccwu.me/vsfm/.

[7] http://www.meshlab.net/.

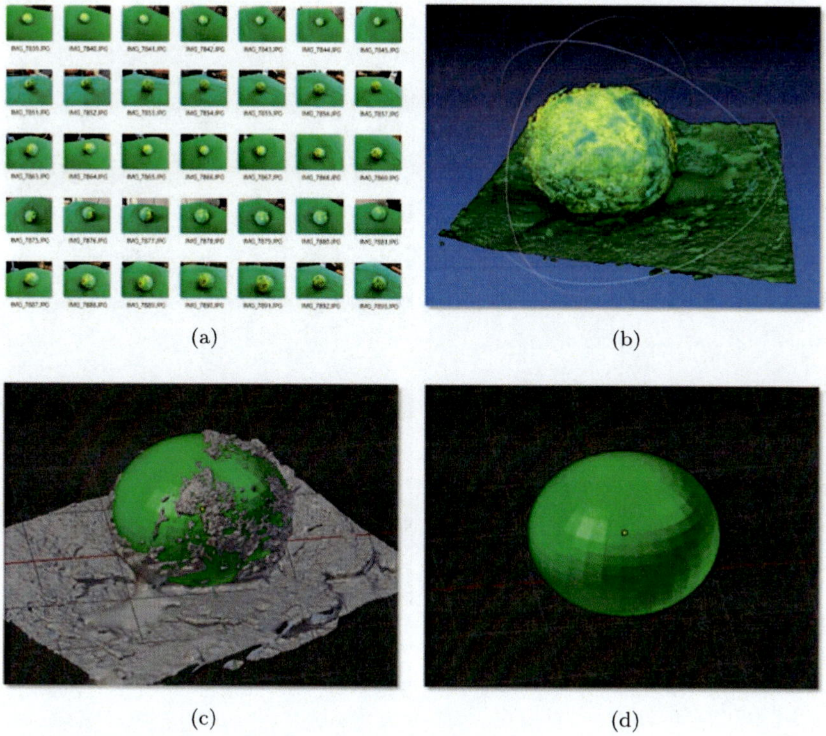

Fig. 12 Measuring of a real dough shape at initial status. **a** pictures with various views for real dough; **b** 3D reconstruction; **c** matching of the reconstructed shapes with a sphere; (d,h) the final matched shapes

8.1 Modelling of a Deformable Object

Even though a model for a deformable object like a pizza dough is designed in Chap. 3.1, the deformable object's properties like viscosity and elasticity vary case by case. Therefore, there is the need to tuning the model's parameters to fit the target object. Figure 12 shows how to measure a real deformable object. During preliminary experiments, a toy clay was used instead of pizza dough for convenience.

The process during the preliminary experiments has been carried out as follows. At first, the measure of the deformable object's surface and the reconstruction of its shape in 3D virtual space, at the initial and deformed statuses, have been carried out. Afterwards, an SPH-based object model is made following the reconstructed shape of the initial and deformed object. Finally, by applying the various actions on the SPH-based object model for the initial object, the best matching parameters for the model are found, which generates the SPH-based object model that is close to the reconstructed shape of the deformed object.

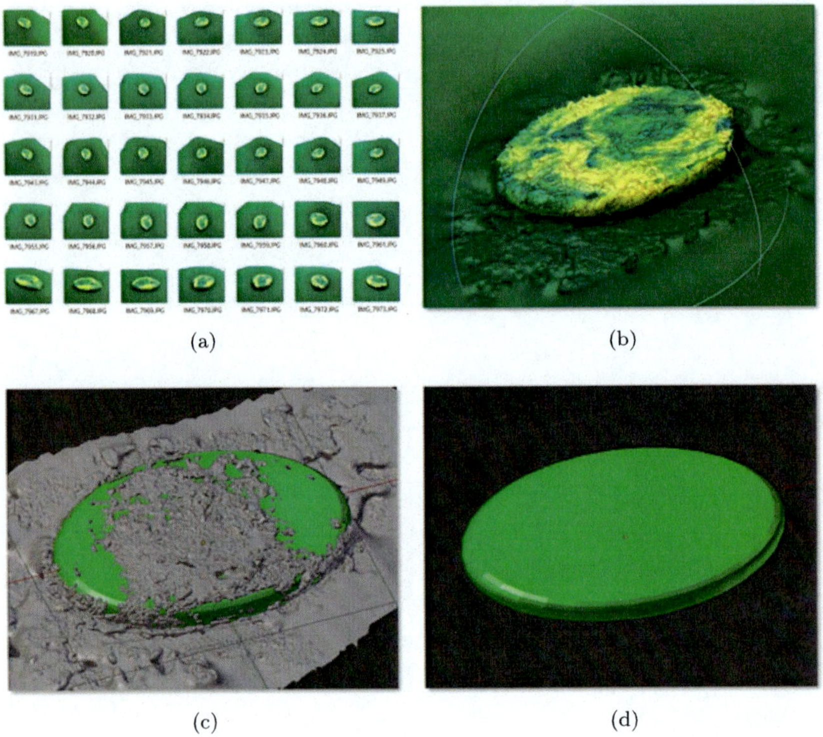

Fig. 13 Measuring of a real dough shape at initial status. **a** pictures with various views for real dough; **b** 3D reconstruction; **c** matching of the reconstructed shapes with a disk; (d,h) the final matched shapes

The reconstruction of a deformable object in the 3D virtual space at the initial status and the deformed status, respectively, are shown in Figs. 12 and 13. The pictures in Fig. 12 deals with a ball shape, representing the pizza dough before any deformation. The pictures in Fig. 12 show a disk shape after the deformation actuated by the rolling pin. To measure the shape of the object, a structure from motion method is used, which gathers some pictures (Figs. 12a and 13a) with various views and generates a 3D model (Figs. 12b and 13b). Usually, the generated 3D model is very rough and too complex. Hence, it needs post-processing to smooth the surface and remove the outliers. A simpler ball mesh-model (Fig. 12c) and cylinder mesh-model (Fig. 13c) are used to match the generated 3D model as close as possible, and the final reconstructed 3D model (Figs. 12d and 13d) are fixed, respectively.

The process of finding the best matching parameters for the deformable object model is shown in Fig. 14. Various deformed shapes are generated with various parameters: among them, we must find the best matching shape and its parameter. The matching process is done offline.

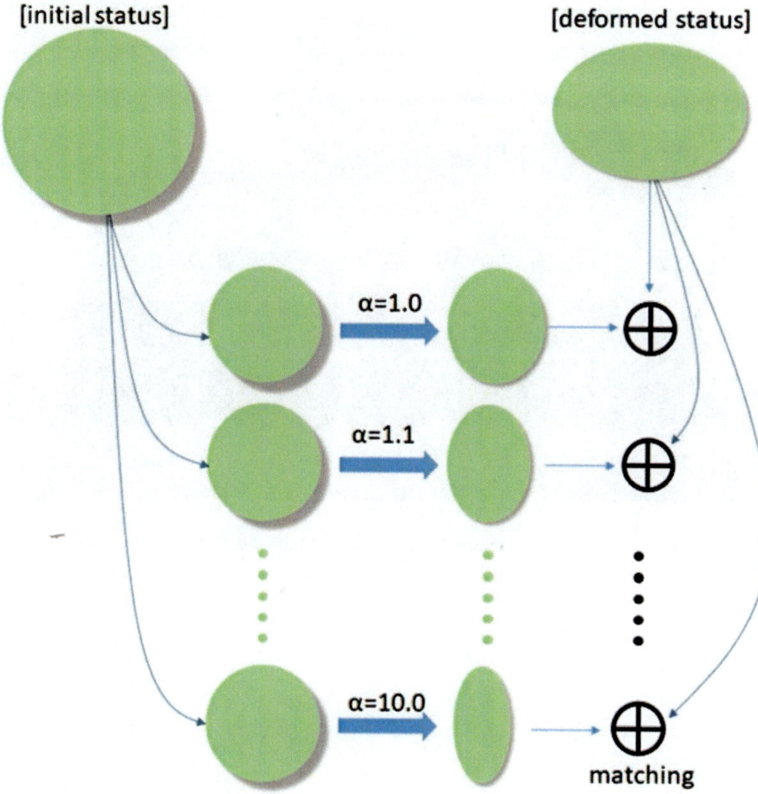

Fig. 14 Schematic diagram for finding the parameters of the deformable object model. α is just an example of a parameter for a SPH-based model: name and values are different depending on the model

8.2 *Pizza Dough Transition Look-up-table*

The transition LUT is used to speed up the planning algorithm, which is generated using various statuses of the pizza dough and various actions. However, it is not easy, and it needs much time to have experiments with real pizza dough. Therefore, using the deformable object model obtained before is more efficient than a real deformable object.

The previous section defines the model for a deformable object and its parameters obtained from real experiments. We had experiments with SPH particle radius of $1.25 \cdot 10^{-3}$ m, a pizza dough density of 1.276 kg/m^3, a solid density of $200 \cdot 10^3$ kg/m^3, an IISPH [40] for an incompressible fluid, and a viscosity coefficient of 250 kg/ms. Base on this model, the transition LUT is made by simulating the deformation from several configurations of the pizza dough and with various actions on the object. In particular, the action set $\mathcal{A}$ consists of seven heights, from $8 \cdot 10^{-3}$ m

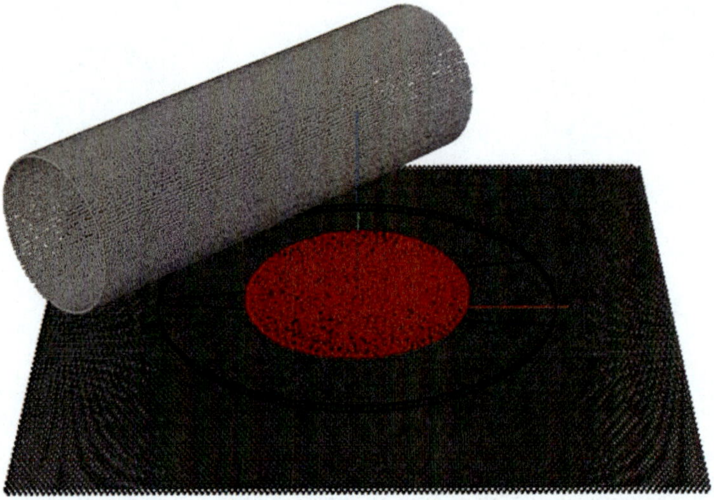

Fig. 15 SPH-based transition simulation for the pizza dough (red object) with a rolling pin (grey object)

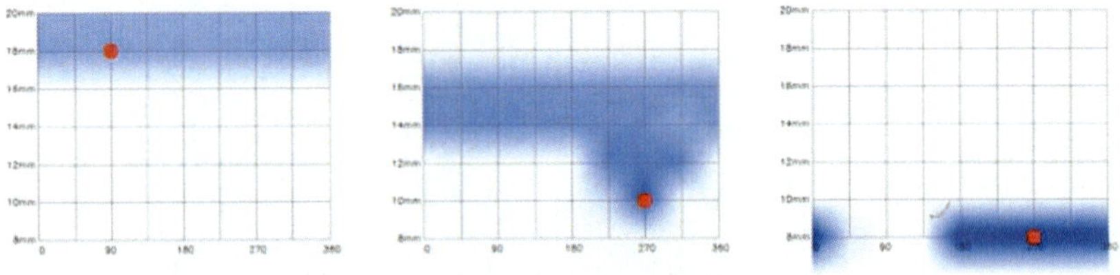

(a) (b) (c)

Fig. 16 LUT method. The x-axis and the y-axis indicate the angle and the height of a rolling pin motion, respectively. The white colour means lower score, while the darker blue colour means higher score. Red ball is the highest score action

to $20 \cdot 10^{-3}$ m spanned each $2 \cdot 10^{-3}$ m, and eight directions, from zero degrees to 180°, spanned each 45° and considering both forward and backward motions.

A simulation is shown in Fig. 15 where the pizza dough is represented by the red particles and the rolling pin by the green ones. An example where the pizza dough is stretched is depicted in Figs. 16 and 17. The former shows the transition LUT for sequential statuses of the pizza dough with height and angular table, while the latter shows the occupancy of the pizza dough for each status.

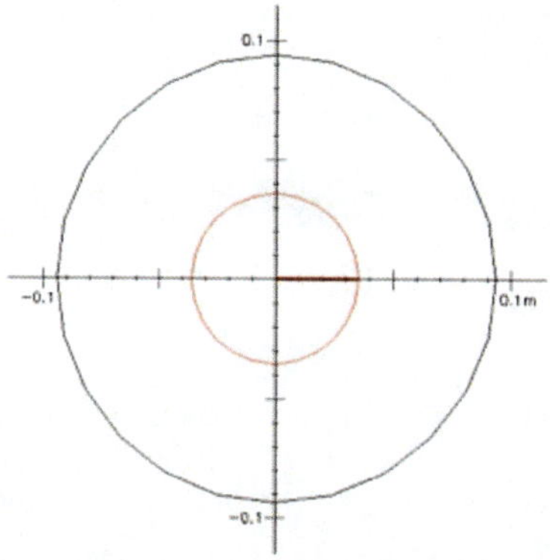

(a) Start shape of a dough placed in center

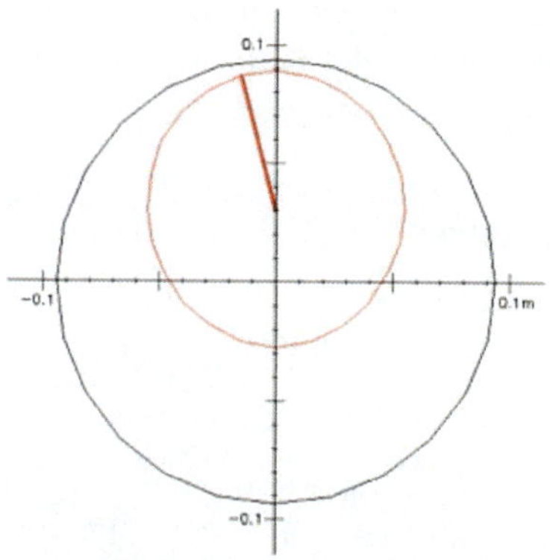

(b) The shape after applying the highest scored action in Fig. 9.15a

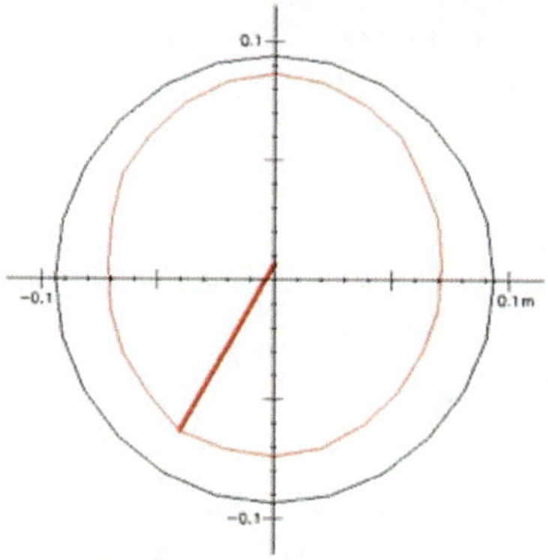

(c) The shape after applying the highest scored action in Fig. 9.15b

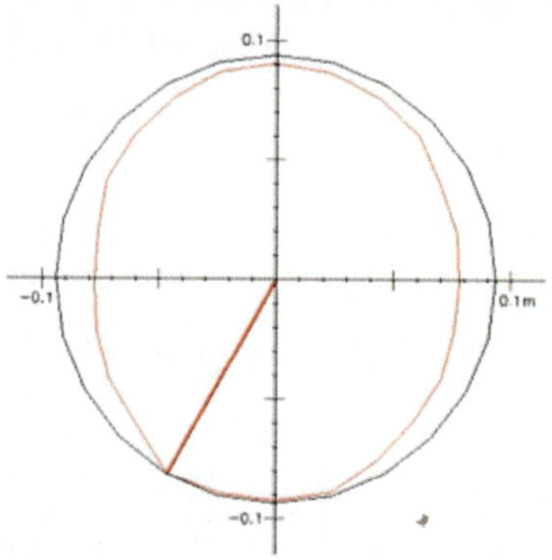

(d) The shape after applying the highest scored action in Fig. 9.15c

Fig. 17 Shapes of the current dough (red) and its target (black). The red line indicates the angle of the first bin in the dough's state

9 Discussion and Conclusion

This chapter explained a planning method to stretch a pizza dough with a rolling pin actuated by a robotic system. Base on the perception chapters, the deformable object is modelled, and the deformation information is used to plan the stretching actions on the pizza dough to reach the desired shape. An object recognition algorithm, a method to model deformable objects with high viscosity, the definition of the status of the pizza dough, the definition of the actions through the rolling pin, an inverse kinematics algorithm for the robot have been integrated to achieve the sought goal. Experiments were carried out to identify the parameters of the pizza dough. The stretching actions were planned offline thanks to a LUT database. Future experiments will definitely validate the proposed approach.

References

1. H. Wieser, Chemistry of gluten proteins. Food Microbiol. **24**(2), 115–119 (2007)
2. F. Van Bockstaele, I. De Leyn, M. Eeckhout, K. Dewettinck, Rheological properties of wheat flour dough and the relationship with bread volume. I. Creep-recovery measurements. Cereal Chem. J. **85**(6), 753–761 (2008)
3. G.M. Campbell, C.D. Rielly, P.J. Fryer, P.A. Sadd, Measurement and interpretation of dough densities. Cereal Chem. **70**(5), 517–521 (1993)
4. M. Charalambides, L. Wanigasooriya, S.M. Goh, J.G. Williams, Viscoelasticity of bread dough, in *2005 SEM Annual Conference and Exposition on Experimental and Applied Mechanics*, pp. 879–888 (2005)
5. J.M. Faubion, R.C. Hoseney, The viscoelastic properties of wheat flour doughs, in *Dough Rheology and Baked Product Texture* (Springer, 1990), pp. 29–66
6. E. Magaña Barajas, B. Ramírez-Wong, P.I. Torres-Chávez, I. Morales-Rosas. Use of the stress-relaxation and dynamic tests to evaluate the viscoelastic properties of dough from soft wheat cultivars, in *Viscoelasticity*, ed. by J. de Vicente (IntechOpen, 2012), pp. 259–272
7. M.R. Amjid, A. Shehzad, S. Hussain, M.A. Shabbir, M.R. Khan, M. Shoaib, A comprehensive review on wheat flour dough rheology. Pak. J. Food Sci. **23**, 105–123 (2013)
8. B.S. Khatkar, A.E. Bell, J.D. Schofield, The dynamic rheological properties of glutens and gluten sub-fractions from wheats of good and poor bread making quality. J. Cereal Sci. **22**(1), 29–44 (1995)
9. S. Uthayakumaran, M. Newberry, M. Keentok, F.L. Stoddard, F. Bekes, Basic rheology of bread dough with modified protein content and glutenin-to-gliadin ratios. Cereal Chem. J. **77**(6), 744–749 (2000)
10. J. Engmann, M.C. Peck, D.I. Wilson, An experimental and theoretical investigation of bread dough sheeting. Food Bioprod. Process. **83**(3), 175–184 (2005)
11. C. Bollain, A. Angioloni, C. Collar, Relationships between dough and bread viscoelastic properties in enzyme supplemented wheat samples. J. Food Eng. **77**, 665–671 (2006)
12. H. Mirsaeedghazi, Z. Eman-Djomeh, S.M.A. Mousavi, Rheometric measurement of dough rheological characteristics and factors affecting it. Int. J. Agric. Biol. **10**(1), 112–119 (2008)
13. S. Sofou, E.B. Muliawan, S.G. Hatzikiriakos, E. Mitsoulis, Rheological characterization and constitutive modeling of bread dough. Rheol. Acta **47**(4), 369–381 (2008)
14. C.R. Beverly, R.I. Tanner, Numerical analysis of three-dimensional Bingham plastic flow. J. Nonnewton. Fluid Mech. **42**(1), 85–115 (1992)
15. E. Mitsoulis, S.G. Hatzikiriakos, Rolling of bread dough: experiments and simulations. Food Bioprod. Process. **87**(2), 124–138 (2009)
16. S. Sofou, E. Mitsoulis, Calendering of pseudoplastic and viscoplastic sheets of finite thickness. J. Plastic Film & Sheet. **20**, 185–222 (2004)
17. S. Tokumoto, S. Hirai, Deformation control of rheological food dough using a forming process model, in *2002 IEEE International Conference on Robotics and Automation*, pp. 1457–1464 (2002)
18. M. Higashimori, K. Yoshimoto, M. Kaneko, Active shaping of an unknown rheological object based on deformation decomposition into elasticity and plasticity, in *2010 IEEE International Conference on Robotics and Automation*, pp. 5120–5126 (2010)
19. A.-M. Cretu, P. Payeur, E.M. Petriu, Neural network mapping and clustering of elastic behavior from tactile and range imaging for virtualized reality applications. IEEE Trans. Instrum. Meas. **57**(9), 1918–1928 (2008)

20. Z. Wang, S. Hirai, Modeling and estimation of rheological properties of food products for manufacturing simulations. J. Food Eng. **102**(2), 136–144 (2011)
21. D. Berenson, Manipulation of deformable objects without modeling and simulating deformation, in *2013 IEEE/RSJ International Conference on Intelligent Robots and Systems*, pp. 4525–4532 (2013)
22. B. Frank, C. Stachniss, N. Abdo, W. Burgard, Efficient motion planning for manipulation robots in environments with deformable objects, in *2011 IEEE/RSJ International Conference on Intelligent Robots and Systems*, pp. 2180–2185 (2011)
23. J. Lang, An acquisition method for interactive deformable models, in *2nd International Conference on Creating, Connecting and Collaborating through Computing*, pp. 165–170 (2004)
24. C. Phillips-Grafflin, D. Berenson, A representation of deformable objects for motion planning with no physical simulation, in *2014 IEEE International Conference on Robotics and Automation*, pp. 98–105 (2014)
25. J. Smolen, A. Patriciu, Deformation planning for robotic soft tissue manipulation, in *2009 Second International Conferences on Advances in Computer-Human Interactions*, pp. 199–204 (2009)
26. R. Hartley, A. Zisserman, *Multiple View Geometry in Computer Vision* (Cambridge University Press, 2004)
27. D.G. Lowe, Object recognition from local scale-invariant features, in *International Conference on Computer Vision*, pp. 1150–1157 (1999)
28. H. Bay, A. Ess, T. Tuytelaars, L. Van Gool, SURF: speeded up robust features, in *Computer Vision and Image Understanding*, pp. 346–359 (2008)
29. E. Rosten, T. Drummond, Machine learning for high-speed corner detection. Eur. Conf. Comput. Vis. **1**, 430–443 (2006)
30. H. Kato, M. Billinghurst, Marker tracking and HMD calibration for a video-based augmented reality conferencing system, in *2nd IEEE and ACM International Workshop on Augmented Reality*, pp. 85–94 (1999)
31. A. Bey, R. Chaine, R. Marc, G. Thibault, S. Akkouche, Reconstruction of consistent 3D CAD models from point cloud data using a priori CAD models. ISPRS - Int. Arch. Photogram. Remote Sens. Spatial Inf. Sci. **3812**, 289–294 (2011)
32. M. Berger, A. Tagliasacchi, L.M. Seversky, P. Alliez, J.A. Levine, A. Sharf, C.T. Silva, State of the art in surface reconstruction from point clouds, in *Eurographics 2014 - State of the Art Reports*, ed. by S. Lefebvre, M. Spagnuolo (The Eurographics Association, 2014)
33. R.C. Gonzalez, R.E. Woods, *Digital Image Processing* (Prentice-Hall, 2006)
34. J. Canny, A computational approach to edge detection. IEEE Trans. Pattern Anal. Mach. Intell. **PAMI-8**(6), 679–698 (1986)
35. D.G. Lowe, Distinctive image features from scale-invariant keypoints. Int. J. Comput. Vis. **60**(2), 91–110 (2004)
36. J. Matas, O. Chum, M. Urban, T. Pajdla, Robust wide-baseline stereo from maximally stable extremal regions. Image Vis. Comput. **22**(10), 761–767 (2004)
37. S. Leutenegger, M. Chli, R.Y. Siegwart, BRISK: Binary Robust Invariant Scalable Keypoints, in *2011 International Conference on Computer Vision*, pp. 2548–2555 (2011)
38. B. Frank, M. Becker, C. Stachniss, W. Burgard, M. Teschner, Efficient path planning for mobile robots in environments with deformable objects, in *2008 IEEE International Conference on Robotics and Automation*, pp. 3737–3742 (2008)
39. B. Siciliano, L. Sciavicco, L. Villani, G. Oriolo, *Robotics: Modelling, Planning and Control.* (Springer, London, UK, 2009)
40. M. Ihmsen, J. Cornelis, B. Solenthaler, C. Horvath, M. Teschner, Implicit incompressible SPH. IEEE Trans. Vis. Comput. Gr. **20**(3), 426–435 (2014)